Optimal Control and System Theory in Dynamic Economic Analysis

A SERIES OF VOLUMES IN

DYNAMIC ECONOMICS: THEORY AND APPLICATIONS

Maurice Wilkinson, *Editor*
Columbia University

PUBLISHED

M. Aoki
University of California, Los Angeles

I. OPTIMAL CONTROL AND SYSTEM THEORY IN DYNAMIC ECONOMIC ANALYSIS

IN PREPARATION

H. Leland
University of California, Berkeley

PRODUCTION THEORY AND FINANCIAL MARKETS

C. Blackorby
Southern Illinois University
D. Primont
University of Massachusetts
R. Russell
University of California, San Diego

FUNCTIONAL STRUCTURE IN ECONOMIC ANALYSIS

G. Rausser
Iowa State University

DYNAMICS OF MARKETING BOARDS: ESTIMATION, PRODUCTION AND CONTROL

Optimal Control and System Theory in Dynamic Economic Analysis

Masanao Aoki
University of California, Los Angeles

AMERICAN ELSEVIER PUBLISHING COMPANY, INC.
52 Vanderbilt Avenue, New York, N. Y. 10017
NORTH-HOLLAND PUBLISHING COMPANY
P.O. Box 211
Amsterdam, The Netherlands
Sole Distributor for all Countries
Except the United States and Canada

Library of Congress Cataloging in Publication Data

Aoki, Masanao.
Optimal control and
system theory in
dynamic economic analysis

(Dynamic economics; 1)
Includes bibliographical references.
1. Economics—Mathematical models. 2. System
theory. I. Title.
HB135.A64 330′.01′82 75-26329
ISBN 0-444-00176-X (American Elsevier)
ISBN 0-7204-8603-3 (North-Holland)

Manufactured in the United States of America

To the Memory of My Father

Contents

PART II DETERMINISTIC DYNAMIC ECONOMIC SYSTEMS

PART III STOCHASTIC DYNAMIC ECONOMIC SYSTEMS

Editor's Preface

This series of monograph and advanced texts is designed to further research on intertemporal decision making and planning. These topics require the combined talents of mathematicians, control engineers and economists to analyze the conceptual problems that derive from the role of time, uncertainty and interdependence of decision makers. The major topics included in the series are dynamic utility (intertemporal preference and portfolio theory) and profit maximization (optimal capital and investment policy), theory of economic planning (public investment, optimal economic growth, centralized and decentralized decision making), optimal control and stabilization (modern control theory, optimal stabilization policy, estimation and control), and related analytical techniques.

The present volume is concerned with the utilization of optimal control and system theory in formulating and analyzing the properties of dynamic economic models. The basic principles and theorems of control and system theory are discussed together with applications to both micro- and macroeconomic models. Many other applications are possible to problems in intertemporal decision making and planning.

Maurice Wilkinson

Preface

Our purpose here is to advance the use of optimal control and system theory in dynamic economic model analysis. To work with economic models described by differential or difference equations, and with some forcing functions, it is crucial if the model is to be dynamic to be able to determine the properties of the model without explicitly solving the system of equations.

We feel that the most important properties of a dynamic economic model are its stability, controllability and observability. Control and system theory can be used for the analysis of these properties of dynamic economic models. We hope that this book will assist economists, management scientists, and other scholars, to utilize the concepts and tools of control and system theory without the necessity of recourse to writings that are exclusively oriented towards engineers.

The problems examined in this book are drawn from both microeconomic and macroeconomic analysis. At the same time, we hope that this book may further help to narrow the communication gap between economists and control and system engineers, by illustrating to the latter, the manner in which economic problems suggest the need for modifying and extending the existing control and system theory. As we shall see in this book, often unconventional and interesting unsolved control problems arise from dynamic economic modeling.

Part I, which is basic to the rest of the book, considers state space, a common form, by which diverse dynamic models are represented. Once a dynamic economic model is given a state space representation, the principles and theorems of control and system theory can be applied to analyze the properties of the model.

Part II analyzes the deterministic economic models and is primarily oriented towards macroeconomics.

Part III mainly considers stochastic price adjustment behavior of individual economic agents. Reference to the table of contents will clearly indicate the particular economic models that are analyzed. The individual sections of Parts I to III may be combined in alternate ways to serve the needs of courses in either economics or engineering departments.

I have benefited greatly from comments, criticisms and suggestions by my colleagues, friends and by students who have taken the course from which this book arose. I would like to mention, in particular, A. Leijonhufvud, R. W. Clower, G. Rausser, T. Takayama, J. Wharton and M. Wilkinson.

Mr. M. K. Nabli and U. Oertel read through most of an earlier draft and have contributed greatly to improve the readability of this book.

The preparation of this book was facilitated greatly by the expert typing of Ms. Jackie Davis and Sherry Dover. Mr. K. Haruki helped with proof-reading.

MASANAO AOKI

Los Angeles, California
May, 1975

PART I
STATE SPACE

1

Dynamic Models, Control and System Theory

1.1 INTRODUCTION

The static economic problem is the allocation of scarce resources at any given moment in time. The dynamic or intertemporal economic problem is the allocation of scarce resources over a finite, or infinite, period of time. Paul Samuelson's (1947) landmark contribution to dynamic theory, which builds upon the earlier work of Frisch (1935), identified a dynamic system as one in which the behavior of variables over time is determined by a set of equations in which the variables at different points in time are involved in an "essential" way. Samuelson's contribution revealed that difference and differential equations can be powerful tools in economics.

During the past thirty to forty years, economists have attempted to develop theories to explain the sequence of decisions of individual economic agents (producers and consumers) that generate the time paths of economic variables. The calculus of variations, and more recently dynamic programming and Pontryagin's Maximum principle, have been of great value in this area, as well as in the theory of growth that deals with highly aggregative economic models (Benavie, 1972; Hadly and Kemp, 1971; Intriligator, 1971). On the other hand, some economists have become greatly interested in describing, or modeling, economic systems with more disaggregated variables or in incorporating into their systems imperfect information. Out of the latter developments, an increasing interest grew in the performance of models out of equilibrium—a concern long familiar to control and systems engineers.

In short, as economists gather more experience with models of varying degrees of complexity, and learn how digital computers can make simulation a more powerful and attractive way of exploring the behavior of complex systems over time, it is reasonable to believe that economists will need to know more of the concepts and tools of control and system theory.

1.1.1 Review of the Beginnings

A brief review of previous attempts to introduce control and system theory to economic analysis may be helpful. In the first half of the 1950's, there was an attempt in England to apply to economic systems the theory of servomechanisms which was developed during World War II. In 1953 Arnold Tustin, professor of electrical engineering, University of Birmingham, wrote a book on this subject entitled *The Mechanisms of Economic Systems*. Its subtitle describes very accurately his intention "An Approach to the Problem of Economic Stabilization From the Point of View of Control Systems Engineering." He viewed economic systems as interdependent dynamic systems and analyzed their time responses to exogenous shocks using the transfer function method which was found to be very successful in designing servomechanisms during and after World War II. He discussed the role of economic models as tools for forecasting and regulation, although he recognized difficulties in estimating parameters due to nonlinearities. He then proceeded to discuss simulation techniques (by means of physical and electronic analogs) of economic system responses in the last part of his book. Except for this last part, which is dated since no large scale electronic digital computers as we now know them were available to him, his book is surprisingly modern in spirit in its attempt to apply control theory to the solutions of some macroeconomic problems. We note, however, that no aspects of decision making by economic agents under imperfect information, such as learning or price expectation formation behavior, were discussed. At about the same time A. W. Phillips began to publish a series of articles applying the theory of servomechanisms to business cycles and stabilization policies of macroeconomic models (1954). These pioneers' efforts, however, did not take hold in England nor in the United States. Phillips is better known now for the Phillips' curve than for his work on business cycles. As for Tustin, his book apparently failed to influence economists, in spite of his acknowledgment of help and encouragement received from young economists such as R. Stone, F. Hahn and E. L. Mills who later became eminent (Tustin, 1953).

It may be instructive to reflect on some of the reasons for this failure. In the author's opinion, there are two major reasons besides the usual difficulty of communication between people with different training. One is technical and the other is circumstantial.

First, we consider the technical reason. We must realize that in the early 1950's nonlinear programming was in its infancy since electronic computers were just coming into general use. At that time no large-scale econometric models were developed. It was not until the 1960's in the United States and Western Europe that the state space theory of dynamic systems were available. Thus, we can see that even though these pioneers had the right idea, their technical tools were not sophisticated enough to tackle the very complex problems associated with the control of economies.

The circumstantial reason is that the interests of mathematically inclined economists in the early 1950's were directed more towards the general competitive equilibrium models as evidenced, for example, by the work of K. J. Arrow and G. Debreu (Arrow and Hahn, 1971; Debreu, 1959). In these general equilibrium models, a set of mostly topological concepts and tools is employed to establish the existence of general equilibria. Sets of prices to support such equilibria, their Pareto efficiency and so on are of interest in these models and thus make them entirely different from those considered by Tustin and Phillips. In these models, problems associated with trades taking place out of equilibrium is avoided by use of the tâtonnement processes of Walras. Stability analysis of tâtonnement (or nontâtonnement) price adjustment schemes and research in growth theory are quite different from the control and system theoretic problems associated with the behavior of economic agents either in or out of equilibrium. Tustin unlike the economic theorists of his time discusses these problems in his book. Tustin was more concerned with the regulation of macroeconomic models. However, he used, it seems to the author, very little space to justify models he used on microeconomic or other economic considerations.

In the 1950's economists were not convinced of the need for acquiring the new tools of control and system theory due to largely circumstantial reasons, also the lack of significant results, and perhaps also to technical data processing difficulties. A total lack of interest existed then on the part of economists for what the control engineer had to offer.

The different attitudes of economists and control engineers towards their models have influenced the tools they use. In economics, many aspects of the models are rather arbitrary since the sheer complexity of real economic phenomena force model builders to adopt many simplifying and sometimes unrealistic assumptions for reasons not entirely justifiable on economic considerations alone. Knowing the arbitrary nature of their models and mistrusting them in some sense, economists do not take their models literally. They are content with establishing qualitative (hence hopefully robust) properties of their models such as existence of optimal decision rules and properties of classes of optimal decision rules such as stationarity and stability. It is fair to say that because of complexities of the phenomena to be modeled, models of disequilibrium adjustment processes rarely exist. Time has played a relatively minor role in some of these economic models.

On the other hand, in control and system theory, developed primarily by engineers, models are assumed as given. They believe in their models much more than economists do in theirs. Even though many engineering models contain some uncertainties in system structures and/or parameter values of systems and environments, engineers are generally willing to analyze their models in detailed quantitative terms. They are interested in construction and implementation of algorithms for optimal decision rules, in addition to establishing some qualitative properties such as stability. Time is usually explicitly incorporated into their models.

1.1.2 Control and System Theory Revisited

Now, in the first half of the 1970's, twenty years after these initial attempts, an interest in applying control and system theory to economic problems has revived, especially in modeling and control of macroeconomic systems (see for example IFAC/IFORS Conference on Dynamic Modeling and Control of National Economies, 1973).

More sophisticated tools to deal with complex dynamic system behavior exist including nonlinear programming, large-scale digital computers, econometric models, and more advanced estimation techniques. Also interest in the Walrasian general competitive equilibrium analysis has been replaced by models in which imperfect and incomplete information, such as search behavior, expectation formation, trading out of equilibrium, can be dealt with. Economists realize that they must not be exclusively concerned with the equilibrium states of the economy, but that they must consider economic systems in disequilibrium that work with imperfect information and without the perfect coordination of all economic activities. A more synpathetic climate seems to be developing among policymakers towards giving these analyses more active roles in the guiding and fine-tuning of the control of economic activities.

In order not to repeat the initial failure of the pioneers in control and system theory, supporters must be careful not to discuss problems that are of no interest to economists. Control and system theorists who choose to work in economics must have a good understanding of economics and must also choose those problems that are of genuine interest to economists. Economists, on the other hand, need to know the limitations of system theory.

1.2 DYNAMIC MODELS IN ECONOMICS

Economic policies cannot be based only on a day-to-day or a month-to-month basis, since obviously current policy decisions have impact not only in the present but also in the future. Economic policymakers are interested in controlling the national economy. This is a dynamic object. Thus, the economic system must be modeled as a dynamic situation, not as a static situation.† The same comments apply to basic economic decision making units, such as households and firms. Their decision making activities are not always modeled as dynamic processes in some text books. We could cite many other macroeconomic and microeconomic problems that could and should be modeled as dynamic systems. Static models do have their own place and usefulness in economics, and certainly economic theory has progressed a great deal with only the use of static models.

For some time, economists have been aware of these considerations. What is not as widely known, perhaps, is that recent developments in system and control theory can be of help to economists in constructing more realistic models of dynamic economic systems. Such models can be understood better quantitatively and also be controlled better.

In this book, we will be primarily interested in those tools that are necessary for the analysis of economic activities and shortrun phenomena that require a truly dynamic approach to be realistic. In some cases, we will incorporate learning or expectation formation behavior of economic agents under the assumption of imperfect information. We will postulate models of household behavior, firms' decision processes and government policies. Their behavior or interactions over time are usually known to us only implicitly, since their behavior is usually given as solutions to a set of optimization problems involving differential equations or difference equations.‡

We must expand static economic models to be dynamic ones. Dynamic models function in an environment which is perhaps changing with time or reacting to decisions made by some economic agents. These dynamic models have enlarged the scope of activities and phenomena that may be

---

†In this book we shall not be concerned so much with the long-run equilibrium motions such as those in growth theory but rather with the disequilibrium behavior of economic models.

‡A mixture of the two types of equations is possible, however, we will not discuss the class of models that are described by systems of difference and differential equations.

usefully describable by economic models in general. We duscuss, for example, the possibility of guiding or nudging some economic variables to follow some specified trajectories over time by the proper manipulation of instruments.

At the same time, however, because dynamic models are more complex we have introduced problems that do not occur in static models. For instance, the problem of instability does not exist in static models nor does the problem of convergence or divergence of some adjustment rules. In dynamic models a foremost concern is to stabilize a model's time behavior in some specific technical sense. For example, can policymakers be certain of finding rules to stabilize the economy? This question is examined in Chapter 5. System and control theories are useful in coping with such complications. The choice of differential or difference equations for a specific time frame is usually dictated by convenience and by the goodness of the approximation. Models described by a set of difference equations are called *discrete time dynamic models*, or *systems*. These are dynamic systems for which time indices vary over a set of discrete values in a given interval or time period. They arise primarily from two types of systems:

(1) Systems that inherently operate in discrete time. A prime example is the digital computer. We shall not be concerned with this class of systems.
(2) Systems that operate continuously in time but are observed or are of interest for only a discrete instant of time.

Many economic systems are of this latter type, since observations can be made only on an hourly, daily, monthly, quarterly, yearly, and so forth, basis. An example is

$$\dot{y}(t) = ay(t) + bx(t) \tag{1.2-1}$$

where all the variables are scalar and the value of the input variable or the control instrument x is held constant over T, a basic time interval. That is, a change in x can occur only at the beginning of each basic time interval T. The period T may be a day or a quarter or any specified time. It is not necessary for T to be constant. T can be variable and changes in x could occur at a set of discrete time instants $t_0 < t_1 < t_2 < \cdots$.

Integrating (1.2-1) over one basic unit of time, we obtain

$$\begin{aligned} y((k+1)T) &= y(kT)e^{aT} + \int_{kT}^{(k+1)T} e^{a[(k+1)T-s]}bx(s)\,ds \\ &= y(kT)e^{aT} + \left(\int_0^T e^{a(T-t)}b\,dt\right)x(kT) \\ &= y(kT)e^{aT} + \frac{1}{a}(e^{aT}-1)bx(kT) \end{aligned} \tag{1.2-2}$$

Let $y_k = y(kT)$, $x_k = x(kT)$, $k = 0, 1, \ldots$. Then (1.2-2) may be rewritten as

$$y_{k+1} = Fy_k + Gx_k \tag{1.2-3}$$

where

$$F = e^{aT} \qquad G = \frac{1}{a}(e^{aT} - 1)b$$

This is the discrete-time model that relates the output values and instrument values at discrete time instants. The output values are observed, and the predicted results compared to the actual values. Here, F and G are constant scalars. Such systems are called *time-invariant* or *constant discrete-time linear dynamic systems*. The procedure for converting differential equations into difference equations is general. It is applicable to time-varying systems and to systems like (1.2-1) where y and x are vectors and a and b are matrices, as long as the dynamics are linear in y and the instruments are held constant over a basic time unit.

This discussion suggests a way of performing stock-flow analysis of a dynamic model described by

$$\dot{\boldsymbol{y}}(t) = \boldsymbol{f}(\boldsymbol{y}(t), \boldsymbol{x}(t), \boldsymbol{e}(t))$$

since

$$\boldsymbol{y}(t + \Delta) - \boldsymbol{y}(t) = \Delta \cdot \boldsymbol{f}(\boldsymbol{y}(t), \boldsymbol{x}(t), \boldsymbol{e}(t)) + \boldsymbol{o}(\Delta)$$

relates the change in the stock variable $\boldsymbol{y}$ in the short-run period of length Δ to that of the flow variable $\boldsymbol{f}(\boldsymbol{y}(t), \boldsymbol{x}(t), \boldsymbol{e}(t))$. We should always keep in mind that a change in Δ results in a *different* period model. Thus, strictly speaking, Δ should be retained in a period model as a parameter although we do not do this. For example, from

$$\dot{\boldsymbol{y}}(t) = A\boldsymbol{y}(t) + \boldsymbol{b}x(t)$$

we derive

$$\boldsymbol{y}_{k+1} = e^{A\Delta}\boldsymbol{y}_k + \Delta\boldsymbol{b}x_k$$

When z denotes a stock variable, its time derivative $\dot{z}$ is a flow variable and $\dot{z}\Delta$ is approximately the change in the stock variable in time period of duration Δ, with error of the order $\boldsymbol{0}(\Delta)$. We must keep in mind that stock variables and flow variables generally obey two distinct sets of constraints.

1.3 DYNAMIC MODELS AND THEIR REPRESENTATION

This book discusses techniques for obtaining the time paths of some variables or for modifying their behavior over time. The variables are governed by systems of differential or difference equations and we will examine how to manipulate these systems. First, however, the system of equations is transformed into a standard representation. Basically, two methods of constructing systems of dynamic equations exist for describing a given system or model. In the one, attention is focused on the so-called instruments that are the inputs to the system, and on specific variables, that are the current and possibly past outputs of the model. Control engineers call this description of the behavior of a dynamic system the *input-output* model. The reduced form or autoregressive moving average form are input-output models.

The other way to construct a model is first to introduce some intermediate variables called state variables. These state variables may not of themselves be of intrinsic interest, but they help to describe the internal configuration or state of the model. These state variables are related by one equation to the instruments, or inputs, and a second equation relates both the state variables and the instruments to the outputs. Control engineers call this the internal or *state-space* model.

These two representations are obviously closely related (Polak and Wong, 1970); in fact, they are equivalent in the sense that under suitable conditions one can be deduced from the other, and vice versa. We will illustrate this equivalence later. The choice of a particular representation is, in general, a matter of convenience. We shall rely primarily on the state-space model because it allows a varying number of variables to be treated uniformly. Also, results from control and system engineering literature will then be applicable to economic problems.

In the case of a model described by a system of differential equations, the state-space representation is of the form

$$\dot{z} = \boldsymbol{f}(z(t), \boldsymbol{x}(t))$$
$$y(t) = \boldsymbol{g}(z(t), \boldsymbol{x}(t))$$

where $\boldsymbol{f}$ and $\boldsymbol{g}$ are vector-valued functions, or in the case of linear systems, it is of the form

$$\begin{aligned} \dot{z} &= Az(t) + Bx(t) \\ y(t) &= Cz(t) + Dx(t) \end{aligned} \tag{1.3-1}$$

where $z(t)$, $x(t)$, and $y(t)$ are usually vectors and A, B, C, and D are usually matrices. The variable $\boldsymbol{y}$ is the output, or measurement, vector; and the variable $\boldsymbol{x}$ denotes the instruments. In some cases, the vector variables may become scalar variables, and the matrices may become vectors or scalars. The state vector $\boldsymbol{z}(t)$ is the intermediate vector relating $\boldsymbol{x}(t)$ to $\boldsymbol{y}(t)$. The matrices A, B, C, and D may change with time or be constant. In the former case, we say the system is time-varying, otherwise the system is time-invariant or constant.

Thus, the state-space representation of a discrete-time linear dynamic system, with respect to a suitable time index, is

$$\begin{aligned} z(t+1) &= Az(t) + Bx(t) \\ y(t) &= Cz(t) + Dx(t) \end{aligned} \tag{1.3-2}$$

We will show that a state-space representation is equivalent to dynamic systems given by second-order or higher order differential equations, or difference equations with more than a single lag. As we have already noted the former can be transformed into the latter, and vice versa, thus we need only consider the one case. In Chapter 3, we will examine standard tests to be applied to see if a state-space model has the desired dynamic properties.

In econometrics, relations and identities relating variables are usually given, initially, by a set of simultaneous equations, called *structural equations*. A model is said to be in structural form. A model is frequently transformed into different forms or representations to suit different applications of the model, such as a reduced or final form. (See, for example, Preston and Wall, 1973, for a more precise discussion.)

We shall discuss how structural, reduced or final form represent representations of dynamic economic models can be put into state-space representations. Also we will consider how final form representation may be obtained from state-space representation.

2
State-Space Models

To be effective a dynamic economic model must have stable adjustment processes by means of a given set of instruments. We shall obtain criteria for models in a standard form to be stable. First, we will show how to put systems of linear differential, or difference, equations into standard form called the state-space form or representation.

2.1 SECOND-ORDER EQUATIONS

2.1.1 Differential Equations

For systems described by second-order differential equations, we can introduce the state vector $z(t)$ with two components $z_1(t)$ and $z_2(t)$ as follows.

Given

$$\ddot{y}(t) = \alpha_1 \dot{y}(t) + \alpha_2 y(t) + \beta x(t) \tag{2.1-1}$$

define†

$$z_1(t) = y(t)$$

$$z_2(t) = \dot{z}_1(t) - \alpha_1 y(t)$$

$$= \dot{z}_1(t) - \alpha_1 z_1(t)$$

or

$$\dot{z}_1(t) = \alpha_1 z_1(t) + z_2(t)$$

†The state-space vectors are defined in such a way that no derivations of $x(t)$ appear in the final state-space equation (2.1-3), since in general the derivatives have numerically undesirable properties that we wish to avoid.

Then

$$\begin{aligned}\dot{z}_2(t) &= \ddot{z}_1(t) - \alpha_1\dot{z}_1(t)\\ &= \ddot{y}(t) - \alpha_1\dot{y}(t)\\ &= \alpha_2 y(t) + \beta x(t)\\ &= \alpha_2 z_1(t) + \beta x(t)\end{aligned}$$

In the vector notation

$$\begin{aligned}\dot{z}(t) &= Az(t) + \boldsymbol{b}x(t)\\ y(t) &= (1\ \ 0)z(t)\end{aligned} \tag{2.1-2}$$

where

$$A = \begin{pmatrix}\alpha_1 & 1\\ \alpha_2 & 0\end{pmatrix} \qquad \boldsymbol{b} = \begin{pmatrix}0\\ \beta\end{pmatrix}$$

where $z(t) = (z_1(t), z_2(t))^t$ and where t denotes transpose. Equation (2.1-2) is a state-space representation of (2.1-1). The initial conditions for (2.1-1) are $y(0)$ and $\dot{y}(0)$, which give the initial condition for (2.1-2) to be $z(0)$.

State-space representations are by no means unique. For example, we could define a two-dimensional vector $\tilde{z}(t)$ by

$$\begin{aligned}\tilde{z}_1(t) &= y(t)\\ \tilde{z}_2(t) &= \dot{\tilde{z}}_1(t) = \dot{y}(t)\end{aligned}$$

Then

$$\begin{aligned}\dot{\tilde{z}}_2(t) &= \ddot{y}(t)\\ &= \alpha_1\tilde{z}_2(t) + \alpha_2\tilde{z}_1(t) + \beta x(t)\end{aligned}$$

In other words, instead of (2.1-2), we have

$$\begin{aligned}\dot{\tilde{z}}(t) &= \tilde{A}\tilde{z}(t) + \tilde{\boldsymbol{b}}x(t)\\ y(t) &= (1\ \ 0)\tilde{z}(t)\end{aligned} \tag{2.1-3}$$

where

$$\tilde{A} = \begin{pmatrix}0 & 1\\ \alpha_2 & \alpha_1\end{pmatrix} \qquad \tilde{\boldsymbol{b}} = \begin{pmatrix}0\\ \beta\end{pmatrix}$$

As a matter of fact, any nonsingular 2×2 matrix T may be used to define a new state vector and a new state-space representation of (2.1-1).

For example, from (2.1-2), let $z = T\bar{z}$. Then

$$\dot{\bar{z}}(t) = \bar{A}\bar{z}(t) + \bar{b}x(t) \qquad (2.1\text{-}4)$$
$$y(t) = [1 \quad 0]\,T\,\bar{z}$$

where $\bar{A} = T^{-1}AT$ and $\bar{b} = T^{-1}b$. Equations (2.1-2) and (2.1-4) are called equivalent representations.

Let T be

$$T = \begin{pmatrix} 1 & 0 \\ -\alpha_1 & 1 \end{pmatrix}$$

Then

$$T^{-1} = \begin{pmatrix} 1 & 0 \\ \alpha_1 & 1 \end{pmatrix}$$

We obtain

$$T^{-1}AT = \begin{pmatrix} 0 & 1 \\ \alpha_2 & \alpha_1 \end{pmatrix}$$

and

$$T^{-1}b = \begin{pmatrix} 0 \\ \beta \end{pmatrix}$$

which show that (2.1-2) and (2.1-3) are indeed equivalent.

2.1.2 Difference Equations

We can perform the analogous transformation for systems of difference equations. Consider a second-order difference equation†

$$y_t = \alpha_1 y_{t-1} + \alpha_2 y_{t-2} + \beta x_{t-1} \qquad (2.1\text{-}5)$$

where the y and x variables are all scalars.

Let us define‡ a two-dimensional vector z_t with components $z_1(t)$ and $z_2(t)$ by

$$z_1(t) = y_t$$
$$z_2(t-1) = z_1(t) - \alpha_1 y_{t-1} - \beta x_{t-1} \qquad (2.1\text{-}6)$$
$$= z_1(t) - \alpha_1 z_1(t-1) - \beta x_{t-1}$$

†Subscripts denote discrete time instants (when no confusion is likely to arise).

‡We choose to avoid lagged instruments appearing in the state-space equation (2.1-7), since such expressions are undesirable for numerical reasons.

or

$$z_1(t) = \alpha_1 z_1(t-1) + z_2(t-1) + \beta x_{t-1}$$

then, from (2.1-6),

$$\begin{aligned} z_2(t) &= z_1(t+1) - \alpha_1 z_1(t) - \beta x_t \\ &= y_{t+1} - \alpha_1 y_t - \beta x_t \\ &= \alpha_2 y_{t-1} \end{aligned}$$

Thus, we have

$$z_2(t) = \alpha_2 z_1(t-1)$$

In the vector notation

$$\begin{aligned} \boldsymbol{z}_t &= A\boldsymbol{z}_{t-1} + \boldsymbol{b}x_{t-1} \\ y_t &= (1 \quad 0)\boldsymbol{z}_t \end{aligned} \tag{2.1-7}$$

where†

$$A = \begin{pmatrix} \alpha_1 & 1 \\ \alpha_2 & 0 \end{pmatrix} \qquad \boldsymbol{b} = \begin{pmatrix} \beta \\ 0 \end{pmatrix}$$

Equation (2.1-7) is a state-space representation of (2.1-5).

As in the differential equation case, any nonsingular matrix T can be used to define a new state vector

$$\boldsymbol{z}_t = T\tilde{\boldsymbol{z}}_t$$

In terms of $\tilde{\boldsymbol{z}}_t$, we have

$$\begin{aligned} \tilde{\boldsymbol{z}}_t &= \tilde{A}\tilde{\boldsymbol{z}}_{t-1} + \tilde{\boldsymbol{b}}x_{t-1} \\ y_t &= (1 \quad 0)T\tilde{\boldsymbol{z}}_t \end{aligned} \tag{2.1-8}$$

where

$$\tilde{A} = T^{-1}AT \quad \text{and} \quad \tilde{\boldsymbol{b}} = T^{-1}\boldsymbol{b}$$

Equations (2.1-7) and (2.1-8) are equivalent.

---

†We write the components of z_t as z_{1t} and z_{2t}, instead of $z_1(t)$ and $z_2(t)$.

Example 1 Let y_t be the change in the national output in the tth period (or quarter)

$$y_t = Y_t - Y_{t-1}$$

where Y_t is the output during the tth quarter. Similarly, let m_t be the change in the money stock and Δr_t be the change in the interest rate.

Often, economists work with linear difference equations such as

$$\begin{aligned} y_t &= a_1 y_{t-1} + a_2 y_{t-2} + a_3 m_{t-1} \\ \Delta r_t &= b_1 y_t + b_2 m_t \end{aligned} \tag{2.1-9}$$

which are the result of regression analysis on such time-series data as the GNP statistics. We will put (2.1-9) into a state-space form.

Let us define

$$\begin{aligned} z_1(t) &= y_t \\ z_2(t-1) &= z_1(t) - a_1 y_{t-1} - a_3 m_{t-1} \end{aligned}$$

A state-space representation of this macroeconomic model is then

$$\begin{aligned} \begin{pmatrix} z_1(t) \\ z_2(t) \end{pmatrix} &= \begin{pmatrix} a_1 & 1 \\ a_2 & 0 \end{pmatrix} \begin{pmatrix} z_1(t-1) \\ z_2(t-1) \end{pmatrix} + \begin{pmatrix} a_3 \\ 0 \end{pmatrix} m_{t-1} \\ \Delta r_t &= (b_1 \quad 0) z(t) + b_2 m_t \end{aligned} \tag{2.1-10}$$

The ultimate use of any model determines which of these two representations or any of other equivalent representations is to be used.

For example, if the time path to exogenously changing money supply is of interest, it may be desirable to have A in the diagonal form. Consider (2.1-10). The eigenvalues are the roots of the characteristic equation

$$\det\begin{pmatrix} \lambda - a_1 & -1 \\ -a_2 & \lambda \end{pmatrix} = \lambda^2 - a_1\lambda - a_2 = 0$$

Denote the eigenvalues as λ_1 and λ_2.

We have

$$\lambda_1 + \lambda_2 = a_1 \quad \text{and} \quad \lambda_1\lambda_2 = -a_2$$

The matrix A can therefore be written as

$$A = \begin{pmatrix} \lambda_1 + \lambda_2 & 1 \\ -\lambda_1\lambda_2 & 0 \end{pmatrix}$$

If the two eigenvalues are distinct, there are two linearly independent eigenvectors, for example,

$$\boldsymbol{u}_1 = \begin{pmatrix} 1 \\ a_2/\lambda_1 \end{pmatrix} = \begin{pmatrix} 1 \\ -\lambda_2 \end{pmatrix} \quad \text{and} \quad \boldsymbol{u}_2 = \begin{pmatrix} 1 \\ a_2/\lambda_2 \end{pmatrix} = \begin{pmatrix} 1 \\ -\lambda_1 \end{pmatrix}$$

are the eigenvectors with eigenvalue λ_1 and λ_2, respectively. Let

$$T = \begin{pmatrix} 1 & 1 \\ -\lambda_2 & -\lambda_1 \end{pmatrix}$$

Verify that $T^{-1}AT$ is indeed equal to the diagonal matrix diag(λ_1, λ_2). Choosing the new state vector to be

$$\begin{aligned} \bar{z}(t) &= T^{-1}z_t = \frac{1}{\lambda_2 - \lambda_1} \begin{pmatrix} -\lambda_1 & -1 \\ \lambda_2 & 1 \end{pmatrix} \begin{pmatrix} z_{1t} \\ z_{2t} \end{pmatrix} \\ &= \frac{1}{\lambda_2 - \lambda_1} \begin{pmatrix} -(\lambda_1 z_{1t} + z_{2t}) \\ \lambda_2 z_{1t} + z_{2t} \end{pmatrix} \end{aligned}$$

the corresponding state-space representation is

$$\bar{z}(t) = \begin{pmatrix} \lambda_1 & 0 \\ 0 & \lambda_2 \end{pmatrix} \bar{z}(t-1) + \frac{a_3}{\lambda_2 - \lambda_1} \begin{pmatrix} -\lambda_1 \\ \lambda_2 \end{pmatrix} m_{t-1} \qquad (2.1\text{-}11)$$

In terms of the original variables, the new state vector is related to y_t and y_{t-1} by

$$\bar{z}_1(t) = -\lambda_1(y_t - \lambda_2 y_{t-1})/(\lambda_2 - \lambda_1) \qquad (2.1\text{-}12)$$

and

$$\bar{z}_2(t) = \lambda_2(y_t - \lambda_1 y_{t-1})/(\lambda_2 - \lambda_1)$$

$$\Delta r_t = b_1(1 \quad 1)\bar{z}(t) + b_2 m_t = b_1 y_t + b_2 m_t$$

Equation (2.1-11) is an equivalent representation of the system (2.1-9) as is equation (2.1-10). Equation (2.1-11) is easier to use to calculate the time path.

$$\bar{z}(t) = A^t \bar{z}(0) + \frac{a_3}{\lambda_2 - \lambda_1} \sum_{j=0}^{t-1} A^{t-1-j} \begin{pmatrix} -\lambda_1 \\ \lambda_2 \end{pmatrix} m_j$$

or in terms of the components

$$\bar{z}_1(t) = \lambda_1{}^t\bar{z}_1(0) - \frac{a_3}{\lambda_2 - \lambda_1} \sum_{j=0}^{t-1-j} \lambda_1{}^{t-j} m_j$$

$$\bar{z}_2(t) = \lambda_2{}^t\bar{z}_2(0) + \frac{a_3}{\lambda_2 - \lambda_1} \sum_{j=0}^{t-1-j} \lambda_2{}^{t-j} m_j$$

where the initial conditions are obtainable from (2.1-12).

Example 2 (Business Cycle) We analyze the business cycle in the state-space form. We use the second-order difference equation describing the business cycle derived from Samuelson (1939).

Consider the saving and investment behavior given by

$$C_t = \alpha Y_{t-1} \tag{2.1-13}$$

$$I_t = \beta (C_t - C_{t-1}) \tag{2.1-14}$$

where

$$Y_t = C_t + I_t + g_t \tag{2.1-15}$$

The consumption C_t during the tth period (say quarter) depends on the output of the $(t-1)$th quarter Y_{t-1}. The investment during the tth quarter I_t is given by the behavioral equation (2.1-14) and the output Y_t is as given by (2.1-15) where g_t is the government expenditure on goods and services during the tth quarter.

Substituting (2.1-13) and (2.1-14) into (2.1-15), we obtain the second-order linear difference equation of output

$$Y_t = \alpha(1 + \beta)Y_{t-1} - \alpha\beta Y_{t-2} + g_t \tag{2.1-16}$$

Unlike (2.1-5), in this equation Y_t is influenced by the instrument (government expenditure) of the same time period. Define

$$z_1(t) = Y_t - g_t \tag{2.1-16$'$}$$

and

$$z_2(t-1) = z_1(t) - \alpha(1 + \beta)Y_{t-1}$$

Then

$$z_1(t) = \alpha(1 + \beta)z_1(t-1) + z_2(t-1) + \alpha(1 + \beta)g_{t-1}$$

and

$$z_2(t) = z_1(t+1) - \alpha(1+\beta)Y_t$$
$$= -\alpha\beta Y_{t-1} = -\alpha\beta z_1(t-1) - \alpha\beta g_{t-1}$$

The state-space representation of the dynamics of business cycle is

$$\mathbf{z}_t = A\mathbf{z}_{t-1} + \boldsymbol{b}g_{t-1} \tag{2.1-17}$$

where

$$A = \begin{pmatrix} \alpha(1+\beta) & 1 \\ -\alpha\beta & 0 \end{pmatrix} \qquad \boldsymbol{b} = \begin{pmatrix} \alpha(1+\beta) \\ -\alpha\beta \end{pmatrix}$$

and the state vector $\mathbf{z}(t)$ is related to the output by

$$Y_t = z_1(t) + g_t \tag{2.1-18}$$

Alternately, let

$$\hat{z}_1(t) = Y_{t-1} \quad \text{and} \quad \hat{z}_2(t) = Y_t \tag{2.1-18$'$}$$

Then instead of (2.1-17), we have a representation

$$Y_t = (0 \quad 1)\hat{\mathbf{z}}_t \tag{2.1-19}$$

$$\hat{\mathbf{z}}_t = \hat{A}\hat{\mathbf{z}}_{t-1} + \hat{\boldsymbol{b}}g_t \tag{2.1-20}$$

where

$$\hat{A} = \begin{pmatrix} 0 & 1 \\ a_2 & a_1 \end{pmatrix} \qquad \hat{\boldsymbol{b}} = \begin{pmatrix} 0 \\ 1 \end{pmatrix}$$

and where

$$a_1 = \alpha(1+\beta) \qquad a_2 = -\alpha\beta$$

Note that $\mathbf{z}_t$ defined by (2.1-16′) excludes g_t while $\hat{\mathbf{z}}_t$ of (2.1-18′) includes g_t. These differences are reflected in (2.1-18) and (2.1-19).

2.2 *n*th ORDER EQUATION

When the equations do not involve derivatives of the instruments or lagged instruments, conversion into state-space form can be carried out very easily as the next two examples show.

Given

$$y^{(n)} + a_1(t)y^{(n-1)} + \cdots + a_n(t)y = x(t) \tag{2.2-1}$$

define an n-vector $z = (z_1 \quad z_2 \quad \cdots \quad z_n)^t$ with $z_1 = y, z_2 = \dot{y}, \ldots, z_n = y^{(n-1)}$.

Then the equation is converted into the first-order differential equation for z

$$\dot{z} = \frac{d}{dt}\begin{pmatrix} z_1 \\ \vdots \\ z_n \end{pmatrix} = A(t)z + \boldsymbol{b}x(t)$$

$$y = (1 \quad 0 \quad \cdots \quad 0)z$$

where

$$A(t) = \begin{bmatrix} 0 & 1 & 0 & \cdots & 0 & 0 \\ & & & \vdots & & \\ 0 & 0 & 0 & \cdots & 0 & 1 \\ -a_n(t) & -a_{n-1}(t) & \cdots & -a_2(t) & -a_1(t) & \end{bmatrix} \quad \text{and} \quad \mathbf{b} = \begin{bmatrix} 0 \\ \vdots \\ 0 \\ 1 \end{bmatrix}$$

Therefore, we can always work with the latter form when a differential equation for a scalar variable is given and the forcing (driving) term does not involve derivatives of the instruments. Note that A or $\boldsymbol{b}$ need not be restricted to the special structure shown, since, with any nonsingular matrix T, the differential equation for $\hat{z}$ where $z = T\hat{z}$, is $\dot{\hat{z}} = T^{-1}AT\hat{z} + T^{-1}\boldsymbol{b}x(t)$ and $T^{-1}AT$ and $T^{-1}\boldsymbol{b}$ may not have the special forms shown above.

Next consider a difference equation with n lags,

$$y_t = \alpha_1(t)y_{t-1} + \alpha_2(t)y_{t-2} + \cdots + \alpha_n(t)y_{t-n} + \beta(t)x_{t-1} \qquad (2.2\text{-}2)$$

Let

$$z_1(t) = y_{t-n+1}, \quad \ldots, \quad z_{n-1}(t) = y_{t-1}, \quad z_n(t) = y_t$$

Then the difference equation involving n lags is put into that with a single lag

$$z_t = A_t z_{t-1} + \boldsymbol{b}_t x_{t-1}$$

$$y_t = (0 \quad 0 \quad \cdots \quad 1)z_t$$

where

$$z_t = [z_1(t) \quad \cdots \quad z_n(t)]^t$$

is an n-dimensional state vector, and where

$$A_t = \begin{bmatrix} 0 & 1 & 0 & \cdots & 0 & 0 \\ 0 & 0 & 1 & \cdots & 0 & 0 \\ & & & \vdots & & \\ 0 & 0 & 0 & \cdots & 0 & 1 \\ \alpha_n(t) & \alpha_{n-1}(t) & \alpha_{n-2}(t) & \cdots & \alpha_2(t) & \alpha_1(t) \end{bmatrix} \qquad \boldsymbol{b}_t = \begin{bmatrix} 0 \\ 0 \\ \vdots \\ 0 \\ \beta(t) \end{bmatrix}$$

Again A_t and $\boldsymbol{b}_t$ may not in general have the special structures shown here. When the right hand side of the differential equations (2.2-1) or the difference equations (2.2-2) contain derivatives or lagged values of the instruments, the simple-minded transformations of this section result in the appearance of derivatives or lagged values of the instruments in the state-space equations.

We extend the procedure introduced in §2.1 for second-order systems to avoid their appearances in the state-space equations. Given

$$\begin{aligned} y^{(n)}(t) &= \alpha_1 y^{(n-1)}(t) + \cdots + \alpha_n y(t) \\ &\quad + \beta_0 x^{(n)}(t) + \beta_1 x^{(n-1)}(t) + \cdots + \beta_n x(t) \end{aligned} \tag{2.2-3}$$

where some of β's may be zero, define an n-dimensional vector $z(t)$ with components $z_1(t), \ldots, z_n(t)$ by

$$\begin{aligned} z_1(t) &= y(t) - \beta_0 x(t) \\ z_2(t) &= \dot{z}_1(t) - \alpha_1 y(t) - \beta_1 x(t) \\ &= \dot{z}_1(t) - \alpha_1 z_1(t) - (\alpha_1 \beta_0 + \beta_1) x(t) \\ &= \dot{y}(t) - \alpha_1 y(t) - \beta_0 \dot{x}(t) - \beta_1 x(t) \\ z_3(t) &= \dot{z}_2(t) - \alpha_2 y(t) - \beta_2 x(t) \\ &= \dot{z}_2(t) - \alpha_2 z_1(t) - (\alpha_2 \beta_0 + \beta_2) x(t) \\ &= \ddot{y}(t) - \alpha_1 \dot{y}(t) - \alpha_2 y(t) - \beta_0 \ddot{x}(t) - \beta_1 \dot{x}(t) - \beta_2 x(t) \\ &\vdots \\ z_n(t) &= \dot{z}_{n-1}(t) - \alpha_{n-1} y(t) - \beta_{n-1} x(t) \\ &= \dot{z}_{n-1}(t) - \alpha_{n-1} z_1(t) - (\alpha_{n-1} \beta_0 + \beta_{n-1}) x(t) \end{aligned}$$

Then

$$\dot{z}_n(t) = \ddot{z}_{n-1}(t) - \alpha_{n-1}\dot{y}(t) - \beta_{n-1}\dot{x}(t) \tag{2.2-4}$$

By definition

$$z_{n-1}(t) = y^{(n-2)}(t) - \alpha_1 y^{(n-3)}(t) - \cdots - \alpha_{n-2}y(t)$$
$$- \beta_0 x^{(n-2)}(t) - \cdots - \beta_{n-2}x(t)$$

Then from (2.2-3) and (2.2-4)

$$\dot{z}_n(t) = y^{(n)}(t) - \alpha_1 y^{(n-1)}(t) - \cdots - \alpha_{n-2}\ddot{y}(t) - \alpha_{n-1}\dot{y}(t)$$
$$- \beta_0 x^{(n)}(t) - \cdots - \beta_{n-2}\ddot{x}(t) - \beta_{n-1}\dot{x}(t)$$
$$= \alpha_n y(t) + \beta_n x(t)$$
$$= \alpha_n z_1(t) + (\alpha_n\beta_0 + \beta_n)x(t)$$

Ther ‶re, collecting these components of $z(t)$, we see that

$$\dot{\boldsymbol{z}}(t) = A\boldsymbol{z}(t) + \boldsymbol{b}x(t)$$
$$y(t) = (1 \quad 0 \quad \ldots \quad 0)\boldsymbol{z}(t) + \beta_0 x(t)$$

where

$$A = \begin{pmatrix} \alpha_1 & 1 & 0 & \cdots & 0 \\ \alpha_2 & 0 & 1 & \cdots & \\ & & & \vdots & \\ \alpha_{n-1} & 0 & 0 & \cdots & 1 \\ \alpha_n & 0 & 0 & \cdots & 0 \end{pmatrix} \qquad \boldsymbol{b} = \begin{pmatrix} \gamma_1 \\ \gamma_2 \\ \vdots \\ \gamma_{n-1} \\ \gamma_n \end{pmatrix} \tag{2.2-5}$$

and where

$$\gamma_i = \alpha_i\beta_0 + \beta_i, \quad i = 1, \ldots, n$$

Consider now a linear difference equation involving n lags (nth order difference equation).

$$y_t = \alpha_1 y_{t-1} + \alpha_2 y_{t-2} + \cdots + \alpha_n y_{t-n}$$
$$+ \beta_0 x_t + \beta_1 x_{t-1} + \cdots + \beta_n x_{t-n} \tag{2.2-6}$$

where the x's and y's are as well as the coefficients α's and β's are all scalars. Some of the β's may be zero. For example, if the right hand side involves x's only up to m lags, $m \leqslant n$, then we have $\beta_{m+1} = \cdots = \beta_n = 0$. Assume $\alpha_n \neq 0$, otherwise the system does not have n lags.

We show that (2.2-6) can be rewritten by a change of variables to be discussed next as a first-order difference equation of an n-dimensional vector. Denote this vector by z. This change of variables permits us to discuss linear difference equations of all orders uniformly as the first-order difference equations of vector variables of various dimensions. This turns out to be very convenient in theoretical discussions.

We now define n scalar variables $z_1, \ldots, z_n$ which will be treated as the n components of the n-dimensional vector z.

Let

$$z_1(t) = y_t - \beta_0 x_t \tag{2.2-7}$$

Next, define

$$\begin{aligned} z_2(t-1) &= z_1(t) - \alpha_1 y_{t-1} - \beta_1 x_{t-1} \\ &= z_1(t) - \alpha_1\{z_1(t-1) + \beta_0 x_{t-1}\} - \beta_1 x_{t-1} \\ &= z_1(t) - \alpha_1 z_1(t-1) - \gamma_1 x_{t-1} \end{aligned} \tag{2.2-8}$$

where (2.2-7) is used to obtain the second line of (2.2-8) from the first and we introduce

$$\gamma_1 = \alpha_1 \beta_0 + \beta_1$$

Rearrange (2.2-8) as

$$z_1(t) = \alpha_1 z_1(t-1) + z_2(t-1) + \gamma_1 x_{t-1}$$

Proceed analogously, i.e., define z_{k+1} by

$$\begin{aligned} z_{k+1}(t-1) &= z_k(t) - \alpha_k y_{t-1} - \beta_k x_{t-1} \\ &= z_k(t) - \alpha_k\{z_1(t-1) + \beta_0 x_{t-1}\} - \beta_k x_{t-1} \\ &= z_k(t) - \alpha_k z_1(t-1) - \gamma_k x_{t-1} \end{aligned}$$

where

$$\gamma_k = \alpha_k \beta_0 + \beta_k, \quad k = 1, 2, \ldots, n-1$$

Rewrite it as

$$z_k(t) = \alpha_k z_1(t-1) + z_{k+1}(t-1) + \gamma_k x_{t-1}$$

Finally, let

$$z_n(t) = \alpha_n z_1(t-1) + \gamma_n x_{t-1}$$

where

$$\gamma_n = \alpha_n \beta_0 + \beta_n$$

Collecting them together into a form of a first-order difference equation for z, we see that (2.2-6) has been transformed into (2.2-9)

$$z_t = A z_{t-1} + b x_{t-1} \tag{2.2-9a}$$

$$y_t = (1 \quad 0 \quad \cdots \quad 0) z_t + \beta_0 x_t \tag{2.2-9b}$$

where

$$A = \begin{pmatrix} \alpha_1 & 1 & 0 & \cdots & 0 \\ \alpha_2 & 0 & 1 & \cdots & 0 \\ & & & \vdots & \\ \alpha_{n-1} & 0 & 0 & \cdots & 1 \\ \alpha_n & 0 & 0 & \cdots & 0 \end{pmatrix}, \quad b = \begin{pmatrix} \gamma_1 \\ \gamma_2 \\ \vdots \\ \gamma_{n-1} \\ \gamma_n \end{pmatrix}$$

and where

$$\gamma_i = \alpha_i \beta_0 + \beta_i, \qquad i = 1, \ldots, n$$

$$z_t = (z_1(t), z_2(t), \ldots, z_n(t))^{\mathrm{t}}$$

Note that (2.2-9b) shows how the vector z_t is related to the original variables y_t and x_t.

Next, we verify that the procedure is reversible. Given (2.2-9), it can be put into the nth order linear difference equation for a scalar variable. We can start from (2.2-9b)

$$y_t = z_1(t) + \beta_0 x_t$$

From (2.2-9a), we see that $z_1(t)$ is expressible in terms of $z_1(t-1)$, $z_2(t-1)$, etc., as

$$\begin{aligned} y_t &= \alpha_1 z_1(t-1) + z_2(t-1) + \gamma_1 x_t + \beta_0 x_t \\ &= \alpha_1 y_{t-1} + \beta_0 x_t + \beta_1 x_{t-1} + z_2(t-1) \end{aligned} \tag{2.2-10}$$

where we use $\gamma_1 = \alpha_1 \beta_0 + \beta_1$.

We know from (2.2-9a) that $z_2(t-1) = \alpha_2 z_1(t-2) + z_3(t-2) + \gamma_2 x_{t-2}$. Then substituting† this into (2.2-10) and recalling that $\gamma_2 = \alpha_2 \beta_0 + \beta_2$,

$$y_t = \alpha_1 y_{t-1} + \beta_0 x_t + \beta_1 x_{t-1} + \alpha_2 y_{t-2} + \beta_2 x_{t-2} + z_3(t-2)$$

Proceeding similarly, we see that

$$y_t = \alpha_1 y_{t-1} + \alpha_2 y_{t-2} + \cdots + \alpha_{n-1} y_{t-n+1} + \beta_0 x_t + \beta_1 x_{t-1} + \cdots + \beta_{n-1} x_{t-n+1} + z_n(t-n+1)$$

Since $z_n(t-n+1) = \alpha_n z_1(t-n) + \gamma_n x_{t-n}$, finally we see that

$$y_t = \alpha_1 y_{t-1} + \cdots + \alpha_{n-1} y_{t-n+1} + \alpha_n y_{t-n} + \beta_0 x_t + \cdots + \beta_{n-1} x_{t-n+1} + \beta_n x_{t-n}$$

This is exactly (2.2-6).

So far, x and y are both scalars. Structural equations for economic models are often represented in a form analogous to (2.2-3) in which $\boldsymbol{x}$ and $\boldsymbol{y}$ are generally vectors,

$$\boldsymbol{y}_t = A_0 \boldsymbol{y}_t + A_1 \boldsymbol{y}_{t-1} + \cdots + A_m \boldsymbol{y}_{t-m} + G_0 \boldsymbol{x}_t + G_1 \boldsymbol{x}_{t-1} + \cdots + G_j \boldsymbol{x}_{t-j} + \boldsymbol{f}_t$$

where $\boldsymbol{y}_t$ is a vector of economic variables, $\boldsymbol{x}_t$ is an instrument vector and $\boldsymbol{f}_t$ is some exogenous vector. Kendrick (1972) and Pindyck (1973), for example, use this form as the basis of their further analysis with $G_0 = 0$, and $\boldsymbol{f}_t = 0$ for all t.

If $I - A_0$ is nonsingular, we obtain the reduced form equation from it

$$\boldsymbol{y}_t = A_1 \boldsymbol{y}_{t-1} + \cdots + A_m \boldsymbol{y}_{t-m} + C_0 \boldsymbol{x}_t + C_1 \boldsymbol{x}_{t-1} + \cdots + C_m \boldsymbol{x}_{t-m} + \boldsymbol{d}_t \tag{2.2-11}$$

where some of C's may be zero and where we redefine $(I - A_0)^{-1} A_i$ as A_i and $C_i = (I - A_0)^{-1} G_i$ and $d_t = (I - A_0)^{-1} f_t$, for convenience.‡ When $(I - A_0)$ is singular, no general formula is available. See, however, Example 4. Suppose $\dim y_t = l$, $\dim x_t = r \leqslant l$. The techniques we introduced above in this section for scalar y and scalar x can be applied to

†This may be done very simply by writing expressions for $z_j(t-1)$, $j = 2, \ldots, n$ on top of each other and cancelling like terms.

‡This transformation complicates covariance calculations when the A's are random.

dynamic systems with vector variable input-output relations such as (2.2-11).

Define m l-dimensional subvectors of the state vector as follows:

$$z_t^{\,1} = y_t - C_0 x_t - d_t \tag{2.2-12}$$

$$z_{t-1}^{\,k+1} = z_t^{\,k} - A_k y_{t-1} - C_k x_{t-1}, \qquad k = 1, 2, \ldots, m-1 \tag{2.2-13}$$

Substitute the expression for y_{t-1} from (2.2-12) into (2.2-13). Then we see that

$$z_{t-1}^{\,k+1} = z_t^{\,k} - A_k z_{t-1}^{\,1} - (A_k C_0 + C_k) x_{t-1} - A_k d_{t-1}$$

or

$$z_t^{\,k} = A_k z_{t-1}^{\,1} + z_{t-1}^{\,k+1} + (A_k C_0 + C_k) x_{t-1} + A_k d_{t-1},$$
$$k = 1, \ldots, m-1 \tag{2.2-14}$$

Eliminating the intermediate z's from (2.2-13), we see that the subvectors z's are related to y_t, x_t and their lagged values as follows

$$\begin{aligned} z_t^{\,1} &= y_t - C_0 x_t - d_t \\ z_{t-1}^{\,2} &= y_t - A_1 y_{t-1} - C_0 x_t - C_1 x_{t-1} - d_t \\ &\vdots \\ z_{t-m+1}^{\,m} &= y_t - A_1 y_{t-1} - \cdots - A_{m-1} y_{t-m+1} \\ &\quad - C_0 x_t - \cdots - C_{m-1} x_{t-m+1} - d_t \end{aligned}$$

From (2.1-11), the last equation may be written as

$$z_{t-m+1}^{\,m} = A_m y_{t-m} + C_n x_{t-m}$$

or

$$\begin{aligned} z_t^{\,m} &= A_m y_{t-1} + C_m x_{t-1} \\ &= A_m z_{t-1}^{\,1} + (A_m C_0 + C_m) x_{t-1} + A_m d_{t-1} \end{aligned} \tag{2.2-15}$$

This equation completes the dynamic relations of (2.2-14). Define

$$z_t = \begin{bmatrix} z_t^{\,1} \\ \vdots \\ z_t^{\,m} \end{bmatrix} : ml\text{-vector}$$

Collecting terms as indicated by (2.2-14) and (2.2-15), a state-space representation is obtained

$$z_{t+1} = Fz_t + Gx_t + \begin{bmatrix} A_1 \\ \vdots \\ A_m \end{bmatrix} d_t$$

$$y_t = Hz_t + C_0 x_t + d_t \tag{2.2-16}$$

where

$$F = \begin{bmatrix} A_1 & I & 0 & \cdots & 0 \\ A_2 & 0 & I & \cdots & 0 \\ & & & \vdots & \\ A_{m-1} & 0 & 0 & \cdots & I \\ A_m & 0 & 0 & \cdots & 0 \end{bmatrix}, \quad G = \begin{bmatrix} \gamma^1 \\ \gamma^2 \\ \vdots \\ \gamma^{m-1} \\ \gamma^m \end{bmatrix}$$

where the $l \times r$ matrices γ^i are defined by

$$\gamma^i = A_i C_0 + C_i, \quad i = 1, \ldots, m$$

and

$$H = (I \quad 0 \quad \cdots \quad 0)$$

The procedure developed in this section for putting differential or difference equations into state-space forms results in the state space with the minimal dimension.† It works for systems with more than one instrument and with exogenous (or disturbance) terms. We illustrate this with an example.

Example 3 Kmenta and Smith (1973) considered a short-run macroeconomic model example of the following type (the coefficients are rounded-off to two digits after the decimal point). The model is given by ‡

$$\begin{aligned} Y_t = {} & (3.07Y_{t-1} - 3.66Y_{t-2} + 2.09Y_{t-3} - 0.56Y_{t-4} + 0.05Y_{t-5}) \\ & + (1.14G_t - 2.53G_{t-1} + 1.38G_{t-2} + 0.59G_{t-3} - 0.79G_{t-4} + 0.18G_{t-5}) \\ & + (0.32M_t - 0.75M_{t-1} + 0.63M_{t-2} - 0.20M_{t-3} + 0.01M_{t-4}) \\ & + (0.05L_t - 0.11L_{t-1} + 0.06L_{t-2} + 0.02L_{t-3} - 0.03L_{t-4} + 0.01L_{t-5}) \\ & + \xi_t \end{aligned}$$

†Compare, for example, Chow (1973).

‡$A_1 = 3.07, \ldots, A_5 = 0.05$, etc., in (2.2-11).

where Y_t is the gross national product in the tth quarter ($ bill.); M_t the money supply (demand deposit plus currency outside banks; $ bill.); G_t the government purchases of goods and services plus net foreign investment ($ bill.); L_t the money supply plus time deposits in commercial banks ($ bill.); and ξ_t the trend factor plus random error.

We define a state vector with components $z^1 \sim z^5$ and the instruments as follows. Let

$$\boldsymbol{x}_t = \begin{bmatrix} G_t \\ M_t \\ L_t \end{bmatrix}$$

Let

$$z_t^{\ 1} = Y_t - 1.14G_t - 0.32M_t - 0.05L_t - \xi_t$$

$$\begin{aligned} z_{t-1}^{\ 2} &= z_t^{\ 1} - 3.07Y_{t-1} + 2.53G_{t-1} + 0.75M_{t-1} + 0.11L_{t-1} \\ &= z_t^{\ 1} - 3.07(z_{t-1}^{\ 1} + 1.14G_{t-1} + 0.32M_{t-1} + 0.05L_{t-1} + \xi_{t-1}) \\ &\quad + 2.53G_{t-1} + 0.75M_{t-1} + 0.11L_{t-1} \\ &= z_t^{\ 1} - 3.07z_{t-1}^{\ 1} - \gamma_{11}G_{t-1} - \gamma_{12}M_{t-1} - \gamma_{13}L_{t-1} - 3.07\xi_{t-1} \end{aligned}$$

where

$$\begin{aligned} \gamma_{11} &= 3.07 \times 1.14 - 2.53 = 0.97 \\ \gamma_{12} &= 3.07 \times 0.32 - 0.75 = 0.23 \\ \gamma_{13} &= 3.07 \times 0.05 - 0.11 = 0.04 \end{aligned}$$

$$\begin{aligned} z_{t-1}^{\ 3} &= z_t^{\ 2} + 3.66Y_{t-1} - 1.38G_{t-1} - 0.63M_{t-1} - 0.06L_{t-1} \\ &= z_t^{\ 2} + 3.66(z_{t-1}^{\ 1} + 1.14G_{t-1} + 0.32M_{t-1} + 0.05L_{t-1} + \xi_{t-1}) \\ &\quad - 1.38G_{t-1} - 0.63M_{t-1} - 0.06L_{t-1} \\ &= z_t^{\ 2} + 3.66z_{t-1}^{\ 1} - \gamma_{21}G_{t-1} - \gamma_{22}M_{t-1} - \gamma_{23}L_{t-1} + 3.66\xi_{t-1} \end{aligned}$$

where

$$\begin{aligned} \gamma_{21} &= -3.66 \times 1.14 + 1.38 = -2.79 \\ \gamma_{22} &= -3.66 \times 0.32 + 0.63 = -0.54 \\ \gamma_{23} &= -3.66 \times 0.05 + 0.06 = -0.12 \end{aligned}$$

$$
\begin{aligned}
z_{t-1}{}^4 &= z_t{}^3 - 2.09\,Y_{t-1} - 0.59\,G_{t-1} + 0.20\,M_{t-1} - 0.02\,L_{t-1} \\
&= z_t{}^3 - 2.09\left(z_{t-1}{}^1 + 1.14\,G_{t-1} + 0.32\,M_{t-1} + 0.05\,L_{t-1} + \xi_{t-1}\right) \\
&\quad - 0.59\,G_{t-1} + 0.20\,M_{t-1} - 0.02\,L_{t-1} \\
&= z_t{}^3 - 2.09\,z_{t-1}{}^1 - \gamma_{31}G_{t-1} - \gamma_{32}M_{t-1} - \gamma_{33}L_{t-1} - 2.09\,\xi_{t-1}
\end{aligned}
$$

where

$$
\begin{aligned}
\gamma_{31} &= 2.09 \times 1.14 + 0.59 = 3.97 \\
\gamma_{32} &= 2.09 \times 0.32 - 0.20 = 0.47 \\
\gamma_{33} &= 2.09 \times 0.05 + 0.02 = 0.12
\end{aligned}
$$

$$
\begin{aligned}
z_{t-1}{}^5 &= z_t{}^4 + 0.56\,Y_{t-1} + 0.75\,G_{t-1} - 0.10\,M_{t-1} + 0.03\,L_{t-1} \\
&= z_t{}^4 + 0.56\left(z_{t-1}{}^1 + 1.14\,G_{t-1} + 0.32\,M_{t-1} + 0.05\,L_{t-1} + \xi_{t-1}\right) \\
&\quad + 0.75\,G_{t-1} - 0.01\,M_{t-1} - 0.03\,L_{t-1} \\
&= z_t{}^4 + 0.56\,z_{t-1}{}^1 - \gamma_{41}G_{t-1} - \gamma_{42}M_{t-1} - \gamma_{43}L_{t-1} + 0.56\,\xi_{t-1}
\end{aligned}
$$

where

$$
\begin{aligned}
\gamma_{41} &= -0.56 \times 1.14 - 0.75 = -1.39 \\
\gamma_{42} &= -0.56 \times 0.32 + 0.01 = -1.78 \\
\gamma_{43} &= -0.56 \times 0.05 + 0.03 = 0
\end{aligned}
$$

Finally,

$$
\begin{aligned}
z_t{}^5 &= 0.05\,Y_{t-1} + 0.18\,G_{t-1} + 0.01\,L_{t-1} \\
&= 0.05\,z_{t-1}{}^1 + \gamma_{51}G_{t-1} + \gamma_{52}M_{t-1} + \gamma_{53}L_{t-1} + 0.05\,\xi_{t-1}
\end{aligned}
$$

where

$$
\begin{aligned}
\gamma_{51} &= 0.05 \times 1.14 + 0.18 = 0.24 \\
\gamma_{52} &= 0.05 \times 0.32 = 0.16 \\
\gamma_{53} &= 0.05 \times 0.05 + 0.01 = 0.01
\end{aligned}
$$

Thus a state-space equation is given by

$$z_t = Az_{t-1} + B\mathbf{x}_{t-1} + f\xi_{t-1}$$

$$Y_t = (1 \quad 0 \quad 0 \quad 0 \quad 0)z_t + 1.14G_t + 0.32M_t + 0.05L_t + \xi_t$$

where

$$A = \begin{bmatrix} 3.07 & 1 & 0 & 0 & 0 \\ -3.66 & 0 & 1 & 0 & 0 \\ 2.09 & 0 & 0 & 1 & 0 \\ -0.56 & 0 & 0 & 0 & 1 \\ 0.05 & 0 & 0 & 0 & 0 \end{bmatrix}, \quad B = \begin{bmatrix} 0.97 & 0.23 & 0.04 \\ -2.79 & -0.54 & -0.12 \\ 3.97 & 0.47 & 0.12 \\ -1.39 & -1.78 & 0.00 \\ 0.24 & 0.16 & 0.01 \end{bmatrix}$$

$$f = \begin{bmatrix} 3.07 \\ -3.66 \\ 2.09 \\ -0.56 \\ 0.05 \end{bmatrix}$$

In obtaining (2.2-11), we assumed $I - A_0$ to be nonsingular. Even when $I - A_0$ is singular, we can sometimes apply the technique of this section to put a set of simultaneous equations into a state-space form. We consider such an example next.

Example 4 (Leontief Dynamic Input-Output Model) The dynamic input-output model is a dynamic extension of the Leontief input-output model, and is of the form

$$\mathbf{q}_t = A\mathbf{q}_t + B(\mathbf{q}_{t+1} - \mathbf{q}_t) + \mathbf{d}_t$$

where $\mathbf{q}_t$ is the vector of gross output level; A the matrix of input coefficients; B the matrix of stock (capital) coefficients; and $\mathbf{d}_t$ the vector of final demands (excluding investments).

The matrices A and B would be time-varying. Here they are assumed to be constant. In an n-sector model, they are $n \times n$ matrices.

Suppose that in some sectors of the economy, no investment takes place. This makes B singular, since some rows of B will be all zero. After renumbering the sectors, we assume, therefore, that B is of the form

$$B = \begin{pmatrix} B_{11} & B_{12} \\ 0 & 0 \end{pmatrix}$$

where B_{11} is a nonsingular $n_1 \times n_1$ matrix.

Partition $\boldsymbol{q}_t$, $\boldsymbol{d}_t$ and A conformably and assume that the matrix

$$R = I - A + B$$

is nonsingular. This assumption is usually satisfied. We wish to derive a state-space representation of this model.

The structure of B implies that $\boldsymbol{q}_{t+1}{}^2$ is given by

$$(I - A_{22})\boldsymbol{q}_{t+1}{}^2 = A_{21}\boldsymbol{q}_{t+1}{}^1 + \boldsymbol{d}_{t+1}{}^2 \tag{E1}$$

The matrix $I - A_{22} = R_{22}$ is nonsingular by assumption; hence, $\boldsymbol{q}_\tau{}^2 = R_{22}{}^{-1}(A_{21}\boldsymbol{q}_\tau{}^1 + \boldsymbol{d}_\tau{}^2)$ for $\tau = t, t+1$ are used to eliminate $\boldsymbol{q}_t{}^2$ and $\boldsymbol{q}_{t+1}{}^2$ from

$$R_{11}\boldsymbol{q}_t{}^1 + R_{12}\boldsymbol{q}_t{}^2 = B_{11}\boldsymbol{q}_{t+1}{}^1 + B_{12}\boldsymbol{q}_{t+1}{}^2 + \boldsymbol{d}_t{}^1 \tag{E2}$$

These two equations can be written together conveniently by use of two matrices

$$T_1 = \begin{pmatrix} I_{n_1} & 0 \\ 0 & 0 \end{pmatrix} \quad \text{and} \quad T_2 = \begin{pmatrix} 0 & 0 \\ 0 & I_{n_2} \end{pmatrix}$$

as

$$[B - T_2(I - A)]\boldsymbol{q}_{t+1} = [B + T_1(I - A)]\boldsymbol{q}_t - T_1\boldsymbol{d}_t - T_2\boldsymbol{d}_{t+1} \tag{E3}$$

The first n_1-components of (E3) is (E2) and the remaining n_2-components are those of (E1).

Now (E3) can be transformed into a state-space form by the method of this section since $B - T_2(I - A)$ is nonsingular. Define

$$z_t = [B - T_2(I - A)]\boldsymbol{q}_t + T_2\boldsymbol{d}_t$$

or

$$\boldsymbol{q}_t = [B - T_2(I - A)]^{-1}(z_t - T_2\boldsymbol{d}_t)$$

From (E3), the state-space dynamic equation is

$$z_{t+1} = Fz_t + G\boldsymbol{d}_t$$

where

$$F = [B + T_1(I - A)][B - T_2(I - A)]^{-1}$$

and where

$$G = -T_1 - [B + T_1(I - A)][B - T_2(I - A)]^{-1}T_2$$

In the above, we see from the structures of T_1, T_2 and B, that

$$F = \begin{pmatrix} F_{11} & F_{12} \\ 0 & 0 \end{pmatrix} \quad \text{and} \quad G = \begin{pmatrix} I & G_{12} \\ 0 & 0 \end{pmatrix}$$

In other words, the last n_2 components are decoupled from the first n_1 components. The model is not controllable as we will see in §3.2 (see also Example 3, §3.2).

Note that in some models the reduced form equation can be obtained directly from the structural equation without the apparatus developed in this section.

Example 5 Let us put into state-space form a macroeconomic model of Sargent and Wallace (1974) that incorporates Lucas' aggregate supply hypothesis (1973a). The money market is assumed to clear instantaneously.

The model is given by a set of three difference equations and an algebraic identity

Supply side: $$y_t = a_1 k_{t-1} + a_2(p_t - \rho_t) \qquad a_1, a_2 > 0 \tag{E1}$$

Demand side: $$y_t = b_1 k_{t-1} + b_2(r_t - \pi_t) + b_3 g_t \qquad b_1 > 0, b_2 < 0 \tag{E2}$$

Portfolio balance: $$m_t = p_t + c_1 y_t + c_2 r_t \qquad c_1 > 0, c_2 < 0 \tag{E3}$$

Investments: $$k_t = d_1 k_{t-1} + d_2(r_t - \pi_t) + d_3 \tau_t \qquad 0 < d_1 \leqslant 1, d_2 < 0 \tag{E4}$$

where ρ_t is the expected price that is treated as exogenously given at the moment; g_t the government expenditure; and m_t and τ_t are the instruments. The variable τ_t may be a proxy for such things as investment tax credits. All the variables are logarithms of the corresponding quantities: $k_t = \ln K_t$, $y_t = \ln Y_t$, and so on; except for the interest rate r_t and π_t which is the expected rate of inflation. To close the model we must specify how π_t and ρ_t are related to the current and past endogenous and exogenous variables. In this example, we leave this unspecified so that we may discuss later different schemes for generating π and ρ. By eliminating r_t from (E2) through (E4), we obtain the structural equation

$$Az_t = Bz_{t-1} + Cx_t + \hat{\alpha}\pi_t + \hat{\beta}\rho_t \tag{E5}$$

where

$$z_t = (y_t \quad p_t \quad k_t)^{\mathrm{t}}, \quad \boldsymbol{x}_t = (m_t \quad g_t \quad \tau_t)^{\mathrm{t}}$$

$$A = \begin{bmatrix} 1 & a_2 & 0 \\ 1 + b_2c_1/c_2 & b_2/c_2 & 0 \\ c_1d_2/c_2 & d_2/c_2 & 1 \end{bmatrix}, \quad B = \begin{bmatrix} 0 & 0 & a_1 \\ 0 & 0 & b_1 \\ 0 & 0 & d_1 \end{bmatrix}$$

$$C = \begin{bmatrix} 0 & 0 & 0 \\ b_2/c_2 & b_3 & 0 \\ d_2/c_2 & 0 & d_3 \end{bmatrix}, \quad \hat{\boldsymbol{\alpha}} = \begin{bmatrix} 0 \\ -b_2 \\ -d_2 \end{bmatrix}, \quad \hat{\boldsymbol{\beta}} = \begin{bmatrix} -a_2 \\ 0 \\ 0 \end{bmatrix}$$

The matrix A is nonsingular if and only if

$$\begin{aligned} \Delta &= a_{22} - a_{21}a_{12} \\ &= (b_2/c_2) + a_2 + (a_2b_2c_1/c_2) \neq 0 \end{aligned} \tag{E6}$$

Assuming that (E6) is satisfied, (E5) is transformed into the reduced form

$$z_t = Fz_{t-1} + Gx_t + \boldsymbol{\alpha}\pi_t + \boldsymbol{\beta}\rho_t \tag{E7}$$

where

$$F = A^{-1}B, \quad G = A^{-1}C, \quad \boldsymbol{\alpha} = A^{-1}\hat{\boldsymbol{\alpha}}$$

and where

$$\boldsymbol{\beta} = A^{-1}\hat{\boldsymbol{\beta}}$$

It is convenient to define 2-dimensional vectors composed of some elements of A

$$\boldsymbol{\gamma} = \begin{pmatrix} a_{22} \\ -a_{21} \end{pmatrix}, \quad \boldsymbol{\delta} = \begin{pmatrix} -a_{12} \\ 1 \end{pmatrix}, \quad \boldsymbol{\mu} = \begin{pmatrix} a_{31} \\ a_{32} \end{pmatrix}$$

Then, after straightforward matrix manipulation, we obtain

$$F = \begin{bmatrix} 0 & 0 & f_{13} \\ 0 & 0 & f_{23} \\ 0 & 0 & f_{33} \end{bmatrix}$$

where

$$\begin{pmatrix} f_{13} \\ f_{23} \end{pmatrix} = \frac{1}{\Delta}(a_1\boldsymbol{\gamma} + b_1\boldsymbol{\delta}) \quad \text{and} \quad f_{33} = d_1 - \frac{\boldsymbol{\mu}^{\mathrm{t}}}{\Delta}(a_1\boldsymbol{\gamma} + b_1\boldsymbol{\delta})$$

and

$$G = \begin{pmatrix} G_{11} & 0 \\ g_{31}\, g_{32} & g_{33} \end{pmatrix}$$

where

$$G_{11} = (c_{21}\delta \quad c_{22}\delta)/\Delta$$

$$g_{31} = c_{31} - \mu^t\delta c_{21}/\Delta, \quad g_{32} = -\mu^t\delta c_{22}/\Delta \quad \text{and} \quad g_{33} = d_3$$

Note that the rank $G = 2$. We also have

$$\alpha = \begin{pmatrix} -a_2\gamma \\ -d_2 + a_2\mu^t\gamma/\Delta \end{pmatrix}$$

Static Expectation Suppose that the rate of inflation π_t is generated as $(P_t - P_{t-1})/P_{t-1}$, and $\rho_t = p_{t-1}$.† From the assumption, then

$$\pi_t = (e^{p_t} - e^{p_{t-1}})/e^{p_{t-1}} = e^{p_t - p_{t-1}} - 1 \cong (p_t - p_{t-1})T$$

to the first order of approximation, where T is the interval between the $(t-1)$th and the tth period.

Since $\pi_t = Te_2{}^t z_t$, where $e_2{}^t = (0 \quad 1 \quad 0)$, the reduced equation (E7) becomes modified as

$$z_t = \hat{F}z_{t-1} + Gx_t \tag{E7$'$}$$

where

$$\hat{F} = F + T\alpha e_2{}^t + \beta e_2{}^t = \begin{bmatrix} 0 & -a_2 & 0 \\ 0 & -a_2T\gamma & (a_1\gamma + b_1\delta)/\Delta \\ 0 & (-d_2 + a_2\mu^t\gamma/\Delta)T & d_1 - \mu^t(a_1\gamma + b_1\delta)/\Delta \end{bmatrix}$$

†If people predict P_t at time $t-1$ by $P_{t-1} + \pi_{t-1}\cdot T$, then $\rho_t \cong p_{t-1} + (p_{t-1} - p_{t-2})T$. Then we must modify the state vector z_t to be $(y_t \quad p_t \quad p_{t-1} \quad k_t)^t$. There is no difficulty with this modification. Then, instead of (E7), the state equation becomes

$$\begin{pmatrix} z_t \\ p_{t-1} \end{pmatrix} = \hat{\hat{F}}\begin{pmatrix} z_{t-1} \\ p_{t-2} \end{pmatrix} + \begin{pmatrix} G \\ 0 \end{pmatrix}x_t$$

where

$$\hat{\hat{F}} = \begin{pmatrix} \hat{F} + \beta(0 \quad 1+T \quad 0) & \beta T \\ 0 \quad 1 \quad 0 & 0 \end{pmatrix}$$

Adaptive Expectation Other schemes for generating π are possible. For example, suppose

$$\pi_t = a_0\pi_{t-1} + b_0 p_t \quad \text{and} \quad \rho_t = p_{t-1} + \pi_{t-1}T \tag{E8}$$

Then we must augment the state vector z_t by π_t. From (E7), we derive the structural equation

$$\begin{pmatrix} I_3 & -\alpha \\ 0 & 1 \end{pmatrix}\begin{pmatrix} z_t \\ \pi_t \end{pmatrix} = \begin{pmatrix} F & 0 \\ b^t & a_0 \end{pmatrix}\begin{pmatrix} z_{t-1} \\ \pi_{t-1} \end{pmatrix} + \begin{pmatrix} G \\ 0 \end{pmatrix}x_t + \begin{pmatrix} \beta \\ 0 \end{pmatrix}(e_2{}^tT)\begin{pmatrix} z_{t-1} \\ \pi_{t-1} \end{pmatrix}$$

where

$$f^t = (0 \quad b_0 \quad 0)$$

and the reduced form equation becomes

$$\begin{pmatrix} z_t \\ \pi_t \end{pmatrix} = \Phi\begin{pmatrix} z_{t-1} \\ \pi_{t-1} \end{pmatrix} + \Psi x_t \tag{E9}$$

where

$$\Phi = \begin{pmatrix} I_3 & -\alpha \\ 0 & 1 \end{pmatrix}^{-1}\left(\begin{pmatrix} F & 0 \\ f^t & a_0 \end{pmatrix} + \begin{pmatrix} \beta \\ 0 \end{pmatrix}(e_2{}^tT)\right)$$

$$= \begin{pmatrix} \hat{F} + \alpha f^t & \beta T + a_0\alpha \\ f^t & a_0 \end{pmatrix}$$

where $\hat{F} = F + \beta e_2{}^t$ and

$$\Psi = \begin{pmatrix} I_3 & - & \alpha \\ 0 & & 1 \end{pmatrix}\begin{pmatrix} G \\ 0 \end{pmatrix} = \begin{pmatrix} G \\ 0 \end{pmatrix}$$

We shall examine this model again in connection with its stability and controllability in stabilization policy discussions.

Pegged Interest Rate Suppose the monetary authority decides to peg r at some specified value, letting m take whatever value is necessary to peg r at this rate. A consequence of this decision is to remove m from the set of discretionary instruments at the disposal of the government. Instead, a rule

$$\begin{aligned} m_t &= p_t + c_1 y_t + c_2 \overline{r}_t \\ &= (c_1 \quad 1 \quad 0)z_t + c_2 \overline{r}_t \end{aligned} \tag{E10}$$

is adopted.

From (E7′)

$$z_t = \hat{F} z_{t-1} + G_1 m_t + (G_2 \quad G_3)\begin{pmatrix} g_t \\ \tau_t \end{pmatrix} \tag{E11}$$

where

$$G_1 = \begin{pmatrix} c_2\delta \\ g_{31} \end{pmatrix}$$

Substitute (E10) into (E11) to obtain

$$z_t = \bar{F} z_{t-1} + (\bar{G}_2 \quad \bar{G}_3)\begin{pmatrix} g_t \\ \tau_t \end{pmatrix}$$

$$I - G_1(c_1 \quad 1 \quad 0) z_t = \hat{F} z_{t-1} + (G_2 \quad G_3)\begin{pmatrix} g_t \\ \tau_t \end{pmatrix}$$

where

$$\begin{aligned} \bar{F} &= I - G_1(c_1 \quad 1 \quad 0)^{-1}\hat{F} \\ &= \hat{F} + \frac{1}{1-d}\begin{pmatrix} c_2\delta \\ g_{31} \end{pmatrix}(0 \quad k \quad l) \end{aligned}$$

where

$$\begin{aligned} d &= (c_1 \quad 1)\delta = 1 - c_1 a_{12} \\ k &= -a_2 T(c_1 \quad 1)\gamma = -a_2 T(c_1 a_{22} - a_{21}) \\ l &= (c_1 \quad 1)(a_1\gamma + b_1\delta)/\Delta = (b_1 d - a_1 k/a_2 T)/\Delta \end{aligned}$$

2.2.1 Solution of Linear Differential Equations

Next to the scalar first-order differential equation, perhaps the simplest differential equation is that of the first-order differential equation with constant coefficient matrix A, $\dot{z} = Az$, and in which A has n linearly independent eigenvectors. Then the solution can be reduced to one of n first-order scalar equations

$$Af^i = \lambda_i f^i, \qquad i = 1, \ldots, n$$

where $f^1, \ldots, f^n$ are linearly independent eigenvectors with eigenvalues $\lambda_1, \ldots, \lambda_n$. Note that the λ's need not all be distinct. The equations can be rewritten together as

$$AT = T\Lambda$$

where $T = (f^1, \ldots, f^n)$ and $\Lambda = \mathrm{diag}(\lambda_1, \ldots, \lambda_n)$ are $n \times n$ matrices. Since the column vectors of T are linearly independent, T^{-1} exists; hence $T^{-1}AT = \mathrm{diag}(\lambda_1, \ldots, \lambda_n)$. Change the variable from z to y by $z = Ty$. Then

$$\dot{y} = T^{-1}ATy$$

or

$$\dot{y}_i = \lambda_i y_i, \qquad i = 1, \ldots, n$$

as asserted.

Their solutions are $y_i(t) = y_i(0)e^{\lambda_i t}$, $i = 1, \ldots, n$. Thus, if the eigenvalues are all negative, then $y_i(t) \to 0$ as $t \to \infty$ for every $y_i(0)$, $i = 1, \ldots, n$. When this happens for every $y_i(0)$† we say $y_i(t)$ is asymptotically stable in the large. (Refer to Chapter 4 for a general discussion of stability of solutions of linear differential or difference equations.)

We now describe a general method for solving linear differential equations with constant coefficients by solving first-order differential equations of an n-dimensional vector $\dot{z} = Az$ when the coefficient matrix A does not necessarily have n linearly independent eigenvectors.

TIME-INVARIANT DIFFERENTIAL EQUATION We first discuss the solution to a homogeneous time-invariant differential equation with no forcing terms, $x(t) \equiv 0$,

$$\dot{z}(t) = Az(t), \quad z(0) = z^0 \neq 0$$

We establish first that $\int_0^t z(s)\,ds$ is bounded from above. Then we show that $z(t)$ is equal to $e^{At}z^0$, where e^{At} is defined as the limit of $(I + At + A^2t^2/2 + \cdots)$ which exists for all finite $t \geqslant 0$.

We can integrate it once to obtain

$$z(t) = z^0 + A\int_0^t z(s)\,ds \tag{2.2-17}$$

†It does here because of the linearity.

Denote by $\| \quad \|$ the norm of vectors defined by

$$\|z(t)\| = (z^t(t)z(t))^{1/2}$$

Similarly we denote the norm† of A by $\|A\|$. It satisfies

$$\|Az(s)\| \leqslant \|A\| \cdot \|z(s)\|$$

From (2.2-17), we obtain

$$\|z(t)\| \leqslant \|z^0\| + \|A\| \int_0^t \|z(s)\|\, ds$$

or

$$\frac{\|A\|\ \|z(t)\|}{\|z^0\| + \|A\| \int_0^t \|z(s)\|\, ds} \leqslant \|A\|$$

Integrate the above to obtain

$$\ln\left(\|z^0\| + \|A\| \int_0^t \|z(s)\|\, ds \right) - \ln\|z^0\| \leqslant \|A\| t$$

or

$$\|z^0\| + \|A\| \int_0^t \|z(s)\|\, ds \leqslant \|z^0\| \exp \|A\| t \tag{2.2-18}$$

This inequality‡ shows that $\int_0^t \|z(s)\|\, ds$, and *a fortiori* $\int_0^t z(s)\, ds$, is bounded above for all finite t.

Denote this bound by M_t. For example, we may take it as

$$M_t = \frac{\|z^0\| \exp \|A\| t}{\|A\|}$$

---

†Also see Bellman (1953, 1960), Desoer (1970), or Polak and Wong (1970). See Appendix A for further detail on functions of matrices.

‡This is a special case of the Bellman–Gronwall inequality (see page 35 of Bellman (1953)).

Since

$$z(s) = z^0 + A\int_0^s z(s')\,ds', \qquad s' \leqslant t$$

we can write (2.2-17) as

$$z(t) = z^0 + A\int_0^t \left(z^0 + A\int_0^s z(s')\,ds'\right) ds$$

$$= (I + At)z^0 + A^2\int_0^t \left(\int_0^s z(s')\,ds'\right) ds$$

Repeating the above, we derive

$$z(t) = (I + At)z^0 + A^2\int_0^t \left(z^0 + A\int_0^s ds'\int_0^{s'} z(s'')\,ds''\right) ds$$

$$= (I + At + A^2t^2/2!)z^0 + A^3\int_0^t ds\int_0^s ds'\int_0^{s'} z(s'')\,ds''$$

By mathematical induction we see that

$$z(t) = (I + At + \cdots + A^k t^k/k!)z^0 + \boldsymbol{r}_k$$

where

$$\boldsymbol{r}_k = A^{k+1}\int_0^t ds_1\int_0^{s_1} ds_2 \cdots \int_0^{s_k} z(s_{k+1})\,ds_{k+1}$$

and $\boldsymbol{r}_k$ is an n-dimensional vector. From (2.2-18) we know that

$$\left\|\int_0^s z(s')\,ds\right\| \leqslant \int_0^s \|z(s')\|\,ds' \leqslant \int_0^t \|z(s')\|\,ds' \leqslant M_t$$

Then

$$\|\boldsymbol{r}_k\| \leqslant \frac{\|A\|^{k+1}t^{k+1}}{(k+1)!} M_t$$

Therefore, we conclude that for any z^0 and for any t fixed

$$\|z(t) - (I + At + \cdots + A^k t^k/k!)z^0\| \to 0 \quad \text{as } k \to \infty$$

or $z(t) = e^{At}z^0$ is the solution of the homogeneous equation, where e^{At} is defined to be the limit of the infinite series $I + At + \cdots$ (see Appendix A for further detail). Noting that $d/dt\, e^{At} = Ae^{At}$, we can verify that

$$z(t) = e^{At}z^0 + \int_0^t e^{A(t-s)}Bx(s)\,ds \tag{2.2-19}$$

is the solution of the inhomogeneous equation $\dot{z} = Az + Bx$, $z(0) = z^0$. The matrix $e^{A(t-s)}$ is called the *transition matrix* and is usually written as $\phi(t - s)$.

2.2.2 Solution of Linear Difference Equations

We can develop solutions of linear difference equations analogously. We give only a brief discussion here. First, let us consider a homogeneous case given by

$$z_{t+1} = A(t)z_t, \quad \text{with } z_{t_0} = z^0$$

and where $A(t)$ is some $n \times n$ matrix. The solution is

$$z_t = \phi_{t,t_0} z_0$$

where†

$$\begin{aligned} \phi_{t_0,t_0} &= I \\ \phi_{t,t_0} &= A(t-1)\phi_{t-1,t_0}, \qquad t > t_0 \\ &= A(t-1)A(t-2)\cdots A(t_0) \end{aligned} \tag{2.2-20}$$

When the matrix $A(t)$ is constant, (2.2-20) becomes simpler $\phi_{t,t_0} = A^{t-t_0}$. This can be verified by mathematical induction in a straightforward way.

For the inhomogeneous case given by

$$z_{t+1} = A(t)z_t + B(t)x_t$$

†For difference equations we have an index set of $I = \{t_0, t_1, t_2, \ldots\}$ over which the subscript t ranges, i.e., $t \in I$. We usually do not explicitly show what I is.

the solution is

$$z_t = \phi_{t, t_0} z_{t_0} + \sum_{s=t_0}^{t-1} \phi_{t, s+1} B(s) x_s \tag{2.2-21}$$

The matrices ϕ_{t, t_0} and $\phi_{t, s}$ in (2.2-20) and (2.2-21) are called *transition matrices* and have the properties (Polak and Wong (1970))

$$\phi_{t_0, t_0} = I \quad \text{and} \quad \phi_{t, s}\phi_{s, t_0} = \phi_{t, t_0}$$

2.3 TRANSFORM OF THE STATE-SPACE EQUATION†

Given a continuous time-invariant linear dynamic system in a state-space representation

$$\dot{z}(t) = Az(t) + Bx(t) \tag{2.3-1}$$

$$y(t) = Cz(t) + Dx(t) \tag{2.3-2}$$

it is sometimes convenient to express these two equations using Laplace transforms to take full advantage of the conveniences associated with the transform method, and to be able to manipulate the expressions involving derivatives algebraically.

For a scalar function of time $f(\ )$ defined on $[0, \infty)$, its Laplace transform is defined by

$$\mathcal{L}(f(t)) = \hat{f}(s) = \int_0^\infty f(t) e^{-st}\, dt$$

if the integral exists for some complex variable s. The symbols ^ or $\mathcal{L}(\ )$ are used to denote Laplace transforms. We see that the integral exists if

$$\int_0^\infty |f(t)| e^{-\sigma t}\, dt < \infty \qquad \text{for some } \sigma > 0$$

The Laplace transform of a vector-valued function of time is a vector whose components are the Laplace transforms of the component functions of time. We make frequent use of the following facts:

†Material in this section are used mostly in Chapters 2 and 5. Thus, the reader, in the first reading of this text, may go directly to §2.6.

(1) If

$$g(t) = \frac{df(t)}{dt}, \qquad 0 \leqslant t$$

then

$$\hat{g}(s) = s\hat{f}(s) - f(0)$$

(2) If

$$g(t) = \int_0^t f(\tau)\, d\tau, \qquad 0 \leqslant t$$

then

$$\hat{g}(s) = \frac{1}{s}\hat{f}(s)$$

Under some mild regularity conditions on f, we have two useful theorems:

Final Value Theorem

$$\lim_{s\to 0} s\hat{f}(s) = \lim_{t\to\infty} f(t)$$

Initial Value Theorem

$$\lim_{s\to\infty} s\hat{f}(s) = \lim_{t\to 0} f(t)$$

Assuming the existence of the Laplace transforms, (2.3-1) and (2.3-2) become

$$s\hat{z}(s) - z(0) = A\hat{z}(s) + B\hat{x}(s) \qquad (2.3\text{-}3)$$

$$\hat{y}(s) = C\hat{z}(s) + D\hat{x}(s) \qquad (2.3\text{-}4)$$

where ˆ indicates the Laplace transform.

Finally, let us recall that the Laplace transform of a convolution integral is the product of the respective Laplace transforms

$$\mathcal{L}\left(\int_0^t f(t-\tau)\, g(\tau)\, d\tau\right) = \hat{f}(s) \cdot \hat{g}(s)$$

when they exist.

Rewrite (2.3-3) as

$$(sI - A)z(s) = \hat{z}(0) + B\hat{x}(s)$$

When s is not equal to an eigenvalue of A, $(sI - A)$ is invertible, since $|sI - A| \neq 0$. Hence, we can solve it for $\hat{z}(s)$

$$\hat{z}(s) = (sI - A)^{-1}(z(0) + B\hat{x}(s)), \tag{2.3-5}$$

when s is not an eigenvalue of A.

We know that

$$\frac{d}{dt} e^{At} = Ae^{At}$$

Let

$$\mathcal{L}(e^{At}) = \hat{F}(s)$$

Then from the above

$$s\hat{F}(s) - I = A\hat{F}(s)$$

or

$$(sI - A)\hat{F}(s) = I$$

So we see that $(sI - A)^{-1}$ is nothing but the Laplace transform of e^{At}. Equation (2.3-5) is the Laplace transform of the solution of the inhomogeneous differential equation (2.2-19). Substituting (2.3-5) into (2.3-4), we obtain the relation that connects the Laplace transform of the instruments to that of the outputs (targets) directly

$$\hat{y}(s) = C(sI - A)^{-1}(z(0) + B\hat{x}(s)) + D\hat{x}(s) \tag{2.3-6}$$

What has been accomplished, in effect, is to eliminate the state vector and to convert the state-space representation into the input–output representation of the system (see also §2.4 and §5.2).

2.3.1 Computation of $(sI - A)^{-1}$

From matrix theory, we know

$$(sI - A)^{-1} = \frac{\operatorname{adj}(sI - A)}{|sI - A|}$$

where adj ( ) denotes the adjoint matrix. The determinant $|sI - A|$ has the same expression as the characteristic equation of A, $|\lambda I - A|$, when s is substituted for λ. Thus it is an nth degree polynomial in s. Let

$$\rho(s) = |sI - A| = s^n + d_1 s^{n-1} + \cdots + d_n$$

From the identity

$$|sI - A| = \text{adj}(sI - A) \cdot (sI - A)$$

we see that the adjoint matrix $\text{adj}(sI - A)$ is a polynomial of degree of at most $n - 1$, which takes the form

$$\text{adj}(sI - A) = B_0 s^{n-1} + B_1 s^{n-2} + \cdots + B_{n-1}$$

where $B_0, \ldots, B_{n-1}$ are $n \times n$ matrices.

Substituting this into the identity and equating like powers of s, we can express $B_0, \ldots, B_{n-1}$ as

$$\begin{aligned} B_0 &= I \\ B_1 &= B_0 A + d_1 I \\ &\vdots \\ B_{n-1} &= B_{n-2} A + d_{n-1} I \\ 0 &= B_{n-1} A + d_n I \end{aligned}$$

If we know $d_1, \ldots, d_n$, then we can determine the B's from these.

The iterative technique for determining d's and B's to compute $(sI - A)^{-1}$ is known variously as the method of Souriau or of Leverrier (see, for example, Fadeeva (1955) or Zadeh-Desoer (1963)).

These identities between the B's and d's may be used to prove the Cayley-Hamilton Theorem for a general matrix A.

Cayley–Hamilton Theorem *For any $n \times n$ matrix A of real numbers, $\rho(A)$ is the $n \times n$ null matrix where*

$$\rho(\lambda) = |\lambda I - A|$$

By back substitution, we have

$$\begin{aligned} 0 &= d_n I + B_{n-1} A \\ &= d_n I + (d_{n-1} I + B_{n-2} A) A \\ &= d_n I + d_{n-1} A + d_{n-2} A^2 + B_{n-3} A^3 \\ &\vdots \\ &= d_n I + d_{n-1} A + \cdots + d_2 A^{n-1} + A^n \\ &= \rho(A) \end{aligned}$$

as was to be proved.

2.3.2 Operator Notation

Expressions such as (2.2-1) and (2.2-3) can be conveniently expressed if the usual conventions involving the differential operator or lag operator are followed.

Let p stand for d/dt so that the process of taking the derivative of $y(t)$ is interpreted as applying an operator p to the function y

$$py(t) = \frac{dy(t)}{dt}$$

Similarly, we define $p^2, \ldots, p^n$ as operators which when applied to $y(t)$ produce higher-order derivatives of y such as

$$p^2y(t) = \frac{d^2y(t)}{dt^2}$$
$$\vdots$$
$$p^ny(t) = \frac{d^ny(t)}{dt^n}$$

Let L be an operator† which when applied to a sequence $\{y_t\}$ produces

$$Ly_t = y_{t-1}$$
$$\vdots$$
$$L^ny_n = y_{t-n}$$

We can formally define polynomials in p or L. For example, let

$$M(p) = p^n + \alpha_1 p^{n-1} + \cdots + \alpha_n$$

be a polynomial in p of degree n, called a polynomial operator. Then, $M(p)y$ is a short-hand notation for $y^{(n)} + \alpha_1 y^{(n-1)} + \cdots + \alpha_n y$. Similarly, $M(L)y_t$ is a short-hand notation for $y_{t-n} + \alpha_1 y_{t-n+1} + \cdots + \alpha_n y_t$.

We note a formal similarity of $M(p)$ with the Laplace transform of $y^{(n)} + \alpha_1 y^{(n-1)} + \cdots + \alpha_n y$, since the latter is $M(s)\hat{y}(s)$ except for terms due to the initial conditions, where $M(s)$ is the same nth order polynomial $M(p)$, with p replaced by s.

†This is known as the lag operator in econometrics.

We can define algebraic operations for polynomial operators. For example, given two polynomial operators in p, $M(p)$ and $N(p)$, their sum $M(p) + N(p)$ and their product $M(p)N(p)$ are also polynomial operators which are defined by

$$[M(p) + N(p)]y = M(p)y + N(p)y$$

and

$$[M(p)N(p)]y = M(p)N(p)y$$

respectively, for any suitably differentiable function y.

Also, we understand that the equation

$$y = \frac{x}{M(p)}$$

means a function $y(t)$ which is a solution to the differential equation $M(p)y = x$. For example $y(t) = (1/p)x(t)$ defines $y(t) = \int_0^t x(\tau)\, d\tau$, since it solves the differential equation $py = x$. The solution to the differential equation $M(p)y = x$ is not uniquely determined unless initial conditions are specified. We shall consider $x/M(p)$ to represent all such solutions. With this convention, we have

$$\frac{1}{M(p)}[M(p)y] = y$$

In other words, we put $M(p)/M(p)$ equal to 1. We then have the relation

$$\frac{1}{N(p)}[M(p)y] = M(p)\left[\frac{1}{N(p)}y\right]$$

and we can write $M(p)/N(p)$ without ambiguity. An expression such as $M(p)/N(p)$, where $M(p)$ and $N(p)$ are polynomial operators, is called a (rational) transfer function in p.

Similarly, we can define a rational transfer function in L, $M(L)/N(L)$. The expression is called a rational lag form in the econometric literature.

2.3.3 Lag Operator (L-transform)

Effects of an exogenous signal (disturbance) appear distributed over time for economic systems. It is natural, therefore, to describe dynamics in terms of distributed lag or the impulse response type description.

Suppose that we have

$$y_t = \sum_{s=0}^{\infty} w_s x_{t-s} + \zeta_t$$

This type of equation involving lags of either finite or infinite order is used extensively in economics since it shows explicitly that past instruments, say, x_{t-3}, are responsible for y_t with weight w_3. In other words, $(w_3/\Sigma_0^{\infty} w_i) \times 100$ percent of y_t is caused by x_{t-3} (ignoring for the moment the effects of ζ_t). We assume that $\Sigma_{i=0}^{\infty}|w_i| < \infty$. The sequence $\{w_i\}$ is known as the impulse response or the weighting sequence (in the system literature).

It can be written concisely by using the lag operator L (introduced in subsection "Operator Notation")

$$L^s x_t = x_{t-s}, \qquad s = 0, 1, \ldots$$

as

$$y_t = w(L)x_t$$

where

$$w(L) = \sum_{s=0}^{\infty} w_s L^s$$

provided the latter is well-defined for some L.

The expression $y_t = M(L)/N(L)x_t$ then represents all solutions of the linear difference equation

$$N(L)y_t = M(L)x_t$$

unless definite initial or boundary conditions are specified.

2.3.4 Discrete-Time (ζ-) Transform

In the engineering literature, it is a common practice to use transforms of sequences in handling sequences rather than the lag operator.

For example, given a sequence $\{y_t, t = 0, 1, \ldots\}$, we define its transform formally by

$$\hat{y}(\zeta) = \sum_{t=0}^{\infty} y_t \zeta^{-t} \tag{2.3-7}$$

usually without specifying explicitly the domain of convergence of ζ.

The symbol ζ plays the role of the place marker. For example, the expression $1/(1+\zeta^{-1})$ will stand for a sequence $y_t = (-1)^t$, $t = 0, 1, \ldots$, since

$$\frac{1}{1+\zeta^{-1}} = 1 - \zeta^{-1} + \zeta^{-2} - \zeta^{-3} + \cdots$$

In the engineering literature, it is common practice to use z rather than ζ in defining the transform (that is why the transform is sometimes called the z-transform (Truxal 1955). To avoid any possible confusion with state-space vectors, which are usually denoted by z in this book, we will use ζ. We shall show shortly that ζ^{-1} corresponds to L.

Assume for convenience that this infinite sum converges in some appropriate region of the complex plane.

Then, the ζ-transform of $\{y_{t-i}\}$ for some fixed i becomes

$$\begin{aligned}\sum_{t=0}^{\infty} y_{t-i}\zeta^{-t} &= \left(\sum_{\tau=-i}^{\infty} y_\tau \zeta^{-\tau}\right)\zeta^{-i} \\ &= \hat{y}(\zeta)\zeta^{-i} + \sum_{\tau=1}^{i} y_{-\tau}\zeta^{-i+\tau}\end{aligned}$$

and

$$\begin{aligned}\sum_{t=0}^{\infty} y_{t+j}\zeta^{-t} &= \left(\sum_{\tau=j}^{\infty} y_\tau \zeta^{-\tau}\right)\zeta^{j} \\ &= \hat{y}(\zeta)\zeta^{j} - \sum_{\tau=0}^{j-1} y_\tau \zeta^{-\tau+j}\end{aligned}$$

Given a difference equation

$$y_t - \sum_{i=1}^{n} \alpha_i y_{t-i} = \sum_{i=0}^{n} \beta_i x_{t-i}$$

its ζ-transform is

$$\begin{aligned}\left(1 - \sum_{i=1}^{n} \alpha_i \zeta^{-i}\right)\hat{y}(\zeta) &= \left(\sum_{i=0}^{n} \beta_i \zeta^{-i}\right)\hat{x}(\zeta) \\ &\quad + \sum_{i=0}^{n} \alpha_i\left(y_{-1}\zeta^{-i+1} + \cdots + y_{-i}\right) \\ &\quad + \sum_{i=0}^{n} \beta_i\left(x_{-1}\zeta^{-i+1} + \cdots + x_{-i}\right)\end{aligned}$$

We write this as a relation between the ζ-transforms of two sequences of real numbers $\{y_t\}$ and $\{x_t\}$

$$\hat{y}(\zeta) = \frac{M(\zeta)}{N(\zeta)}\hat{x}(\zeta) + \frac{1}{N(\zeta)}\left\{\sum_{i=0}^{n}\alpha_i(y_{-i}\zeta^{-i+1} + \cdots + y_{-i}) + \sum_{i=0}^{n}\beta_i(x_{-i}\zeta^{-i+1} + \cdots + x_{-i})\right\}$$

where

$$M(\zeta) = \sum_{i=0}^{n}\beta_i\zeta^{-i}$$

and

$$N(\zeta) = 1 - \sum_{i=1}^{n}\alpha_i\zeta^{-i}$$

We see that ζ^{-1} corresponds to L, where L is the lag operator introduced earlier, since

$$y_t - \sum_{i=1}^{n}\alpha_i y_{t-i} = M(L)y_t$$

We shall also refer to $M(\zeta)/N(\zeta)$ as the transfer function.

When sequences of vectors are given, their ζ-transforms are vectors each component of which is the ζ-transform of its component sequence, and we speak of transfer matrices instead of transfer functions to relate two such ζ-transforms.

For example, given $\mathbf{y}_{t+1} = A\mathbf{y}_t + B\mathbf{x}_t$, its ζ-transform is

$$\zeta\hat{\mathbf{y}}(\zeta) - \zeta\mathbf{y}_0 = A\hat{\mathbf{y}}(\zeta) + B\hat{\mathbf{x}}(\zeta)$$

or solving this for $\hat{y}(\zeta)$, we obtain

$$\hat{\mathbf{y}}(\zeta) = (\zeta I - A)^{-1}B\hat{\mathbf{x}}(\zeta) + (\zeta I - A)^{-1}\zeta\mathbf{y}_0$$

The (input–output) transfer matrix is $(\zeta I - A)^{-1}B$.

The ζ-transforms have properties analogous to the Laplace transforms. For example, products of two ζ-transforms correspond to the ζ-transform of the convolution-like summation, i.e., let

$$x(\zeta) = \sum_{i=0}^{\infty}x_i\zeta^{-i}, \quad \hat{y}(\zeta) = \sum_{j=0}^{\infty}y_j\zeta^{-j}$$

Then

$$\hat{x}(\zeta)\cdot\hat{y}(\zeta) = \hat{w}(\zeta)$$

where

$$w_i = \sum_{j=0}^{i} x_{i-j}y_j, \qquad i = 0, 1, \ldots$$

We can extend the summation in the above to

$$w_i = \sum_{j=0}^{\infty} x_{i-j}y_j$$

if we define $x_{i-j} = 0$ or $i - j < 0$. See, for example, Truxal (1955) or Zadeh and Desoer (1963) for further details of the properties of the ζ-transform.

2.4 TARGET-INSTRUMENT PAIR REPRESENTATION†

2.4.1 Transfer Function

Suppose the relation between the scalar variables y and x is given by

$$y^{(n)} = \alpha_1 y^{(n-1)} + \cdots + \alpha_n y + \beta_n x + \beta_{n-1}\dot{x} + \cdots + \beta_{n-m}x^{(m)} \tag{2.4-1}$$

Take the Laplace transforms of both sides to obtain

$$\begin{aligned}(s^n - \alpha_1 s^{n-1} - \cdots - \alpha_n)\hat{y}(s) &= (\beta_{n-m}s^m + \cdots + \beta_n)\hat{x}(s)\\ &\quad + \{\text{terms due to initial conditions},\\ &\qquad x(0), \ldots, x^{m-1}(0), y(0), \ldots, y^{(n-1)}(0)\}\end{aligned}$$

†The results of this section are used mainly in Chapter 5. Thus the reader, in the first reading of this text, may go directly to §2.5.

The ratio of $\hat{y}(s)/\hat{x}(s)$ when the initial conditions are all zero is known as the *transfer function* between the instrument (or input) and the target (or output) and is denoted by

$$h(s) = \frac{\hat{y}(s)}{\hat{x}(s)} = \frac{\beta_{n-m}s^m + \cdots + \beta_n}{s^n - \alpha_1 s^{n-1} - \cdots - \alpha_n}, \qquad m \leqslant n \tag{2.4-2}$$

We know that (2.4-1) can be put into a state-space representation

$$\begin{aligned} \dot{z} &= Az + Bx \\ y &= Cz + Dx \end{aligned} \tag{2.4-2$'$}$$

When x and y are scalars as in (2.4-1), then B is a column vector, C is a row vector and D is a scalar. We can represent the dynamics of (2.4-2′) by a transfer function. Take the Laplace transform of (2.4-2′) to obtain

$$\begin{aligned} (sI - A)\hat{z}(s) &= B\hat{x}(s) + z(0) \\ \hat{y}(s) &= C\hat{z}(s) + D\hat{x}(s) \end{aligned}$$

Eliminating $\hat{z}(s)$ and assuming that $z(0) = 0$, we see that

$$h(s) = C(sI - A)^{-1}B + D \tag{2.4-3}$$

We see that $D = 0$ unless $n = m$ and $\beta_0 \neq 0$. Equating (2.4-2) with (2.4-3), we establish that

$$\frac{\beta_{n-m}s^m + \cdots + \beta_n}{s^n - \alpha_1 s^{n-1} - \cdots - \alpha_n} = C(sI - A)^{-1}B + D \tag{2.4-4}$$

where $D = 0$, if $m < n$. We factor the denominator as $(s - s_1) \cdots (s - s_n)$, where the s_i for $i = 1, \ldots, n$ are called poles of the transfer function.

From the right-hand side of (2.4-3), we see that the poles of the transfer function are the eigenvalues of A since

$$(sI - A)^{-1} = \operatorname{adj}(sI - A)/|sI - A| \tag{2.4-3$'$}$$

Conversely, suppose a transfer function

$$h(s) = \frac{\beta_0 s^n + \beta_1 s^{n-1} + \cdots + \beta_n}{s^n - \alpha_1 s^{n-1} - \cdots - \alpha_n}$$

is given. Assume that any common factor of both the numerator and the denominator have been cancelled out so that they are relatively prime (i.e., have no factors in common). We next derive a state-space representation of the system whose transfer function is $h(s)$ to show the equivalence of these two representations.

Consider the case $m = n$, that is,

$$y^{(n)} = \alpha_1 y^{(n-1)} + \cdots + \alpha_n y + \beta_0 x^{(n)} + \cdots + \beta_n x$$

Its Laplace transform is

$$(s^n - \alpha_1 s^{n-1} - \cdots - \alpha_n)\hat{y}(s) = (\beta_0 s^n + \cdots + \beta_n)\hat{x}(s) + \hat{w}(s)$$

where

$$\begin{aligned}\hat{w}(s) = {} & s^{n-1}(y(0) - \beta_0 x(0)) \\ & + s^{n-2}(\dot{y}(0) - \beta_0 \dot{x}(0) - \alpha_1 y(0) - \beta_1 x(0)) + \cdots + \\ & \left[y^{(n-1)}(0) - \beta_0 x^{(n-1)}(0) - \cdots - \alpha_{n-1} y(0) - \beta_{n-1} x(0)\right] \qquad (2.4\text{-}5)\end{aligned}$$

Thus, this equation would yield $h(s)$ as the transfer function provided the zero initial vector of the state-space representation corresponds to

$$\begin{aligned} y(0) - \beta_0 x(0) &= 0 \\ \dot{y}(0) - \beta_0 \dot{x}(0) - \alpha_1 y(0) - \beta_1 x(0) &= 0 \\ &\vdots \\ y^{(n-1)}(0) - \beta_0 x^{(n-1)}(0) - \cdots - \alpha_{n-1} y(0) - \beta_{n-1} x(0) &= 0 \end{aligned}$$

Thus, we define n components of a state-space vector z by

$$\begin{aligned} z_1(t) &= y(t) - \beta_0 x(t), && \text{so that } z_1(0) = 0 \\ z_2(t) &= \dot{z}_1(t) - \alpha_1 y(t) - \beta_1 x(t), && \text{so that } z_2(0) = 0 \\ &\vdots \\ z_n(t) &= \dot{z}_{n-1}(t) - \alpha_{n-1} y(t) - \beta_{n-1} x(t), && \text{so that } z_n(0) = 0 \end{aligned}$$

Finally,

$$z_n(t) = \alpha_n y(t) + \beta_n x(t)$$

Note that this is precisely the way the state-space vector was introduced in (2.2-10).

Eliminating $y(t)$ from the above n equations, we obtain

$$\dot{z}(t) = Az(t) + Bx(t)$$

$$y(t) = (1 \quad 0 \quad \cdots \quad 0)z(t) + \beta_0 x(t)$$

where

$$A = \begin{bmatrix} \alpha_1 & 1 & 0 & \cdots & 0 \\ \alpha_2 & 0 & 1 & \cdots & 0 \\ & & & \vdots & \\ \alpha_{n-1} & 0 & 0 & \cdots & 1 \\ \alpha_n & 0 & 0 & \cdots & 0 \end{bmatrix}, \quad \boldsymbol{b} = \begin{bmatrix} \gamma_1 \\ \gamma_2 \\ \vdots \\ \gamma_{n-1} \\ \gamma_n \end{bmatrix}, \quad \gamma_i = \alpha_i \beta_0 + \beta_i,$$

$i = 1, \ldots, n$

2.4.2 State-Space Equation via Partial Fraction Expansion

We can use the partial fraction expansion of the transfer function to obtain another equivalent state-space representation of (2.4-1).

Suppose A has n distinct eigenvalues. Then

$$|sI - A| = \Pi_{i=1}^{n}(s - \lambda_i)$$

From (2.4-3) and (2.4-3′), we know the denominator of $h(s)$ is $|sI - A|$, assuming there are no common factors in the numerator and the denominator of $h(s)$. We assume that $D = 0$. Let us expand $h(s)$ into the partial fractions.

$$h(s) = \frac{N(s)}{\Pi(s - \lambda_i)} = \sum_{i=1}^{n} \frac{\gamma_i}{s - \lambda_i}$$

where from (2.4-4), $N(s) = C \operatorname{adj}(sI - A)B$ and

$$\gamma_i = \lim_{s \to \lambda_i} \frac{(s - \lambda_i)N(s)}{\Pi(s - \lambda_j)} = \frac{N(\lambda_i)}{\prod_{j \neq i} (\lambda_i - \lambda_j)}$$

We see then that by (2.4-2)

$$\hat{y}(s) = \sum_{i=1}^{n} \hat{z}_i(s)$$

where

$$\hat{z}_i(s) = \frac{\gamma_i}{s - \lambda_i} \hat{x}(s)$$

or in the time domain equation

$$\dot{z}_i = \lambda_i z_i + \gamma_i x(t)$$

from which we can obtain a state-space representation of the system with the transfer function $h(s)$,

$$\dot{z} = \begin{pmatrix} \dot{z}_1 \\ \vdots \\ z_n \end{pmatrix} = \begin{bmatrix} \lambda_1 & 0 & \cdots & 0 & 0 \\ 0 & \lambda_2 & \cdots & 0 & 0 \\ & & \vdots & & \\ 0 & 0 & \cdots & \lambda_{n-1} & 0 \\ 0 & 0 & \cdots & 0 & \lambda_n \end{bmatrix} \begin{pmatrix} z_1 \\ \vdots \\ z_n \end{pmatrix} + \begin{pmatrix} \gamma_1 \\ \vdots \\ \gamma_n \end{pmatrix} x,$$

$$y = (1 \quad 1 \quad \cdots \quad 1) z$$

The initial condition for z can be determined by

$$z_i(0) = w_i, \qquad i = 1, \ldots, n$$

where from the initial conditions (2.4-5)

$$\frac{\hat{w}(s)}{s^n - \alpha_1 s^{n-1} - \cdots - \alpha_n} = \sum_{i=1}^{n} \frac{w_i}{s - \lambda_i}$$

This idea can obviously be extended to systems with multiple eigenvalues.

2.4.3 Closed Loop Systems

A closed loop control system is a system in which the instrument x is generated as a function of both y the actual output and y^* the desired output or target, so that $x = g(y, y^*)$. Different functions g generally affect the system quite differently. Consider a dynamic system with, for simplicity, $\dim y = \dim x = 1$. Then the transfer function $h(s)$ is a rational function in s, i.e., a ratio of two polynomials of degree at most n, where $\dim z = n$,

$$\hat{y}(s) = h(s)\hat{x}(s) + (\text{initial condition term}) \tag{2.4-6}$$

Now suppose that the instrument x is generated as a sum of the actual target variable being fedback with a transfer function $G(s)$ and an exogenous signal,

$$\hat{x}(s) = \hat{v}(s) - kG(s)\hat{y}(s)$$

where $v(t)$ is exogenously given, and where k is a scalar parameter (called gain). The second term means that it is generated by a differential equation with the forcing term $y(t)$ with the transfer function $kG(s)$. Then (2.4-6) becomes

$$\hat{y} = T(s)\hat{v}(s) \tag{2.4-7}$$

where

$$T(s) = h(s)/[1 + kh(s)G(s)]$$

is the equation relating the exogenous variable to the target variable.

Note that $T(s)$ is a rational function of s. Now, suppose the degrees of the numerators of G and h are at most equal to those of the denominators. Let

$$h(s) = N_1(s)/D_1(s) \quad \text{and} \quad G(s) = N_2(s)/D_2(s)$$

Then

$$T(s) = \frac{N_1(s)/D_1(s)}{1 + (kN_1(s)N_2(s)/D_1(s)D_2(s))} = \frac{N_1(s)D_2(s)}{D_1(s)D_2(s) + kN_1(s)N_2(s)} \tag{2.4-8}$$

is a rational function in s, hence it has at most a finite number of singularities, called poles.

First, we cancel common factors from the numerator and the denominator of T, if any. Suppose the denominator of T has poles at s_i with the multiplicity m_i, $i = 1, \ldots, n$, where n is the degree of the denominator of T. We can expand $T(s)$ into partial fractions and invert the Laplace transform expression to obtain $y(t)$. Therefore, we see that $y(t) \to 0$ as $t \to \infty$ if and only if the roots s_i $i = 1, \ldots, N$, all lie in the left half of the complex k-plane. When $N_1(s)N_2(s)$ and $D_1(s)D_2(s)$ have no common factors for $\mathrm{Re}(s) \geqslant 0$, we see from (2.4-8) that the poles of T are either given by $1 + kG(s)h(s) = 0$ or are poles of $h(s)$. (See any textbook (Truxal, 1955) on classical control systems or servomechanisms for details of stability analysis by graphical means, such as the Nyquist diagram or the Bode diagram.)

Later, we will use one graphical technique called the root locus technique. This is a method of graphically plotting the poles s_i as a function of the scalar parameter k (see Appendix D).

2.4.4 Transfer Matrix

We can generalize the idea of transfer functions and introduce the transfer matrix as the matrix relating $\hat{y}(s)$ to $\hat{x}(s)$ when initial conditions are all zero, i.e.,

$$\hat{y}(s) = H(s)\hat{x}(s)$$

where

$$H(s) = C(sI - A)^{-1}B + D$$

By means of the transfer matrices, we may compute discrepancies between the desired and actual target levels and other related variables, and evaluate a particular instrument generation policy in this way. The ideas illustrated in §5.2 are thus generalizable to vector-valued instruments and target variables.

We have seen in §2.4.3, one example of generating the instrument by feedback

$$x = k(y^* - y)$$

where y^* and y are the desired and actual target values.

When this instrument is applied to the economic model

$$\dot{z} = Az + Bx$$

$$y = Cz + Dx$$

then x can be eliminated from the above, and we obtain an expression for y and y^*. In control terminology this is known as closing the loop. The resulting system is called a closed-loop system. Whenever we adopt some transformation rule for the instrument, or for the equations for x involving the actual output y, then the system is called a closed-loop system.

The analysis of consequences of "closing the loop" is most conveniently carried out in the Laplace transform domain. Since

$$\hat{y}(s) = C(sI - A)^{-1}(z(0) + B\hat{x}(s)) + D\hat{x}(s)$$

if $x(t)$ is generated as some function of y and an exogenously given target y^* and if the Laplace transforms are related by†

$$\hat{x}(s) = G_1(s)\hat{y}(s) + G_2(s)\hat{y}^*(s)$$

then the actual target will be related to the desired target by

$$(I - HG_1)\hat{y} = HG_2\hat{y}^* + C(sI - A)^{-1}z(0)$$

Assuming that $I - HG_1$ is invertible, except possibly at some finite member of isolated points in the s-plane, we obtain

$$\hat{y} = (I - HG_1)^{-1}HG_2\hat{y}^* + C(sI - A)^{-1}z(0)$$

Then the discrepancy of the actual and the desired target values is $\hat{y}^*(s) - y(s)$.

If $\lim_{s\to 0} s(\hat{y}^*(s) - y(s)) = 0$, then by the final value theorem the actual target value will eventually reach the desired target value.

We will examine the asymptotic behavior of closed-loop dynamic systems in §3.4.

2.5 BILINEAR DYNAMIC SYSTEMS

There is a simple nonlinear dynamic equation that appears naturally in economic systems. With one scalar instrument variable x, it may be written as

$$\dot{z}(t) = (A + Bx_t)z(t) + cx_t$$

†We assume that $G_1(s)$ and $G_2(s)$ are stable so that we may ignore $y(0)$ and $y^*(0)$.

or as

$$z_{t+1} = (A + Bx_t)z_t + cx_t$$

where A and B are (possibly time-varying) matrices, and z and c are n-dimensional vectors. If $B = 0$, then the above reduce to the usual linear differential and difference equations, respectively. With $B \neq 0$, the instrument x_t multiplies the state vector z_t, making the systems nonlinear.

When there are several instruments, $x_t^1, \ldots, x_t^k$, we encounter nonlinear equations that are generalizations of the above

$$\dot{z}(t) = \left(A + \sum_{i=1}^{k} x_t^i B_i\right) z(t) + \sum_{i=1}^{k} C_i x_t^i$$

or

$$z_{t+1} = \left(A + \sum_{i=1}^{k} x_t^i B_i\right) z_t + \sum_{i=1}^{k} C_i x_t^i$$

These systems are called bilinear systems, since they are linear (actually affine) in z for x held fixed or in x with z held fixed, see, for example, Rink and Mohler (1968), Tarn *et al.* (1973).

In §3.1 we introduce a two-sector economy composed of the public and private sectors as an example of an uncontrollable dynamic system. The perturbed growth path of the economy is shown there composed of two components $\zeta_t = \xi_t + \eta_t$, where $\{\eta_t\}$ corresponds to the growth path initiated by the initial redistribution of the capital goods between the two sectors. A linearized approximation to η_t can be shown to be governed by a bilinear difference equation with two instruments, i.e., $k = 2$ in the above. We refer the reader to §3.5 for details.

Another example of a deterministic bilinear system is an industry composed of several firms, each of which adjusts its output rate in response to the market-clearing price. Such behavior may arise when the firms produce homogeneous nonstorable goods. See §7.1 and §11.2 for details. Here, we merely note that the adjustment steps of the individual firms are the control instruments of the firms, and the dynamic equation for output rate then turns out to be bilinear. The controllability of such a system was investigated by Tarn *et al.* (1973).

2.6 LINEARIZED APPROXIMATIONS OF NONLINEAR DYNAMIC SYSTEMS ABOUT REFERENCE PATH

So far we have discussed only linear dynamic systems, generally economic systems are nonlinear. We can sometimes deal with nonlinear dynamic equations directly, but very often we must resort to some approximations.

If we can treat nonlinear systems as perturbations on linear dynamic systems, then we can generally say more about the local behavior of such nonlinear systems. Very often, in order to apply the results available in linear systems and control theories, we may construct approximations to nonlinear dynamic models that are locally valid, i.e., in some small neighborhoods about the equilibrium points or about some specified reference time path. The technique of such expansions about the equilibrium points or reference functions of time are useful not only for nonlinear systems but also for linear systems. We often make use of such approximations.

There remains, however, the question of how such reference time paths *are* chosen and also of how they *should* be chosen. A reference path usually represents a compromise among the competing or conflicting goals of the policymakers or societies. The reference paths may be given implicitly as solutions to some intertemporal optimization problems. Usually, there are technical (or numerical) difficulties to be considered.† Here, we will proceed on the assumption that the reference time paths have somehow been chosen.

Suppose we have a linear dynamic system expressed as

$$\dot{y} = Ay + c \tag{2.6-1}$$

where c is a constant vector. Suppose that y^e is such that $Ay^e + c = 0$. Denote the deviation of y from y^e by z, i.e., $z = y - y^e$. Then (2.6-1) can be rewritten for z as

$$\dot{z} = A(z + y^e) + c = Az \tag{2.6-2}$$

since y^e is a constant vector. Equation (2.6-2) expresses the dynamic behavior of the system (2.6-1) about the equilibrium point. Equation (2.6-2) represents the same system as (2.6-1) with respect to the new coordinate system, the origin of which is now at y^e.

We can perform similar transformations on nonlinear dynamic systems. Let $\bar{y}(t)$ be the path corresponding to $v(t)$ for the system, governed by

$$\dot{\bar{y}}(t) = F(t, \bar{y}(t), v(t)), \quad t_0 \leqslant t \leqslant T$$

Namely, $\bar{y}(t)$ is the reference time path associated with the reference instrument time path $v(t)$. Let $\eta(t)$ be the path corresponding to $v(t) +$

†This is almost always the case with economic models with random coefficients (see, for example, Cooper and Fisher (1975)).

$\boldsymbol{x}(t)$, so that

$$\dot{\boldsymbol{\eta}}(t) = \boldsymbol{F}(t, \boldsymbol{\eta}(t), \boldsymbol{v}(t) + \boldsymbol{x}(t)), \qquad t_0 \leqslant t \leqslant T; \boldsymbol{\eta}(t_0) = \bar{\boldsymbol{y}}(t_0)$$

Define $\boldsymbol{y}(t)$ to be the difference $\boldsymbol{\eta}(t) - \bar{\boldsymbol{y}}(t)$. Then, it is governed by

$$\begin{aligned}\dot{\boldsymbol{y}}(t) &= \dot{\boldsymbol{\eta}}(t) - \dot{\bar{\boldsymbol{y}}}(t) \\ &= \boldsymbol{F}(t, \bar{\boldsymbol{y}}(t) + \boldsymbol{y}(t), \boldsymbol{v}(t) + \boldsymbol{x}(t)) - \boldsymbol{F}(t, \bar{\boldsymbol{y}}(t), \boldsymbol{v}(t)) \\ &\qquad t_0 \leqslant t_0 \leqslant T; \quad \boldsymbol{y}(t_0) = 0\end{aligned}$$

Define the right-hand side as $\boldsymbol{h}(t, \boldsymbol{y}(t), \boldsymbol{x}(t))$, suppressing the dependence on $\bar{\boldsymbol{y}}$ and $\boldsymbol{v}$ since they are known functions of time. Then we have

$$\dot{\boldsymbol{y}}(t) = \boldsymbol{h}(t, \boldsymbol{y}(t), \boldsymbol{x}(t)), \qquad t_0 \leqslant t \leqslant T \tag{2.6-3}$$

with

$$\boldsymbol{h}(t, 0, 0) = 0, \qquad t_0 \leqslant t \leqslant T$$

The initial condition $\boldsymbol{y}(t_0)$ need not be zero if $\boldsymbol{\eta}(t_0)$ is taken to be different from $\bar{\boldsymbol{y}}(t_0)$. The nonlinear differential equation (2.6-3) describes the motion of a control system about the reference path $\bar{\boldsymbol{y}}(t)$. Typically in this kind of approximation, $\boldsymbol{x}(t)$ is small in some sense, e.g. $\|\boldsymbol{x}(t)\| < \sigma$ for all t in $[t_0, T]$ for some $T > t_0$. If $\boldsymbol{y}(t)$ remains small also,† then we may approximate (2.6-3) by linear equations. We obtain the associated linear

---

†We need some conditions to rule out the possibility that the time path $\boldsymbol{y}$ diverges to ∞ in a finite time interval. One common condition is to require that $\boldsymbol{y}'\boldsymbol{h}(t, \boldsymbol{y}, \boldsymbol{x}) < c(1 + \|\boldsymbol{y}\|^2)$ for some c and for all t, $\boldsymbol{y}$ and $\boldsymbol{x}$ in some compact set in R^m, $\dim \boldsymbol{y} = m$. Any other conditions to restrict the path to a compact subset of (t, z) space will do. To see this, note that from (2.6-3) we have

$$\boldsymbol{y}'\dot{\boldsymbol{y}} \leqslant c(1 + \|\boldsymbol{y}\|^2)$$

or

$$\frac{d}{dt}\|\boldsymbol{y}\|^2 \leqslant 2c(1 + \|\boldsymbol{y}\|^2)$$

or integrating the above

$$\|\boldsymbol{y}(t)\| \leqslant (1 + \|\boldsymbol{y}^0\|^2)e^{2c(t-t_0)}, \qquad t_0 \leqslant t \leqslant T$$

Thus, once $T \geqslant t_0$ is chosen, then for $t_0 \leqslant t \leqslant T$ $\boldsymbol{y}(t)$ is in the compact region

$$D(t) = \left\{(t, \boldsymbol{y}) | t_0 \leqslant t \leqslant T; \|\boldsymbol{y}\|^2 \leqslant (1 + \|\boldsymbol{y}^0\|^2)e^{2cT}\right\}$$

differential equation of (2.6-3) by expanding $\boldsymbol{h}$ in a Taylor series, retaining linear terms only,†

$$\dot{y}(t) = A(t)y(t) + B(t)x(t) \tag{2.6-4}$$

where

$$A(t) = \frac{\partial h}{\partial \boldsymbol{y}}(t, 0, 0), \quad B(t) = \frac{\partial h}{\partial \boldsymbol{x}}(t, 0, 0)$$

In the above, we might think of $\boldsymbol{v}$ as the instruments of the policymakers, $\boldsymbol{F}$ as representing dynamics of a macroeconomic model and the reference time path $\bar{y}(t)$ as an outcome of a simulation in which the instruments were manipulated according to a program $\bar{\boldsymbol{x}}(t)$ over $t_0 \leqslant t \leqslant T$.

Policymakers are interested in learning the results of different manipulations of the instruments. The perturbation analysis described here is a particular type of comparative study (known as sensitivity analysis in the engineering literature) by which policymakers evaluate various ways to alter the time paths of the instruments in the future without re-running the simulation, which for large models could be expensive.

Analogous developments can be carried out to obtain the linearized difference equation approximation to a system described by a set of n linear difference equations. Rather than repeat the derivation, we illustrate the linearization procedure in Example 1. A more detailed example of the linearization of an economy is discussed in §3.4.

Example 1 We now obtain the linearized approximation of the nonlinear difference equations system that arises in the two-sector growth model of Arrow and Kurz (1970).

Consider a two-sector economy. Let us choose a quarter as the basic unit of time, then let K_t^1 and K_t^2 denote the capital stocks in Sectors 1 and 2 at the beginning of the tth quarter. We assume that in Sector 1, the private sector, the capital stock depreciates at the rate $1 - \delta$ and the amount of new capital formation is sY_t, where s is the propensity to save of the private sector,

$$K_{t+1}^1 = (1 - \delta)K_t^1 + sY_t$$

Y_t is the output of the whole economy given by

$$Y_t = F(K_t^1, K_t^2, L_t)$$

†Similar derivations are possible for nonlinear difference equations.

and the labor force grows at the rate of

$$L_t = L_0 e^{nt}$$

The capital stock of Sector 2, the government sector, is governed by

$$K_{t+1}{}^2 = (1 - \delta) K_t^2 + G_t$$

where G_t represents the government expenditures.

We work with per capita variables denoted by lower-case letters

$$k_t^i = K_t^i / L_t = K_t^i / L_0 e^{nt}, \qquad i = 1, 2$$

$$y_t = Y_t / L_0 e^{nt}$$

$$g_t = G_t / L_0 e^{nt}$$

Assume a Cobb–Douglas production function

$$Y_t = \gamma (K_t^1 + K_t^2)^\alpha L_t^{1-\alpha}$$

then

$$y_t = \gamma (k_t^1 + k_t^2)^\alpha$$

Choose

$$\begin{pmatrix} k_t^1 \\ k_t^2 \end{pmatrix}$$

as the state vector $\boldsymbol{k}_t$ of this model. Then the growth of the per capita capital stocks† is governed by

$$\boldsymbol{k}_{t+1} = \begin{pmatrix} \beta & 0 \\ 0 & \beta \end{pmatrix} \boldsymbol{k}_t + \begin{pmatrix} 0 \\ e^{-n} \end{pmatrix} g_t + \begin{pmatrix} \hat{s} y_t \\ 0 \end{pmatrix} \tag{2.6-5}$$

where

$$\beta = (1 - \delta) e^{-n}, \qquad \hat{s} = e^{-n} s$$

Since y_t, the output per capita, is a nonlinear function of $\boldsymbol{k}_t$, let us derive

†More precisely, effective per capita. We abbreviate it and use the simpler per capita term since no confusion is likely to result.

its linearized equation about the equilibrium state vector $\boldsymbol{k}_*$. Assume g_t is equal to a constant g_* for ease of illustration.

The equilibrium per capita capital stocks are k_*^1 and k_*^2 and from (2.6-5) must satisfy the relations

$$k_*^1 = \beta k_*^1 + \hat{s}\gamma\left(k_*^1 + k_*^2\right)^\alpha \tag{2.6-6}$$

$$k_*^2 = \beta k_*^2 + e^{-n} g_* \tag{2.6-7}$$

where $\hat{s}$ and g_* are assumed to be exogenously given in this example. From (2.6-7), the equilibrium government sector per capita capital stock is obtained as

$$k_*^2 = e^{-n} g_* / (1 - \beta)$$

Substituting this into (2.6-6), the equilibrium private per capita capital stock is obtained,

$$(1 - \beta) k_*^1 = \bar{s}\left\{k_*^1 + e^{-n} g_* / (1 - \beta)\right\}^\alpha \tag{2.6-8}$$

where

$$\bar{s} = \hat{s}\gamma = e^{-n}\gamma s$$

In (2.6-8), k_*^1 can be determined as the intersection of the line $y = (1 - \beta)x$ and the curve $y = \bar{s}(x + e^{-n} g_*/(1 - \beta))^\alpha$.

Since the equilibrium capital stock levels are determined in the two sectors, we may expand $\hat{s}y_t(k_t^1, k_t^2)$ into a Taylor series about $\boldsymbol{k}_*$ retaining terms up to the first-order term

$$\bar{s}\left(k_t^1 + k_t^2\right)^\alpha = \bar{s}\left(k_*^1 + k_*^2\right)^\alpha + \alpha\bar{s}\left(k_*^1 + k_*^2\right)^{\alpha - 1}\left(k_t^1 - k_*^1 + k_t^2 - k_*^2\right) + \cdots$$

where from (2.6-6), we have

$$\bar{s}\left(k_*^1 + k_*^2\right)^{\alpha - 1} = \frac{(1 - \beta)k_*^1}{k_*^1 + k_*^2}$$

Thus,

$$\bar{s}\left(k_t^1 + k_t^2\right)^\alpha = (1 - \beta)k_*^1 + \frac{\alpha(1 - \beta)k_*^1}{k_*^1 + k_*^2} \times \left\{\left(k_t^1 - k_*^1\right) + \left(k_t^2 - k_*^2\right)\right\} + \cdots \tag{2.6-9}$$

Substitute this expansion into (2.6-5) to obtain the linearized state-space equation for (2.6-5).

Let

$$z_t^i = \boldsymbol{k}_t^i - \boldsymbol{k}_*^i, \qquad i = 1, 2$$

From (2.6-6), (2.6-7) and (2.6-9), retaining terms up to linear ones in z's,

$$z_{t+1}^1 = \beta z_t^1 + \frac{\alpha(1-\beta)k_*^1}{k_*^1 + k_*^2}\left(z_t^1 + z_t^2\right)$$

$$z_{t+1}^2 = \beta z_t^2 + e^{-n}(g_t - g_*)$$

or putting them into the vector–matrix form, the linearized dynamic equation which approximates (2.6-5) is given by

$$z_{t+1} = A z_t + \boldsymbol{b} x_t \tag{2.6-10}$$

where

$$z_t = \begin{pmatrix} z_t^1 \\ z_t^2 \end{pmatrix}, \quad A = \begin{pmatrix} a_{11} & a_{12} \\ 0 & a_{22} \end{pmatrix}, \quad \boldsymbol{b} = \begin{pmatrix} 0 \\ e^{-n} \end{pmatrix}$$

$$x_t = (g_t - g_*)$$

and where

$$a_{11} = \beta + \frac{\alpha(1-\beta)k_*^1}{k_*^1 + k_*^2}, \quad a_{12} = \frac{\alpha(1-\beta)k_*^1}{k_*^1 + k_*^2}, \quad a_{22} = \beta$$

If the propensity to save deviates from a constant s postulated in deriving (2.6-10) and if we want to incorporate the possibility of the saving rate s at the tth quarter s_t differing from s_* (which is taken to be long run average rate), then this can easily be done by expanding $\hat{s}_t y_t$, where $\hat{s}_t = e^{-n} s_t$, using the per capita production function

$$f(k_1, k_2) = F(k_1/L, k_2/L, 1)$$

$$\hat{s}_t y_t = \hat{s}_* f\left(k_*^1, k_*^2\right) + \hat{s}_*\left[\frac{\partial f}{\partial k^1}\left(k^1 - k_*^1\right) + \frac{\partial f}{\partial k^2}\left(k^2 - k_*^2\right)\right]$$

$$+ (\hat{s}_t - \hat{s}_*) f\left(k_*^1, k_*^2\right) + \cdots$$

The state-space equation becomes

$$z_{t+1} = Az_t + \boldsymbol{b}x_t + d_t \tag{2.6-11}$$

where

$$d_t = \frac{(1-\beta)k_*^1}{\left(k_*^1 + k_*^2\right)s_*}\left(s_t - s_*\right)$$

is the additional term due to $s_t \neq s_*$.

If the government wishes to use (2.6-11) as a macroeconomic model to guide the capital stock formation in the two sectors, then d_t acts as disturbance. Whether it may be treated as exogenous depends on matters that we do not discuss here.

Example 2 We consider a very simple-minded macroeconomic model, and we derive a linearized state-space dynamic equation. The model is a two-variable version of a macroeconomic model considered by Bergstrom (1967).

In this model, demand for money is given by

$$M^d = aY^\alpha r^{-\beta}, \qquad \alpha > \beta > 0 \tag{2.6-12}$$

where M^d is the demand for money, Y is the output and r is the interest rate. They are all functions of time t. The time argument is suppressed for ease of notation. The interest rate is changing with time according to

$$\dot{r}/r = h \ln M^d/M^s \tag{2.6-13}$$

where M^s is the supply of money. We say more about M^s later.

The real output rate changes according to

$$\dot{Y} = \lambda(C + \dot{K} + G - Y) \tag{2.6-14}$$

where C is the rate of consumption, $\dot{K}$ is the rate of new capital formation, G is the rate of government expenditure. Price is assumed to be constant in the short run.

To close the model in a simple way, we assume the following relations hold

$$\begin{aligned} C &= (1-s)Y \\ \dot{K} &= bY - dr \\ G &= \nu Y \end{aligned} \tag{2.6-15}$$

where s, b and d are constant. The variable ν is an instrument and hence can vary with time.

We wish to derive a state-space differential equation to describe the (short-run) dynamics of the model. Let

$$w_1 = \ln r \tag{2.6-16}$$

and

$$w_2 = \ln Y$$

From (2.6-12), (2.6-13) and (2.6-15)

$$\begin{aligned} \dot{w}_1 &= h\left[\ln a + \alpha w_2 - \beta w_1 - \ln M^s\right] \\ \dot{w}_2 &= \lambda\left[(-s + b) - de^{w_1 - w_2} + \nu\right] \end{aligned} \tag{2.6-17}$$

We consider a steady growth equilibrium corresponding to

$$\begin{aligned} M^s &= M^* e^{mt} \\ \nu &= \nu^* \end{aligned} \tag{2.6-17'}$$

Particular solutions of (2.6-17) with M^s and ν as specified are obtained by substituting

$$w_1 = \ln \overline{w}_1 + \mu_1 t$$

$$w_2 = \ln \overline{w}_2 + \mu_2 t$$

into (2.6-17) and solving for the $\overline{w}$'s and μ's.

From the second equation in (2.6-17), we immediately see that $\mu_1 = \mu_2$. Also from (2.6-17) we obtain

$$\mu_1 = \mu_2 = (\alpha - \beta)/m$$

Denote this common value by μ. We obtain

$$\overline{w}_1 = \eta_1 \overline{w}_2$$

$$\overline{w}_2 = \exp\left[(\eta_2 - \beta \ln \eta_1)/(\alpha - \beta)\right]$$

where

$$\eta_1 = \frac{-s + b + \nu^* - (\mu/\lambda)}{d}$$

$$\eta_2 = (\mu/h) - \ln(a/M^*)$$

We now define the state vector

$$z_1 = \ln \frac{r}{\overline{w}_1 e^{\mu t}} = w_1 - \ln \overline{w}_1 - \mu t$$
$$z_2 = \ln \frac{Y}{\overline{w}_2 e^{\mu t}} = w_2 - \ln \overline{w}_2 - \mu t \tag{2.6-18}$$

To examine the effect of M^s and ν deviating from $M^* e^{mt}$ and ν^* we consider the instruments given by

$$x_1 = \ln \frac{M^s}{M^* e^{mt}} = \ln M^s - \ln M^* - mt$$
$$x_2 = \nu - \nu^*$$

In other words, the actual money supply is expressible as

$$M^s = (M^* e^{mt}) e^{x_1}$$

and the government expenditure as

$$\nu = \nu^* + x_2$$

so that the actual amount of money supplies is $M^* e^{mt}$ times e^{x_1}. This factor can be less than 1 or greater than 1 depending on whether $x_1 < 0$ or $x_1 > 0$.

Similarly, x_2 measures the deviation of the actual government expenditure rate from the equilibrium growth rate.

The differential equations for the state-space variables are obtained from (2.6-17) and (2.6-18)

$$\dot{z}_1 = h(-\beta z_1 + \alpha z_2) - h x_1$$
$$\dot{z}_2 = -\lambda d \eta_1 (e^{z_1 - z_2} - 1) + \lambda x_2 \tag{2.6-19}$$

Equations (2.6-19) are the differential equations governing the deviations from the equilibrium path. The linearized dynamic model then is given by

$$\dot{z}_1 = h[-\beta z_1 + \alpha z_2] - h x_1$$
$$\dot{z}_2 = -\delta z_1 + \delta z_2 + \lambda x_2 \tag{2.6-20}$$

where $\delta = \lambda d \eta_1$.

3

Properties of Dynamic Systems: Controllability and Observability

When we select a dynamic system to use for modelling economic phenomena, an institution or individual economic agent's decision making processes, we must be aware of the broad implications of choosing one rather than another dynamic system. In other words, we must know the qualitative differences in the descriptive capabilities of the different kinds of systems used in a model.

It would be ridiculous to construct and laboriously to discuss the dynamic behavior of a macroeconomic model, for example, for a certain policy implication study, if the particular dynamic model chosen were incapable of achieving the desired policy objectives. The dynamic model might be incapable due to some inherent limitations of the set of differential or difference equations used to represent the macroeconomic system regardless of what instruments were employed. We wish to know without solving the equations explicitly whether a model possesses the necessary properties. For instance, for a stability property, this wish has been answered partially by Lyapunov (LaSalle and Lefschetz, 1961). The other properties of interest are called controllability and observability.

In the modelling of dynamic systems, whether it be a continuous time model using a system of differential equations or a discrete time model composed of a system of difference equations, in order to use the model effectively, we must have firm understanding of the qualitative properties that broadly prescribe the potential usefulness of the model. We should note that dynamic systems do have some properties that are absent in systems of algebraic equations.

There are only a few qualitative properties of systems that broadly categorize a model's possible responses and capabilities. The existing literature on questions of stabilization policies of macroeconomic systems is fairly large. Two other most important properties of dynamic systems are controllability, or reachability, and observability, or reconstructability.

All three of these properties are intimately related. For example, we show later that if a dynamic model has the proper controllability and observability properties, then under certain technical conditions we can find a stabilization policy for the dynamic model. There are many other connections that will become clear in this text. Properties such as controllability, observability and stabilizability are useful primarily because their presence tells us in a simple way the qualitative capabilities of the dynamic system, and their absence alerts us to the limitations of the system's performance.

Controllability is an existence condition of a dynamic policy. We say that a deterministic dynamic system $\dot{z} = \boldsymbol{f}(t, z, \boldsymbol{x})$ is (completely) controllable at time t_0 if for each pair of states z_0 and z_1, there exists a feasible instrument vector $\boldsymbol{x}(\cdot)$ on some finite interval $t_0 \leqslant t \leqslant t_1$ such that the system moves from the state z_0 at time t_0, to z_1 at time t_1.

Loosely speaking, the controllability of a model has to do with the effectiveness of the control instruments in influencing or modifying dynamic behavior. Problems associated with the existence and the design of policy instruments to achieve a set of targets were considered, for example, by Tinbergen for a static economic system (Tinbergen, 1955). When the assumption of a static economic model is replaced by that of a dynamic one and also the model is treated as a control system, then feasible or attainable targets mean that these targets can be achieved by instruments satisfying all the constraints imposed from economic, political and other reasons. In other words, the concept of controllability arises naturally when we examine the geometry of the set of time paths attainable by a dynamic system using feasible controls (i.e., controls satisfying all the constraints imposed on them). Thus, there is no exaggeration when we say that controllability ranks equally with stability and observability in its importance in discussing capabilities of dynamic economic models.

For dynamic models, it is also relevant to ask if targets can be tracked once they are attained, i.e., if the (subsets of) model outputs or state vector can be made to follow targets if their values change with time and also if there is a policy that accomplishes the above by means of stable instruments. Another important question is whether some outputs or targets can be changed while the rest of the target values remain unchanged. (This question is known as the decoupling problem in control literature, Falb and Wolovich, 1967.) This turns out to be a generalized version of the so-called question of assignment problem, Mundell 1962, 1968; subsets of instruments are assigned to subsets of targets in a one-to-one way so that instruments in one subset influence only a limited number of targets to achieve noninteraction of some groupings of targets. For linear dynamic systems at least, all these questions are closely connected with the controllability property, as we shall see later.

The importance of controllability also derives from the fact that it appears as technical conditions in many optimization questions. For example, it is known (and will be discussed later) that a certain canonical or standard state-space representation is guaranteed to exist for controllable systems. Controllability is sufficient also for the existence of optimal solutions to certain quadratic programming problems that arise in the regulation of linear dynamic systems.

Observability is equally important. As a matter of fact, this concept is dual (in a way to be specified later) to that of controllability. More directly, the observability property has to do with the ability to uncover or recover unobservable systems' data from a set of observed data. Such a property is clearly important in giving an operational definition to variables that are not directly available to model builders. By deterministic observability we refer to an existence condition of an estimator for a quantity or a value that is not accessible to direct measurements. In defining the state (vector) observability, we have a dynamic equation $\dot{z} = \boldsymbol{f}(t, z, \boldsymbol{x})$ and an observation or output equation $\boldsymbol{y} = \boldsymbol{h}(t, z, \boldsymbol{x})$. This combined system is said to be observable at time t_0 if for each feasible instrument time path over some time interval $t_0 \leqslant t \leqslant t_1$, the observation record over the same time interval uniquely determines $z(t_0)$. It also implied a rather unexpected result. For example, later in discussing a particular monetary disequilibrium model, observability will be shown to exclude clearings of a subset of markets.

Although we do not have as extensive or as close connections between the controllability or observability property and the other desired properties described above for nonlinear dynamic systems as in linear ones, these properties are still very important in the intertemporal optimization or stabilization considerations of nonlinear dynamic systems. It is expected that some analogous, perhaps weaker, conditions also appear as technical conditions in examining nonlinear dynamic systems as well. In other words, the study of controllability and observability is expected to be a necessary preliminary step in discussing topics such as stabilization or optimization of nonlinear dynamic economic models.

3.1 UNCONTROLLABLE DYNAMIC SYSTEMS—AN EXAMPLE

Before we turn to formal definitions and discussions of controllability, we discuss one example that illustrates some consequences of choosing a noncontrollable system as a model.† This is Example 1 of §2.6, slightly modified by the tax rates introduced as the instruments of the government.

This model of a national economy consists of two sectors: the public and the private sector, each with its own capital stock.

Suppose the government balances the budget so that we have the accounting identities

$$\begin{aligned} K_p(t+1) &= K_p(t) + S_p(t) \\ K_g(t+1) &= K_g(t) + X_t \end{aligned} \tag{3.1-1}$$

where $K_p(t)$ is the private sector capital stock, $K_g(t)$ is the government capital stock, $S_p(t)$ is the cumulative saving over one unit period by the private sector and X_t is the total tax receipts of the government. The stock variables are defined at the beginning of periods (measured in some convenient unit such as a month, or a year). It is possible to assume that the government uses two taxes as the instruments for control, one for savings u_t^s and one for consumption u_t^c where the private sector saves s_t of the disposable income. However, for simplicity of presentation, assume that these two taxes are the same, $u_t^c = u_t^s = x_t$.‡

Therefore, we have the behavioral relation for S_p

$$S_p(t) = s_t(1 - x_t)Y_t \tag{3.1-2}$$

since

$$X_t = x_t Y_t$$

and consumption is given by

$$C_t = (1 - x_t)(1 - s_t)Y_t \tag{3.1-3}$$

†This example is taken from Aoki (1974a). The model is described in (Arrow and Kurz, 1970).

‡See Aoki (1974a) for the two tax cases.

where Y_t is national output that is assumed to be given by a production function

$$Y_t = F\big(K_p(t), K_g(t), (1+\delta)^t L_t\big)$$

where L_t is the labor force for period t, and where $1 + \delta$ is the technical progress factor, $\delta > 0$, that is assumed to be labor-augmenting. The labor supply is assumed to grow at a given proportional rate n

$$L_t = (1+n)^t L_0$$

Define the lower-case variables by multiplying the upper-case variables by $(1+\gamma)^{-t} L_0^{-1}$

$$k_p(t) = (1+\gamma)^{-t} K_p(t)/L_0$$

$$y_t = (1+\gamma)^{-t} Y_t/L_0, \quad \text{etc.}$$

where

$$1 + \gamma = (1+\delta)(1+n)$$

The discrete time model given by (3.1-1)-(3.1-3) can be expressed as

$$\begin{aligned}(1+\gamma)k_p(t+1) &= k_p(t) + s_t(1-x_t)y_t \\ (1+\gamma)k_g(t+1) &= k_g(t) + x_t y_t\end{aligned} \tag{3.1-4}$$

where assuming the production function to be homogeneous of degree one,

$$y_t = f(k_p(t), k_g(t)) = (1+\gamma)^{-t} F\big(K_p(t), K_g(t), (1+\gamma)^t L_0\big)/L_0$$

We assume the existence of a balanced growth path for (3.1-4).† Define the state vector of the system by

$$\boldsymbol{k}_t = \begin{pmatrix} k_p(t) \\ k_g(t) \end{pmatrix} \in R^2$$

$$f(\boldsymbol{k}_t) = f(k_p(t), k_g(t))$$

†Alternatively we may assume that f is concave $f_p(0,0) = f_g(0,0) = \infty$, $f_p, f_g \to 0$ as $\|(k_p, k_g)\| \to \infty$ or some suitable extension of the Okamoto–Inada condition, or we may assume that $f(k_p, k_g)$ as a function of $k_p(k_g)$ satisfies the Inada condition for each fixed $k_p(k_g)$, where f_p is $\partial f/\partial k_p$.

We note the economic interpretation of some of the terms in (3.1-4):

$$\bar{w}_t = s_t f(k_t) \quad \left\{ \begin{array}{l} \text{before-tax saving} \\ \text{rate (per ef-} \\ \text{fective worker)} \end{array} \right.$$

$$w_t = (1 - x_t)\bar{w}_t \quad \left\{ \begin{array}{l} \text{actual saving} \\ \text{rate (per ef-} \\ \text{fective worker)} \end{array} \right. \tag{3.1-5}$$

We avoid possible ambiguity in timing by assuming that at the beginning of each period, the government announces its tax rates x_t, then the consumer chooses his saving rate s_t.

We illustrate the results for a specific production function

$$f(k_p, k_g) = A(k_p + k_g)^\alpha, \qquad 0 < \alpha < 1 \tag{3.1-6}$$

that assumes that the public and private capital stocks are lumped together with equal weights. A slightly more general function can be similarly treated

$$f(k_p, k_g) = A\left[\lambda k_p + (1 - \lambda)k_g\right]^\alpha, \qquad 0 < \alpha < 1, 0 < \lambda < 1$$

where different weights are assigned to the capital stocks. We carry out the analysis for (3.1-6).

To exhibit the instrument explicitly, put the difference equation governing the economy (3.1-4) into the form

$$\boldsymbol{k}_{t+1} = \boldsymbol{g}(\boldsymbol{k}_t, s_t) + x_t \boldsymbol{h}(\boldsymbol{k}_t, s_t) \tag{3.1-4'}$$

where

$$\boldsymbol{g}(\boldsymbol{k}_t, s_t) = \left[\boldsymbol{k}_t + s_t \begin{pmatrix} 1 \\ 0 \end{pmatrix} f(\boldsymbol{k}_t)\right] / (1 + \gamma)$$

$$\boldsymbol{h}(\boldsymbol{k}_t, s_t) = \begin{pmatrix} -s_t \\ 1 \end{pmatrix} f(\boldsymbol{k}_t) / (1 + \gamma)$$

Before we move to the discussion of dynamic behavior we note the equilibrium state of (3.1-4′).

Let $\boldsymbol{k}^*$ be the equilibrium solution of (3.1-4′) with constant tax rate x^* and saving ratio s^*. Then $\boldsymbol{k}^*$ is given by

$$\boldsymbol{k}^* = \hat{\boldsymbol{s}} f(k^*)$$

where

$$\hat{\boldsymbol{s}} = \frac{1}{\gamma}\begin{pmatrix} s^*(1 - x^*) \\ x^* \end{pmatrix}$$

Suppose that at time t_0 (that is taken to be zero without loss of generality), the government modifies the tax rate to $\{x^* + v_t\}$, where $v_t = 0$ when $t < 0$.

The new tax rate would be expected to influence the consumer behavior, so that $\{s^*\}$ changes to $\{s^* + \sigma_t\}$, $\sigma_t = 0$ for $t < 0$, $\sigma_t \neq 0$ for $t > 0$. Although the relationship between $\{v_t\}$ and $\{\sigma_t\}$ is too complicated to be modeled realistically since it involves considerations such as the government's anticipation of the expected change in the saving behavior by the consumer and so on, we make a simple assumption that s_t is a function of the current tax rate† and the current output or equivalently $\boldsymbol{k}_t$, i.e.,

$$s_t = s(x_t, \boldsymbol{k}_t)$$

Let $\{\boldsymbol{k}^* + \zeta_t\}$ be the resultant perturbed trajectory, i.e., let ζ_t be generated by

$$\begin{aligned}\zeta_{t+1} = {} & \boldsymbol{g}(\boldsymbol{k}^* + \zeta_t, s^* + \sigma_t) - \boldsymbol{g}(\boldsymbol{k}^*, s^*) \\ & + \boldsymbol{h}(\boldsymbol{k}^* + \zeta_t, s^* + \sigma_t)(x^* + v_t) - \boldsymbol{h}(\boldsymbol{k}^*, s^*)x^* \end{aligned} \tag{3.1-7}$$

where

$$v_t, \sigma_t \text{ and } \zeta_t \text{ are all zero for } t < 0,$$

where by assumption we have

$$\sigma_t \cong \frac{\partial s_t}{\partial x_t} v_t + \frac{\partial s_t}{\partial \boldsymbol{k}_t} \zeta_t \tag{3.1-8}$$

†More generally, s_t may be treated as a part of the state vector so that s_t depends on past tax rates as well as on the current tax rate, without changing the analysis of this example in any substantial way.

where we define in view of (3.1-8)

$$\hat{\sigma}_t = \frac{\partial s}{\partial x_t} v_t + \frac{\partial s}{\partial \boldsymbol{k}_t} \boldsymbol{\zeta}_t \tag{3.1-8'}$$

We now assume that $|v_t|$, the magnitude of the perturbation in the tax rate, hence the resultant $\|\boldsymbol{\zeta}_t\|$ and $|\sigma_t|$, are small over some finite time interval (in this case $t = 0, 1$, suffices). Taking (3.1-8′) into account, (3.1-7) gives rise to (3.1-9), the associated linearized dynamic system of (3.1-4′), that is also the equation that the Jacobian matrix satisfies. Using z_t for $\boldsymbol{\zeta}_t$, it is given by

$$z_{t+1} = Az_t + \boldsymbol{b}v_t \tag{3.1-9}$$

with

$$A = F + \boldsymbol{h} \nabla s$$

and

$$\boldsymbol{b} = \boldsymbol{g} + \boldsymbol{h} \epsilon s^* / x^*$$

where

$$F = \frac{\partial \boldsymbol{g}}{\partial \boldsymbol{\zeta}} (k^*, s^*) + x^* \frac{\partial \boldsymbol{h}}{\partial \boldsymbol{\zeta}} (k^*, s^*)$$

$$= \left(I + \begin{pmatrix} s^*(1 - x^*) \\ x^* \end{pmatrix} \nabla f(\boldsymbol{k}^*) \right) \Big/ (1 + \gamma)$$

$$\boldsymbol{g} = \begin{pmatrix} -s^* \\ 1 \end{pmatrix} f(\boldsymbol{k}^*) / (1 + \gamma)$$

$$\boldsymbol{h} = \frac{\partial \boldsymbol{g}}{\partial s} (\boldsymbol{k}^*, s^*) + \frac{\partial \boldsymbol{h}}{\partial s} (\boldsymbol{k}^*, s^*) x^*$$

$$= \begin{pmatrix} 1 - x^* \\ 0 \end{pmatrix} f(\boldsymbol{k}^*) / (1 + \gamma)$$

$$\hat{\sigma} = \epsilon s^* v_t / x^* + \nabla s_t z_t$$

and where the tax elasticity of the saving rate is

$$\epsilon = \frac{x}{s^*} \frac{\partial s_t}{\partial x_t}$$

Suppose $z_0 = 0$. With $v_0 = v_1 = 0$, etc., the capital stocks stay at $\boldsymbol{k}^*$. Suppose the government wishes to modify the ratio of k_g to k_p from that of $\boldsymbol{k}^*$ by having $v_0 \neq 0$, $v_1 \neq 0$, etc. Then from (3.1-9),

$$z_1 = \boldsymbol{b}v_0$$

$$z_2 = \boldsymbol{b}v_1 + A\boldsymbol{b}v_0$$

These are the deviation of the $\boldsymbol{k}$'s from that implied by the policy of no tax change.

We next show that there are a number of plausible parameter value combinations for which the vector $A\boldsymbol{b}$ is proportional to $\boldsymbol{b}$, hence $\boldsymbol{b}v_1 + A\boldsymbol{b}v_0$ is proportional to $\boldsymbol{b}$. This means that the capital stocks can deviate from $\boldsymbol{k}^*$ only in the direction specified by $\boldsymbol{b}$. Because of the Cayley–Hamilton theorem, then, z_t, $t > 0$, stays on the linear manifold spanned by $\boldsymbol{b}$ and cannot escape from it no matter what v's are employed. The only possible modification in the composition of the capital stocks is such that

$$\Delta k_p = -s\Delta k_g$$

No other change is possible by this model. This is the direct result of "unfortunate" choice of the model with the parameter values that make $A\boldsymbol{b}$ proportional to $\boldsymbol{b}$. Such a model is clearly not controllable.

One simple possibility to make $A\boldsymbol{b}$ proportional to $\boldsymbol{b}$ is to assume that $\nabla s = \rho \nabla f$ for some constant ρ. Then ϵ^* is given by†

$$\epsilon^* = \rho f(\boldsymbol{k}^*) + s^*/(1 - x^*) \qquad (3.1\text{-}10)$$

Another is to assume that $s_g = 0$ and $\epsilon^* = x^*/(1 - x^*)$.

†If increase in the private capital stock decreases the saving rate then $\rho > 0$ since $\partial f/\partial k_p < 0$. Then (3.1-10) holds only if $\epsilon^* > 0$, i.e., if the increase of tax rate induces increased saving.

3.2 CONTROLLABILITY OF LINEAR DYNAMIC ECONOMIC SYSTEMS

3.2.1 Complete Controllability of State Vector (State-Space Controllability)

Given an economic system and a set of instruments (also called control vector) of specified type generally varying with time (such as piece-wise continuous in time t or differentiable), can the economic system in some initial state $z(t_0)$ be brought to a desired target state z in some finite time using admissible, i.e., feasible instrument values without violating any constraints? This is a natural and very important question to ask about the set of instruments.

Loosely speaking, if the answer is affirmative, we say that the system is controllable with respect to the admissible set of instrument values. We say controllability is complete if the system in an arbitrary initial state and the initial time can be brought to an arbitrarily specified target state in a finite time.†

Therefore, we must select a set of instruments so that the system is uniformly completely controllable. We now give precise definitions of controllability and characterize controllable systems, first for linear systems and later for nonlinear systems. As usual, available results on controllability of general nonlinear systems are scanty.

LINEAR DIFFERENTIAL SYSTEMS Consider

$$\dot{z}(t) = A(t)z(t) + B(t)x(t), \qquad z(t_0) = z_0 \tag{3.2-1}$$

where $B(t)$ is an $n \times r$ matrix.

Denote by $\phi(t, t_0)$ its $n \times n$ fundamental solution matrix, i.e.,

$$\frac{d\phi(t, t_0)}{dt} = A(t)\phi(t, t_0), \qquad \phi(t_0, t_0) = I_n$$

Then the solution of (3.2-1) at time $t_1 > t_0$ is expressible as

$$z(t_1) = \phi(t_1, t_0)z_0 + \int_{t_0}^{t_1} \phi(t_1, t)B(t)x(t)\, dt, \qquad t_1 \geqslant t_0$$

†The use of these adjectives differs slightly from author to author and there are many variations on the basic idea of controllability. See for example Hermes and LaSalle, 1969.

Rewriting it†

$$0 = \phi(t_1, t_0)(z_0 - \phi(t_0, t_1)z(t_1)) + \int_{t_0}^{t_1} \phi(t_1, t)B(t)x(t)\, dt \quad (3.2\text{-}2)$$

we note that the "initial" state $z_0 - z(t_1)\phi(t_0, t_1)$ is now transferred to the origin at time t_1. We see that the existence of the control vector over $[t_0, t_1]$ to transfer z_0 to $z(t_1)$ is equivalent to the existence of the same control vector over the same time interval to transfer the initial state

$$z_0 - z(t_1)\phi(t_0, t_1)$$

to the origin.

Thus, we say the system (3.2-1) is completely controllable if the system can be transferred from any initial state at any initial time t_0 to the origin in some finite time $t_1 - t_0$.

Equation (3.2-2) can also be written as

$$\xi = \int_{t_0}^{t_1} \phi(t_1, s)B(s)x(s)\, ds \quad (3.2\text{-}3)$$

where ξ is an arbitrary n-vector since z_0 and $z(t_1)$ are arbitrary. When A and B are constant matrices, we have

$$\phi(t_1, t_0) = e^{A(t_1 - t_0)}$$

and (3.2-3) becomes

$$\xi = \int_{t_0}^{t_1} e^{A(t_1 - t)} Bx(t)\, dt \quad (3.2\text{-}3')$$

Equation (3.2-3) (or (3.2-3′)) clearly shows that the controllability is about the range space of the linear mapping $L : R^r \to R^n$, defined by

$$L(x) = \int_{t_0}^{t_1} \phi(t_1, s)B(s)x(s)\, ds \quad (3.2\text{-}4)$$

The system (3.2-1) is uniformly completely controllable if and only if there is a t_1 such that the range space of (3.2-4) is the entire R^n for any $t_0 < t_1$. If the system is not completely controllable, the range space of $L(\cdot)$ is confined to some subspace of R^n.

---

†Recall the relation $\phi(t, t_0) = \phi(t, t_1)\phi(t_1, t_0)$ analogous to the one developed in §2.2.2.

Theorem 1 *The n-dimensional linear time-invariant system*

$$\dot{z}(t) = Az(t) + Bx(t)$$

is completely controllable if and only if the column vectors of the controllability matrix

$$\langle A|B\rangle \stackrel{\text{def}}{=} (B, AB, \ldots, A^{n-1}B): n \times nr \text{ matrix}$$

span the n-dimensional space R^n.

Proof By the Cayley–Hamilton Theorem e^{At} is a polynomial in A of at most $n-1$ degree

$$e^{At} = \alpha_0(t)I + \alpha_1(t)A + \cdots + \alpha_{n-1}(t)A^{n-1}$$

where $\alpha_i(t)$ is a scalar function of t, $i = 1, \ldots, n-1$. (See Appendix A.) Therefore,

$$\int_{t_0}^{t_1} e^{A(t_1-s)}Bx(s)\,ds = B\int_{t_0}^{t_1}\alpha_0(t_1-s)x(s)\,ds + AB\int_{t_0}^{t_1}\alpha_1(t_1-s)x(s)\,ds$$

$$+ \cdots + A^{n-1}B\int_{t_0}^{t}\alpha_{n-1}(t_1-s)x(s)\,ds$$

hence ξ is in the linear subspace spanned by the column vectors of $B, AB, \ldots, A^{n-1}B$.

Thus, if span $\langle A|B\rangle$ is not R^n, the system is not completely controllable. To prove the implication in the other direction, suppose rank$\langle A|B\rangle = n$. Then an x vector given by $x(s) = B^{t}e^{A^{t}(t_1-s)}\eta$ will satisfy (3.2-3′) where η is the solution of

$$\left[\int_{t_0}^{t_1} e^{A(t_1-s)}BB^{t}e^{A^{t}(t_1-s)}\,ds\right]\eta = \xi$$

or

$$\eta = W(t_0, t_1)^{-1}\xi$$

where

$$W(t_0, t_1) = \int_{t_0}^{t_1} e^{A(t_1-s)}BB^{t}e^{A^{t}(t_1-s)}\,ds$$

is nonsingular (by hypothesis span $\langle A|B\rangle = R^n$). To see this last point, suppose $W(t_0, t_1)$ is singular. Then there exists a nonzero vector ζ such that $\zeta^t W(t_0, t_1)\zeta = 0$ or

$$\int_{t_0}^{t_1} \| \zeta^t e^{A(t_1 - s)} B \|^2 \, ds = 0$$

This means that $\zeta^t e^{A(t_1-s)}B = 0$ for all $t_0 \leqslant s \leqslant t_1$ for some $\zeta \neq 0$. But this contradicts the hypothesis.

In one part of the proof of sufficiency of the controllability rank condition for time-invariant systems, we showed that there exists a vector $\zeta \neq 0$ such that $e^{-As}B\zeta \not\equiv 0$. Let $Y(s) = e^{-As}B$. Then

$$\begin{aligned} Y(0) &= B \\ \dot{Y}(0) &= (-A)B \\ Y^{(j)}(0) &= (-A)^j B \\ &\vdots \\ Y^{(n-1)}(0) &= (-A)^{n-1} B \end{aligned}$$

so that the rank condition may be expressed as

$$\begin{aligned} n &= \operatorname{rank}(B \quad AB \quad \dots \quad A^{n-1} \quad B) \\ &= \operatorname{rank}\left[Y(0), \dot{Y}(0), \dots, Y^{(n-1)}(0) \right] \end{aligned}$$

The latter form has its natural generalization to the time varying systems. The rank tests for time varying systems is discussed later (see Theorem 5).

We say (A, B) is a completely controllable pair when $\langle A|B\rangle$ spans R^n. Any nonsingular change of coordinates does not destroy the controllability property.

Example 1 (Controllable Linear Dynamic Systems) Let us see if the linear macroeconomic model (2.6-20) that is obtained by linearizing (2.6-17) about a time path corresponding to the instrument (2.6-17a) (from Example 2 of Chap. 2) is completely controllable or not. From (2.6-20) we see that the model is governed by the differential equation

$$\dot{z} = Az + Bx$$

where

$$B = \begin{pmatrix} -h & 0 \\ 0 & \lambda \end{pmatrix}$$

Since rank $B = 2$, provided $\lambda h \neq 0$, and since dim $z = 2$, B alone spans the two-dimensional state space. The model is therefore completely controllable.

From some concepts of local controllability of nonlinear dynamic systems, see §3.4 and references in Aoki, 1974a.

A slight modification of this proof also proves the next theorem.

Theorem 2 *The system (3.2-1) is completely controllable if and only if*

$$W(t_0, t_1) = \int_{t_0}^{t_1} \phi(t_1, s)B(s)B^{t}(s)\phi^{t}(t_1, s)\, ds$$

is positive definite for some $t_1 > t_0$.

Proof If $W(t_0, t_1) > 0$, then for any $\xi \in R^n$, there exists ζ such that $W(t_0, t_1)\zeta = \xi$. Thus, $\boldsymbol{x}^t = \zeta^t\phi(t_1, s)B(s)$ satisfies (3.2-3), hence the system is controllable.

If $W(t_0, t_1)$ is singular, then there exists $\zeta \neq 0$ such that $\zeta^t W(t_0, t_1)\zeta = 0$ or

$$\zeta^t\phi(t_1, s)B(s) = 0 \qquad \text{for all } t_0 \leqslant s \leqslant t_1$$

Then for any control vector $\boldsymbol{x}$

$$\zeta^t \int_{t_0}^{t_1} \phi(t_1, s)B(s)\boldsymbol{x}(s)\, ds = 0$$

or the range of $\int_{t_0}^{t_1}\phi(t_1, s)B(s)\boldsymbol{x}(s)\, ds$ is orthogonal to the subspace spanned by ζ or the range is a subspace of R^n of dimension at most $n - 1$. Hence (3.2-1) is not completely controllable.

From $\zeta^t W(t_0, t_1)\zeta = \int_{t_0}^{t_1}\|\zeta^t\phi(t_1, s)B(s)\|^2\, ds$ we see that $W(t_0, t_1) > 0$ if and only if row vectors of $\phi(t_1, s)B(s)$ are linearly independent for all $t_0 \leqslant s \leqslant t_1$. Thus Theorem 2 can be restated as follows.

Theorem 3 *System (3.2-1) is uniformly completely controllable over $[t_0, t_1]$ if and only if the row vectors of $\phi(t_1, s)B(s)$ are linearly independent for any $t_0 \leqslant s \leqslant t_1$.*

Writing $\phi(t_1, s)$ *as* $\phi(t_1, t_0)\phi(t_0, s) = \phi(t_1, t_0)\phi(s, t_0)^{-1}$ *and noting that* $\phi(t_1, t_0)$ *is nonsingular, we see that linear independence of the row vectors of* $\phi(s, t_0)^{-1}B(s)$ *is also a necessary and sufficient condition.*

The criteria of controllability for time-varying systems so far discussed require that the fundamental solution to the homogeneous part of (3.2-1) be known.

We next give an algebraic characterization of controllability that does not require this knowledge. We follow Silverman and Meadows, 1969.

Let

$$\Theta(t) = \phi(t, t_0)^{-1}B(t)$$

Then by arguing inductively (assuming the indicated derivatives all exist)†

$$\frac{d^k}{dt^k}\Theta(t) = \phi(t, t_0)^{-1}P_k(t), \qquad k = 0, 1, \ldots, n-1$$

where

$$P_{k+1}(t) = -A(t)P_k(t) + \dot{P}_k(t), \qquad P_0(t) = B(t)$$

Thus,

$$\left(\Theta(t), \dot{\Theta}(t), \ldots, \Theta^{(n-1)}(t)\right) = \phi(t, t_0)^{-1}Q(t)$$

where

$$Q(t) = \left[P_0(t), \ldots, P_{n-1}(t)\right]$$

Since $\phi(t, t_0)^{-1}$ is nonsingular, the range space of the left-hand side equals that of $Q(t)$.

Note that when $\phi(t, t_0) = e^{A(t-t_0)}$, the matrix $Q(t)$ becomes $\langle -A|B\rangle$, the controllability matrix generated by $-A$ and B whose range space is the same as that of $\langle A|B\rangle$.

We state the following theorem.

Theorem 4 *System* (3.2-1) *is uniformly completely controllable on every subinterval of* $[t_0, t_1]$ *if and only if* $Q(t)$ *does not have rank less than n on any subinterval of* (t_0, t_1).

---

†Assume that $A(t)$ has $(n-2)$ continuous derivatives and that $B(t)$ has $(n-1)$ continuous derivatives for all $t \in (t_0, \infty)$.

For a proof to Theorem 4 see Silverman and Meadows, Theorem 4 or Lemma 1 and 3.

Denote the operation of $-A(t) + d/dt$ by Γ. Then

$$P_{k+1}(t) = \Gamma P_k$$
$$= \Gamma^{k+1} B$$

or

$$\frac{d^k}{dt^k}\Theta(t) = \phi(t, t_0)^{-1}\Gamma^k B$$

or

$$\left(\Theta(t_1), \dot{\Theta}(t_2), \ldots, \Theta^{(n-1)}(t_n)\right)$$
$$= \left(\phi(t_1, t_0)^{-1} P_0(t_1), \ldots, \phi(t_n, t_0)^{-1} P_n(t_n)\right)$$

Theorem 5 *The system is controllable at time t_0 if and only if there exist times $t_1, \ldots, t_n \geqslant t_0$ for which rank $[P_0(t_1), \ldots, P_{n-1}(t_n)] = n$.*

For proof to Theorem 5 see (Hermes-LaSalle, Theorem 19.2, 1969.)

LINEAR DIFFERENCE SYSTEMS Controllability of discrete-time linear dynamic systems can be discussed in an entirely analogous way. We therefore merely summarize some useful results for later use.

Time-Invariant Systems: We can interpret the controllability rank condition for discrete-time dynamic systems more readily.

Suppose the dynamic equation is given by

$$z_{t+1} = Az_t + Bx_t$$

Then since $z_t = A^t z_0 + \sum_{i=0}^{t-1} A^{t-1-i} Bx_i$, z_t is expressed as functions of the initial state and the past values of the instruments as

$$z_t = A^t z_0 + Q_t x_{0,t}$$

where

$$Q_t = \left[A^{t-1}B, A^{t-2}B, \ldots, AB, B\right] : n \times tr \text{ matrix} \qquad (3.2\text{-}5)$$

and where the past instrument values are stacked to form a vector

$$x_{0,t} = \begin{bmatrix} x_0 \\ x_1 \\ \vdots \\ x_{t-1} \end{bmatrix} : tr\text{-dimensional vector}$$

The system is said to be completely controllable if any initial state can be transferred to any state in some finite time. Controllability thus concerns the existence of solutions for the algebraic equation

$$z^d - A^t z_0 = Q_t x_{0,t}$$

where z^d is the desired state vector to be reached at the tth time instant. We can take $t \leqslant n$ without loss of generality.

The system is by definition completely controllable if and only if

$$\xi = Q_t x_{0,t}$$

has solution for arbitrary $\xi \in R^n$ for some $t \leqslant n$. It has a solution only if ξ is in the range space of Q_t. Since ξ is arbitrary, the range space $= R^n$ or rank $Q_t = n$ is necessary.

If rank $Q_t = n$, then the sequence of instruments given by

$$x_{0,t} = Q_t^{\mathrm{t}}(Q_t Q_t^{\mathrm{t}})^{-1}\xi$$

solves the equation.

We have established a criterion of controllability

$$\text{rank } \langle A|B \rangle = n$$

The rank condition can be equivalently stated as $Q_n Q_n^{\mathrm{t}}$ is nonsingular.

Time-Varying Systems: When the dynamic equations are time-varying as in

$$z_{t+1} = A_t z_t + B_t x_t$$

then

$$z_{t+s} = \phi_{t+s,t} z_t + Q_{t+s,t} x_{t,t+s}$$

where

$$\phi_{t+s,t} = A_{t+s-1} \cdots A_{t+1}A_t$$

$$Q_{t+s,t} = \left[\phi_{t+s,t+1}B_t\phi_{t+s,t+2}B_{t+1} \cdots \phi_{t+s,t+s-1}B_{t+s-2}B_{t+s-1}\right]$$

and

$$x_{t,t+s} = \begin{bmatrix} x_t \\ x_{t+1} \\ \vdots \\ x_{t+s-1} \end{bmatrix}$$

The controllability question becomes equivalent to the existence of solutions, for t fixed, of

$$\xi = Q_{t+\tau,t}x_{t,t+\tau}$$

for some $\tau > 0$.

Namely, a time-varying dynamic system is controllable at time t if and only if

$$\operatorname{rank}(Q_{t+n-1,t}) = n$$

or equivalently $Q_{t+n-1,t}Q_{t+n-1,t}^{\mathrm{t}}$ is nonsingular.

Example 2 Examine the system (2.6-10) for controllability. Here $n = 2$. The rank condition is

$$\operatorname{rank}(b \quad Ab) = \operatorname{rank}\begin{pmatrix} 0 & a_{12}e^{-n} \\ e^{-n} & a_{22}e^{-n} \end{pmatrix}$$

where a_{12} and a_{22} are as given in (2.6-10). Since $a_{12} \neq 0$, the rank condition is met, hence the system is completely controllable with an instrument (the government expenditure). This means, of course, that any point near $(k_*{}^1, k_*{}^2)$ can be realized by appropriate government expenditure, and any relative composition of the capital stocks of the two sectors is achievable.

Example 3 (Controllability of Dynamic Input-Output Model) We now examine controllability of a Leontief dynamic input-output model, which we introduced in Example 4 of §2.2. The model is described by (2.2-17). Its state-space representation is (2.2-22)

$$z_{t+1} = \begin{pmatrix} F_{11} & F_{12} \\ 0 & 0 \end{pmatrix} z_t + \begin{pmatrix} G_{11} & G_{12} \\ 0 & 0 \end{pmatrix} x_t$$

Since

$$\begin{pmatrix} F_{11} & F_{12} \\ 0 & 0 \end{pmatrix}^k \begin{pmatrix} G_{11} & G_{12} \\ 0 & 0 \end{pmatrix} = \begin{pmatrix} X & X \\ 0 & 0 \end{pmatrix}, \qquad k = 1, 2, \ldots, n-1$$

where X stands for some nonzero submatrices, the controllability matrix is not full rank, i.e., the system is not controllable.

The same is true when the dynamic matrix is

$$\begin{pmatrix} F_{11} & F_{12} \\ 0 & F_{22} \end{pmatrix}$$

We can actually construct a dynamic system of dimension less than n that is controllable. For this, define

$$s_t = (I_n \quad 0)z_t$$

where n_1 is the rank of B matrix in (2.2-17). This operation is known as aggregation and s_t is the aggregated state vector of z_t. It is governed by the dynamic equation

$$s_{t+1} = F_{11}s_t + (G_{11} \quad G_{12})x_t$$

See Aoki, 1971, for further details of obtaining dynamics of the aggregated state vector. We can readily see that this system is controllable.

Controllability with Positive Controls The notion and criteria of controllability that we have developed so far assumes unconstrained control variables, i.e., control variables can take on any finite values. In economics, variables are usually constrained to be positive. When we discuss the small perturbation v of tax rates from the existing ones x, then it may be legitimate to assume that v can take on small positive as well as negative values so long as $0 < x < 1$. However, if $x = 0$ or $x = 1$ then v is constrained to be positive and negative, respectively.

It is therefore of some interest to see how the criterion of controllability in Theorem 1 must be modified when control variables are constrained to be positive. A typical result is the next theorem due to Saperstone and Yorke (1970).

Theorem 1A *Let each component of x take its value on* [0, 1]. *Consider the system $\dot{z} = F(z, x)$ where $F(0, 0) = 0$. Let $A = \partial F/\partial z(0, 0)$ and $B = \partial F/\partial x(0, 0)$. Suppose*

(i) *no eigenvalue of A is real and*
(ii) rank $\langle A|B \rangle = \dim z$

Then the system is locally controllable.

We note that if the dimension of z is odd, then the system is *not* locally controllable since a matrix of odd order must have at least one real eigenvalue. The system must then be oscillatory for it to be controllable with positive controls. This has an important implication on the time to reach any desired point. The time path connecting an initial point to a desired point exhibits oscillatory behavior, and the time required to reach any desired point may be considerably longer with sign-constrained control.

Example 4 Let us see if the macromodel of Example 5, §2.2 is controllable. Suppose that m(logarithm of the money stocks) is the only instrument. For simplicity, assume that

$$\pi_t \cong T(p_t - p_{t-1})$$

Then from (2.2-E7′) the dynamic equation becomes

$$z_t = \hat{F} z_{t-1} + \boldsymbol{h} x_t$$

where

$$\boldsymbol{h} = \begin{pmatrix} c_{21}\boldsymbol{\delta} \\ c_{31} - c_{21}\boldsymbol{\lambda}^{\mathrm{t}}\boldsymbol{\delta}/\Delta \end{pmatrix} \quad : 3 \times 1$$

We have established that $\hat{F} = F + T\boldsymbol{\alpha}\boldsymbol{e}_2^{\mathrm{t}}$ where

$$F = \begin{bmatrix} 0 & 0 & f_{13} \\ 0 & 0 & f_{23} \\ 0 & 0 & f_{33} \end{bmatrix}, \qquad \begin{pmatrix} f_{13} \\ f_{23} \end{pmatrix} = \frac{1}{\Delta}(a_1\gamma + b_1\boldsymbol{\delta}),$$

$$f_{33} = d_1 - \lambda^{\mathrm{t}}/[(a_1\gamma + b_1\boldsymbol{\delta})/\Delta]$$

We must compute rank $(\boldsymbol{h} \quad \hat{F}\boldsymbol{h} \quad \hat{F}^2\boldsymbol{h}^2)$, where

$$\hat{F}\boldsymbol{h} = F\boldsymbol{h} + T(\boldsymbol{e}_2^{\mathrm{t}}\boldsymbol{h})\boldsymbol{\alpha}$$

$$\hat{F}^2\boldsymbol{h} = F^2\boldsymbol{h} + T(\boldsymbol{e}_2^{\mathrm{t}}\boldsymbol{h})\boldsymbol{\alpha} + T^2(\boldsymbol{e}_2^{\mathrm{t}}\boldsymbol{h})(\boldsymbol{e}_2^{\mathrm{t}}\boldsymbol{\alpha})\boldsymbol{\alpha}$$

We note that

$$\operatorname{rank}(\boldsymbol{h} \quad \hat{F}\boldsymbol{h} \quad \hat{F}^2\boldsymbol{h}) = 3$$

if and only if the vectors $\boldsymbol{\alpha}$, $\boldsymbol{h}$ and $F\boldsymbol{h}$ are linearly independent.

3.2.2 Output Controllability

Instead of requiring that a state vector be transferable from any point to any other point in R^n, we sometimes only require that an output vector $\boldsymbol{y}$ can be taken from any initial point to any other point.† This is known as complete output controllability. The vector $\boldsymbol{y}$ is related to the state vector z by

$$y = Cz + Dx$$

Thus, instead of (3.2-2) we use

$$0 = C\left(\phi(t_1, t_0)(z_0 - \phi(t_0, t_1)z(t_1)) + \int_{t_0}^{t_1}\phi(t_1, t)B(t)\boldsymbol{x}(t)\,dt\right)$$
$$+ D\boldsymbol{x}(t_1)$$

We have

Theorem 1′ *The dynamic system*

$$\dot{z} = Az + Bx$$
$$y = Cz + Dx$$

where $\boldsymbol{y}$ *is an m-vector, is completely output controllable if and only if the column vectors of*

$$C\langle A|B\rangle + D = \left[C(B, AB, \ldots, A^{n-1}B), D\right] : m \times (r+1)n \text{ matrix}$$

span the m-dimensional space R^m.

In case of linear discrete-time systems, from

$$y_t - CA^t z_0 = CQ_t x_{0,t} + Dx_t$$

where Q_t is defined by (3.2-5), we conclude that the criterion of complete output controllability is that

$$\text{rank}(C\langle A|B\rangle, D) = m$$

†We must still make sure to exclude such possibilities as some components of $z(t)$ going to infinity or violating some constraints imposed on the models from economic considerations.

3.2.3 Perfect Output Controllability

In practice,† policymakers would be interested not only in achieving desired target values at a point in time, but also in keeping targets on some desired time paths once achieved.‡ For example, they would be interested not only in achieving a certain inflation rate or level of foreign reserves but also in keeping them at some designated levels or changing them with time in some prescribed manner. This type of target variable manipulation requires more than the condition that dynamic systems are completely controllable.§

Since it is important that instruments have this added capability we describe the conditions necessary for policymakers to be able to make the target variables follow desired time trajectories (including stationary values). This condition is known as perfect output controllability or functional reproducibility (Basile and Marro, 1971), (Brockett and Mesarovic, 1965). It generalizes the original Tinbergen condition, which is the condition for the existence of instrument variables to achieve assigned (or desired) fixed target variable values in static macroeconomic models (Tinbergen, 1955). Thus perfect output controllability may be considered to be a proper dynamic generalization of Tinbergen's condition in the theory of policy.

We consider economic models described by differential equations and give conditions for constant (time-invariant) linear economic models under the heading Perfect Output Controllability. Under suitable conditions, perfect output controllability of nonlinear economic models can be deduced from the corresponding linear ones. This is described under the heading Nonlinear Macroeconomic Model.

PERFECT OUTPUT CONTROLLABILITY Conditions for following desired time paths have been studied in control theory, using either an algebraic approach or a geometric approach. We summarize them here using the algebraic approach first and comment on the geometric approach later. These conditions are known as functional reproducibility in (Basile and Marro, 1971) and as perfect output controllability in (Brockett and Mesarovic, 1965). As will be seen in this section when the target is a scalar

---

†This section is based on (Aoki 1975a).

‡An alternative is not to attempt to follow some specified time path exactly but to minimize discrepancies between the actual and desired time path by using an error function. This approach is discussed in Chapters 5 and 9.

§See for example Preston (1974).

variable, i.e., $m = 1$, the condition reduces to that of complete controllability studied in §3.2.1.

Suppose the target vector $\boldsymbol{y}$ is related to the state vector z by $\boldsymbol{y} = Cz$, i.e., assume $D = 0$.

Let $\boldsymbol{f}^d(t)$ be a desired time trajectory of the target vector. Define the error vector when the instrument vector $\boldsymbol{x}$ is utilized

$$\boldsymbol{E}(t; \boldsymbol{x}) = \boldsymbol{f}^d(t) - C\phi(t; \boldsymbol{x}, z^0) \tag{3.2-6}$$

where

$$\phi(t; \boldsymbol{x}, z^0) = e^{At}z^0 + e^{At}\int_0^t e^{-As}B\boldsymbol{x}(s)\, ds \tag{3.2-7}$$

The question is: For any $\eta > 0$ and $\tau > 0$, do there exist $\delta(\eta, \tau) > 0$ and $\boldsymbol{x}(t)$, $T \leqslant t \leqslant T + \tau$ such that $\|\boldsymbol{E}(t; \boldsymbol{x})\| < \delta(\eta, \tau)$ for $T \leqslant t \leqslant T + \tau$, for any z^0 such that $\|z^0\| \leqslant \eta$? Here $\|\cdot\|$ is a suitablly defined norm, e.g. denoting the kth derivative of $\boldsymbol{E}$ with respect to time by $\boldsymbol{E}^{(k)}$,

$$\|\boldsymbol{E}(\boldsymbol{x})\| = \max_{0 \leqslant k \leqslant n} \sup_{0 \leqslant t < \infty} |\boldsymbol{E}^{(k)}(t; \boldsymbol{x})|$$

Let

$$\boldsymbol{F}(t) = \boldsymbol{f}^d(t) - Ce^{At}z^0$$

Then by using the argument s to denote the Laplace transform, $\boldsymbol{E}(\boldsymbol{x}) = 0$ implies (recall the discussions in §2.3, in particular (2.3-5)),

$$\boldsymbol{F}(s) = N(s)\boldsymbol{x}(s), \qquad \text{where } N(s) = C(sI - A)^{-1}B : m \times r \tag{3.2-8}$$

If the $m \times r$ matrix $N(s)$ has rank m, then there exists $\boldsymbol{g}(s)$ that satisfies the above, namely

$$\boldsymbol{x}(s) = N^T(s)\left(N(s)N(s)^T\right)^{-1}F(s)$$

Define the $mn \times (2n - 1)r$ matrix M_n is defined by

$$M_n = \begin{bmatrix} CB & CAB & \cdots & CA^{2n-2}B \\ 0 & CB & \cdots & CA^{2n-3}B \\ \vdots & & & \vdots \\ 0 & CB & \cdots & CA^{n-1}B \end{bmatrix} \tag{3.2-9}$$

Now, it is a fact that rank $N = m$ if and only if rank $M_n = mn$.† We have

Proposition 1 *$C(sI - A)^{-1}B$ is of rank m if and only if the matrix M_n is of rank mn. For a proof see (Brockett and Mesarovic, 1965). Related and perhaps simpler criteria are given by (Sain and Massey, 1969), (Dorato, 1969) and (Wolovich, 1974).*

Corollary 1 *The system with dim $\boldsymbol{y} = m$, dim $\boldsymbol{x} = r$ and dim $\boldsymbol{z} = n$ is perfectly output controllable only if $m \leqslant (2 - 1/n)r$. Rewrite this inequality as $m \leqslant r + (1 - 1/n)r$. We see then that if $r = 1$, then $m = 1$ is necessary for perfect output controllability. Note also that when $m = 1$, rank $M_n = n$. Comparing this with Theorem 1′ of §3.2.2 (with $D = 0$), we see that the distinction between the output controllability and perfect output controllability disappears for a scalar target variable.*

Therefore, in the context of linear dynamic macroeconomic models, the concept of perfect output controllability generalizes the original condition of Tinbergen involving the number of instruments and targets.

In the above, we have shown how the target vector can be made to follow a specified time path by suitable manipulation of the instrument vector and its derivatives. In the time domain description, the condition implies that $\boldsymbol{x}$ and its derivatives are adjusted to match $\boldsymbol{y}$ and its deriva-

†The outline of the proof is as follows: The rank of N is m if and only if there exists no m-dimensional row vector $K(s)$ such that

$$K(s)N(s) \equiv 0.$$

Since $N(s)$ is rational in s, we can take $K(s)$ to be, without loss of generality, of the form,

$$K_0 + K_1 s + \cdots + K_{q-1}s^{q-1}$$

for some q. From (3.2-8), the existence of such a $K(s)$ is equivalent to

$$K(s)F(s) = 0$$

i.e., $F(t)$ satisfies a differential equation of the form

$$K_{q-1}F^{(q-1)}(t) + \cdots + K_0 F(t) \equiv 0$$

Expressing the derivatives of F in terms of those of $\boldsymbol{x}$'s we see that such a $K(s)$ does not exist if and only if

$$\text{rank } M_q = mq$$

where M_q is defined analogously to M_n. See (Brockett and Mesarovic, 1965) for the exact expression. Now, it is easy to show

$$\text{rank } M_q = \text{rank } M_n = mn \quad \text{for } q \geqslant n$$

tives with f^d and its derivatives over the time interval $[T, T + \tau]$, for some $\tau > 0$.

When D is nonzero, M_n of (3.2-9) is replaced by rank $\overline{M}_n = (m + 1)m$, where

$$\overline{M}_n = \begin{bmatrix} D & CB & CAB & \cdots & CA^{n-1}B & CA^{n-1}B & \cdots & 0 \\ 0 & D & CB & \cdots & CA^{n-3}B & CA^{n-2}B & \cdots & 0 \\ & & & \vdots & & & \vdots & \\ 0 & 0 & D & \cdots & 0 & D & \cdots & CA^{n-2}B \\ 0 & 0 & 0 & \cdots & D & CB & \cdots & CA^{n-1}B \end{bmatrix}$$

when $r = m$.

In some applications, it is easier to characterize f^d not in terms of its functional and derivative values but by the requirement that it lies in some subspace S of the state space R^n. For example, the policymaker may desire that the target variables maintain a certain fixed relation among their components, thus confining the target variables to some manifold or subspace in R^n.

The perfect output controllability condition may be stated in terms of the subspace S. We say S is (A, B)-invariant if

$$AS \subset S + \mathcal{R}(B)$$

where $\mathcal{R}(B)$ is the range space of the matrix B.

Proposition 2 *The target vector y, having reached a point in S at time T can be made to remain in S from T on if and only if S is (A, B)-invariant.* For a proof see (Wonham and Morse, 1970).

NONLINEAR MACROECONOMIC MODEL Suppose now that a macroeconomic model is described by a nonlinear control equation

$$\begin{aligned} \dot{z} &= f(z, x) = Az + Bx + Q(z, x) \\ y &= Cz \end{aligned} \qquad (3.2\text{-}10)$$

with $Q(0, 0) = 0$.

We assume the linearized system is output controllable. Then A may be taken to be stable without loss of generality (Wonham, 1967). We have the following Proposition.

Proposition 3 *The nonlinear economic system* (3.2-10) *is perfectly output controllable if the linearized system is perfectly output controllable.* (*See* (*Brockett and Mesarovic*, 1965).)

3.2.4 Transformation into Phase-Canonical Form

Given a time-invariant linear control system, it is sometimes convenient to change the state vector so that the matrix A is in a standard form, called the phase canonical form.† We construct such a transformation of the state vector for

$$\dot{z} = Az + \boldsymbol{b}x \tag{3.2-10}$$

where $\boldsymbol{b}$ is an n-vector and x is scalar.

Theorem 6 *Suppose* (3.2-10) *is completely controllable. Let*

$$z(t) = Tw(t)$$

where

$$T = PM$$

with

$$P = (\boldsymbol{b} \quad A\boldsymbol{b} \quad \dots \quad A^{n-1}\boldsymbol{b})$$

and

$$M = \begin{bmatrix} \alpha_1 & \alpha_2 & \cdots & \alpha_{n-1} & 1 \\ \alpha_2 & \alpha_3 & \cdots & 1 & 0 \\ & & \vdots & & \\ 1 & 0 & \cdots & 0 & 0 \end{bmatrix}$$

where the α*'s are the coefficients of the characteristic polynomial of* A

$$\det(sI - A) = \sum_{i=0}^{n} \alpha_i s^i, \qquad \alpha_n = 1$$

Then

$$\dot{w} = Gw + fx \tag{3.2-11}$$

†This form is used to discuss stabilizability conditions of macroeconomic models in §5.1.

with

$$G = \begin{bmatrix} 0 & 1 & \cdots & 0 & 0 \\ & & \vdots & & \\ 0 & 0 & \cdots & 0 & 1 \\ -\alpha_0 & -\alpha_1 & \cdots & -\alpha_{n-2} & -\alpha_{n-1} \end{bmatrix}, \qquad f = \begin{bmatrix} 0 \\ \vdots \\ 0 \\ \vdots \\ 1 \end{bmatrix}$$

The matrix G is said to be the phase canonical form.

Proof Since P is nonsingular by the assumption of complete controllability and det $M = 1$, T is nonsingular.

Thus, take the n-column vectors of T as the new basis of R^n.

By direct calculation, we see that

$$A\boldsymbol{t}_i = \boldsymbol{t}_{i-1} - \alpha_{i-1}\boldsymbol{t}_n, \qquad i = 2, \ldots, n$$

where $\boldsymbol{t}_i$ is the ith column vector of T since from $T = PM$ we have

$$\boldsymbol{t}_1 = \alpha_1 \boldsymbol{b} + \alpha_2 A\boldsymbol{b} + \cdots + A^{n-1}\boldsymbol{b}$$

$$\boldsymbol{t}_2 = \alpha_2 \boldsymbol{b} + \alpha_3 A\boldsymbol{b} + \cdots + A^{n-2}\boldsymbol{b}$$

$$\vdots$$

$$\boldsymbol{t}_{n-1} = \alpha_{n-1}\boldsymbol{b} + A\boldsymbol{b}$$

$$\boldsymbol{t}_n = \boldsymbol{b}$$

For $i = 1$, we have

$$A\boldsymbol{t}_1 = -\alpha_0 \boldsymbol{t}_n$$

since

$$\alpha_0 I + \alpha_1 A + \cdots + \alpha_{n-1}A^{n-1} + A^n = 0$$

by the Cayley–Hamilton Theorem.

Thus

$$AT = TG$$

Also with respect to this new basis

$$\boldsymbol{b} = T \begin{bmatrix} 0 \\ \vdots \\ 0 \\ 1 \end{bmatrix}$$

To see that no such nonsingular transformation exists if (3.2-10) is not completely controllable, note that:

(i) The system in the phase canonical form is complete controllable (compute $(\boldsymbol{f} \quad G\boldsymbol{f} \ldots \quad G^{n-1}\boldsymbol{f})$) and;
(ii) Any nonsingular transformation preserves the complete controllability.

The phase canonical form is sometimes called the controllable canonical form since the transformation matrix T is given as the product of the controllability matrix and M matrix.

We can put the system $\dot{z} = Az + Bx, y = Cz$ in which x and y are scalars into another form called the observable canonical form by the change of variable $z = S\boldsymbol{s}$ such that $\dot{\boldsymbol{s}} = \overline{A}\boldsymbol{s} + \overline{B}x, y = \overline{C}\boldsymbol{s}$ where $\overline{C} = (0 \quad \ldots \quad 0 \quad 1)$ where S is the same matrix postmultiplied by the observability matrix.

3.2.5 Discussion

We have discussed various aspects of controllability in the preceding several subsections. One final caution remains to be mentioned. You may have noted and wondered about the almost total lack of any constraints on the instruments in our discussion on controllability. In other words, the various concepts of controllability do not take account of magnitude or other constraints that must usually be imposed on the instrument vectors. To be economically meaningful, instrument vectors must satisfy various constraints, e.g., on the magnitude or rate of change. Therefore, in so far as the concepts of controllability are defined independently of these practically important constraints, whatever implications we may draw about controllable systems must be regarded as idealizations that provide benchmarks of possible model performances but nothing more.

When we impose magnitude and rate of change constraints on the instruments, for example, we may not be able to change the "state" of the economy in an arbitrary manner in any finite time intervals. We can usually obtain only local controllability, or reachability, results. See §3.4. We consider controllability of stochastic systems in §3.6.3.

3.3 CLASSIFICATION OF DYNAMIC SYSTEMS BY TRACKING CAPABILITY

We can classify dynamic models by their capability to follow or track exogenously given functions of time. This classification scheme is useful since we prefer, *ceteris paribus*, models that can follow or reproduce more complicated functions of time (which are Laplace-transformable) to those which cannot. Even if the system is not perfectly output controllable, so that it cannot reproduce arbitrary (Laplace-transformable) functions of time, it is useful to characterize dynamic models by the types of time paths they are capable of following: Some dynamic systems may be able eventually to attain any level of values after a sudden change in target levels, while they may not be able to follow target levels changing linearly with time or quadratically and so forth.

Suppose a system is represented in the state-space form

$$\dot{z} = Az + \boldsymbol{b}x$$
$$y = \boldsymbol{c}z \qquad (3.3\text{-}1)$$

where x is the instrument variable, y is the scalar target variable, and where $\boldsymbol{b}$ is a column vector and $\boldsymbol{c}$ is a row vector.

Suppose a desired path to be followed by the model is given by $y^*(t)$. It could be a desired growth path of GNP or a time history desired for the interest rate and so on.

Let us denote the discrepancy between the desired and actual target values called the error of tracking in control literature by

$$e = y^* - y \qquad (3.3\text{-}2)$$

We now classify systems according to the type of y^* for which $e(t) \to 0$, as $t \to \infty$ when systems are controlled by the linear constant feedback law (3.3-4) below. It is convenient to illustrate the concept when y, y^* and x in (3.3-1) and (3.3-2) are all scalar variables.

Denote the Laplace transform by $\hat{}$. From (3.3-1)

$$\hat{y}(s) = c(sI - A)^{-1}(\boldsymbol{b}\hat{x} + z(0))$$

We can ignore the contribution from $z(0)$ if A is a stable matrix since

$$e^{At}z(0) \to 0 \text{ as } t \to \infty$$

or

$$\lim_{s\to 0} s(sI - A)^{-1}z(0) = 0$$

i.e., the contribution to $y(t)$ from $z(0)$ vanishes as $t \to \infty$. From (3.3-2),

$$\begin{aligned}\hat{e}(s) &= \hat{y}^*(s) - \hat{y}(s) \\ &= \hat{y}^*(s) - \boldsymbol{c}(sI - A)^{-1}\boldsymbol{b}\hat{x}(s) \qquad (3.3\text{-}3)\end{aligned}$$

Suppose $x(t)$ is generated as†

$$x(t) = k(y^*(t) - y(t)) \qquad (3.3\text{-}4)$$

or

$$\hat{x}(s) = k(\hat{y}^*(s) - \hat{y}(s))$$

Then from (3.3-3)

$$\hat{e}(s) = \frac{\hat{y}^*(s)}{1 + kh(s)}$$

where $h(s) = \boldsymbol{c}(sI - A)^{-1}\boldsymbol{b}$ is the transfer function.

Suppose it is desired to follow a sudden exogenous change of finite magnitude α in the output. Then

$$\hat{y}^*(s) = \frac{\alpha}{s} \quad \text{since } y^*(t) = 0 \quad \text{for } t < 0 \text{ and } y^*(t) = \alpha$$

for $t > 0$.

Denote the error by $e_0(t)$.

From the final value theorem

$$\begin{aligned}\lim_{t\to\infty} e_0(t) &= \lim_{s\to 0} s\hat{e}_0(s) \\ &= \lim_{s\to 0} \frac{\alpha}{1 + kh(s)}\end{aligned}$$

Therefore $e_0(t) \to 0$, as $t \to \infty$ if and only if $\lim_{s\to 0}|h(s)| = \infty$. If $|h(0)| \neq \infty$, then

$$\frac{e_0(\infty)}{\alpha} = \frac{1}{1 + kh(0)}$$

is the steady state error that the system experiences in catching up with a sudden step change, i.e. the target never reaches the new level.

---

†This type of equation for generating instruments is mentioned in §2.4.3.

If $e_0(t) \to 0$, then the system is capable of eventually catching up with a step change in the target value.

Denote the error e by $e_1(t)$ when $y^* = \alpha t$, $t \geqslant 0$. Since

$$\lim_{t\to\infty} e_1(t) = \lim_{s\to 0} s\,\frac{\alpha/s^2}{1 + kh(s)}$$

$$= \lim_{s\to 0} \frac{\alpha}{s(1 + kh(s))}$$

$$= \lim_{s\to 0} \frac{\alpha}{ksh(s)}$$

the system is capable of following a constant growth path of the target variable without steady state error if $\lim_{s\to 0} sh(s) = \infty$, or if the Laurent series expansion of $h(s)$ about $s = 0$ is of the form

$$h(s) = h_1/s^2 + h_2/s + \cdots$$

Similarly, the steady state error $e_2(t)$ corresponding to $y^* = \alpha t^2/2$, $t \geqslant 0$, is such that

$$\lim_{t\to\infty} e_2(t) = \lim_{s\to 0} \frac{1}{s^2 h(s)}$$

when

$$y^*(t) = \begin{cases} \alpha t^2/2, & t \geqslant 0 \\ 0, & t < 0 \end{cases}$$

We see, therefore, that one way of classifying control systems is by this ability to follow without error more and more complicated functions of time. By definition

$$h(s) = \boldsymbol{c}\,\mathrm{adj}(sI - A)\boldsymbol{b}/|sI - A|$$

We can summarize the above results conveniently as follows:

If:	*Then y can follow*:
$\lim_{s\to 0} \|h(s)\| = 0$	level change in y^*
$\lim_{s\to 0} \|sh(s)\| = 0$	linearly changing y^*
$\lim_{s\to 0} \|s^2 h(s)\| = 0$	quadratically changing y^*

If A is nonsingular, then

$$h(0) = -\boldsymbol{c}A^{-1}\boldsymbol{b}$$

is finite, thus $e_0(t) \nrightarrow 0$ as $t \to \infty$.

If A is singular, then

$$h(0) = \lim_{s \to 0} \boldsymbol{c}\ \mathrm{adj}(sI - A)\boldsymbol{b}/|sI - A|$$

From the Cayley–Hamilton Theorem, we can express A^n as

$$A^n = \sum_{i=1}^{m} \alpha_i A^{n-i}$$

where the α's are determined by the characteristic equation

$$|sI - A| = s^n - \alpha_1 s^{n-1} - \cdots - \alpha_n$$

We know that

$$\mathrm{adj}(sI - A) = s^{n-1}I + s^{n-2}F_1 + \cdots + sF_{n-2} + F_{n-1}$$

where

$$F_k = A^k - \alpha_1 A^{k-1} - \cdots - \alpha_k I, \quad k = 1, \ldots, n-1$$

By assumption $|A| = 0$, hence $\alpha_n = 0$.

Thus, if $\boldsymbol{c}F_{n-1}\boldsymbol{b} \neq 0$, then $|h(s)| \to \infty$ as $|s| \to 0$ hence $e_0(t) \to 0$ as $t \to \infty$. Now,

$$\begin{aligned} \boldsymbol{c}F_{n-1}\boldsymbol{b} &= \boldsymbol{c}\left(A^{n-1} - \alpha_1 A^{n-2} - \cdots - \alpha_n I\right)\boldsymbol{b} \\ &= \boldsymbol{c}\left[A^{n-1}\boldsymbol{b} - \alpha_1 A^{n-2}\boldsymbol{b} - \cdots - \alpha_n \boldsymbol{b}\right] \end{aligned}$$

If $\boldsymbol{b}, A\boldsymbol{b}, \ldots, A^{n-1}\boldsymbol{b}$ are linearly independent, then $\boldsymbol{c}F_{n-1}\boldsymbol{b} \neq 0$ since any linear combination of $\boldsymbol{b}, A\boldsymbol{b}, \ldots, A^{n-1}\boldsymbol{b}$ does not lie in the $(n-1)$-dimension subspace orthogonal to $\boldsymbol{c}$. We note that this condition for $e_0(t) \to 0$ as $t \to \infty$ is precisely the condition for $(A, \boldsymbol{b})$ to be a controllable pair. We have shown:

Proposition *If the control system* (3.3-1) *is completely controllable, then any exogenous change in the desired target value can eventually be attained by the system by suitably adjusting the instrument. When* $\boldsymbol{x}$, $\boldsymbol{y}$ *and* $\boldsymbol{y}^*$ *are* *vectors, we have a transfer matrix $H(s) = C(sI - A)^{-1}B$, and the error vector $E(t)$ instead of the transfer function $h(s)$, and the scalar error variable $e(t)$. Its Laplace transform is related to $\hat{y}^*(s)$ by*

$$\hat{E}(s) = \left[I - H(s)(I + KH(s))^{-1}K\right]\hat{y}^*(s)$$

We can discuss various sufficient conditions on $H(s)$ to force some subset of the error vector components to vanish as $t \to \infty$ as above, even though they are not as simple.

3.4 LOCAL CONTROLLABILITY OF NONLINEAR DISCRETE TIME DYNAMIC SYSTEMS

We next state what we can about local controllability of nonlinear dynamic systems, since no useful results about global controllability of general nonlinar dynamic systems are available. We carry out our investigation of discrete-time nonlinear dynamic systems. Analogous developments of the concept of local controllability may be carried out for continuous time nonlinear systems.

Let $\boldsymbol{\phi}(t; t_0, z_0, \boldsymbol{x}, \boldsymbol{\theta})$ be the solution of the difference equation

$$z_{t+1} = \boldsymbol{f}(z_t, \boldsymbol{x}_t, \boldsymbol{\theta}), \qquad t = t_0, t_0 + 1, t_0 + 2, \ldots, z_{t_0} = z_0$$

where $\boldsymbol{x}_t$ denotes the control vector at time t and $\boldsymbol{\theta}$ denotes the exogenously given variable or known (time-varying) parameter vector.† (Unique solutions exist for the type of dynamic systems treated here.) The vector z_t is the state vector. We say control variables (or controls) are admissible if control variables satisfy all constraints imposed on them, i.e., if controls are feasible. We first give the definition for local controllability about a point in the state space. This definition will be useful in discussing local controllability about the equilibrium point of the dynamic equation such as on a balanced growth path.

A dynamic system, with $\boldsymbol{\theta}$ specified, is said to be locally controllable at a point ξ in the state space if there exists an open neighborhood of ξ, $N(\xi)$, such that any point in the neighborhood can be reached by admissible conrols for that realization of the exogenously given variables.

Suppose a system is locally controllable at ξ in the above sense. Let $z(t_0) = \xi$. Then, for every ξ' in $N(\xi)$, there exists $\boldsymbol{x}_s$, $t_0 \leqslant s \leqslant t$, $\boldsymbol{x}_s$ admis-

†The subscript t does not mean the partial derivative with respect to time. Time is sometimes carried as an argument $\boldsymbol{x}(t)$ in the text to avoid double subscripts.

sible, such that

$$\xi' = \phi(t; t_0, \xi, \boldsymbol{x}, \boldsymbol{\theta})$$

for some finite $t \geqslant t_0$. A dynamic system may be locally controllable for some $\boldsymbol{\theta}$ and not locally controllable for some other $\boldsymbol{\theta}$ values.

The definition does not impose any condition on how short $t - t_0$ should be.†

We note that this definition does not require the path leading from ξ to ξ' be contained in $N(\xi)$. For economic systems, it may be important to require that the path remains in some neighborhood $N(\xi)$. This requirement is related to the stability concept. Thus, it may be desired to combine these two concepts. The concept of local-local controllability of Haynes and Hermes may be used for this purpose (1970). We do not pursue this here, however.

We remark that controllability is a purely dynamic concept. It is independent of any utility or objective function, although it does not exclude it. We therefore discuss controllability of a dynamic system rather than that of a policy as Arrow and Kurz (1970) do. When an objective function is introduced, it may be possible to eliminate states in some subset of the neighborhood as clearly inferior. Then we may modify the definition of local controllability accordingly to read "... every point in $S \cap N(\xi)$ ..." rather than "... every point in $N(\xi)$ ..." for some subset S in R^n.

Very often we are interested in the controllability of every point along a specified path. This is the next definition of controllability.

We say a system with $\boldsymbol{\theta}$ specified is locally controllable along a path

$$\phi(t; t_0, \boldsymbol{x}, \boldsymbol{\theta}), \qquad t_0 \leqslant t \leqslant T$$

if the system is locally controllable at every point on the path, i.e. for each t, there is a neighborhood of $\phi(t; t_0, \boldsymbol{x}, \boldsymbol{\theta})$ such that every point in it can be reached by some admissible control $\boldsymbol{x}$.

Locally-local controllability along a path may be analogously defined.

We say that a system is locally controllable in the large in a set of state vectors S if it is locally controllable at every point in S.

Finally, we show the relation of local controllability with the Jacobian matrix associated with nonlinear dynamic systems.

†The controllability concept may be defined with $t - t_0$ specified. This, however, is more stringent than the concept discussed here and of less intrinsic economic interest.

Consider a nonlinear dynamic system given by

$$\xi_{t+1} = \boldsymbol{f}(\xi_t, \boldsymbol{v}_t, t), \qquad t = 0, 1, \ldots \tag{3.4-1}$$

where $\xi_t \in R^n$, $\boldsymbol{v}_t \in R^m$ for all t and f is continually differentiable in its arguments. The vector $\boldsymbol{v}$ is the vector of instruments.

A system of particular interest is the bilinear system introduced in §2.5. Its dynamics are governed by

$$\xi_{t+1} = \boldsymbol{\alpha}(\xi_t, t) + \boldsymbol{\beta}(\xi_t, t)\boldsymbol{v}_t \tag{3.4-2}$$

where $\boldsymbol{\beta}(\xi_t, t)$ is an $n \times m$ matrix.

Let a sequence of controls $\{\boldsymbol{v}_t\}$ and the corresponding path $\{y_t\}$ be given, i.e.,

$$y_{t+1} = \boldsymbol{f}(y_t, \boldsymbol{v}_t)$$

Let $\boldsymbol{x}_t(\boldsymbol{\theta})$, $\boldsymbol{\theta} \in R^n$ be a family of controls such that $\boldsymbol{x}_t(0) = \boldsymbol{v}_t$, $\partial \boldsymbol{x}_t(0)/\partial \boldsymbol{\theta}$ exists. Denote by $\{\xi_t(\boldsymbol{\theta})\}$ the trajectory corresponding to $\{\boldsymbol{x}_t(\boldsymbol{\theta})\}$. Let $\xi_0(0) = y_0$.

For each $t \geqslant 0$, $\xi_t(0)$ is viewed as a map $\boldsymbol{\theta} \mapsto \xi_t(\theta)$ with $0 \mapsto y_t$. Let Z_t be the Jacobian matrix $\partial \xi_t(\theta)/\partial \boldsymbol{\theta}|_0$. It satisfies

$$Z_{t+1} = A_t Z_t + C_t \left.\frac{\partial \boldsymbol{x}_t}{\partial \boldsymbol{\theta}}\right|_0$$

where

$$A_t = \frac{\partial f}{\partial \xi}(y_t, \boldsymbol{v}_t, t), \qquad C_t = \frac{\partial f}{\partial \boldsymbol{v}}(y_t, \boldsymbol{v}_t, t)$$

When specialized to the system of (3.4-2), the Jacobian matrix satisfies

$$Z_{t+1} = A_t Z_t + C_t \frac{\partial \boldsymbol{x}_t}{\partial \boldsymbol{\theta}} \tag{3.4-3}$$

where

$$A_t = \frac{\partial \boldsymbol{\alpha}}{\partial \xi}(y_t, t) + \sum_{i=1}^{m} B_t^i v_t^i$$

$$B_t^i = \frac{\partial \boldsymbol{\beta}_i}{\partial \xi}(y_t, t)$$

$$C_t = \boldsymbol{\beta}(y_t, t)$$

We have a discrete time system counterpart to the theorem due to Kalman (1960) (see also Weiss, 1970).

Theorem *A necessary and sufficient condition that there exists an* $m \times n$ *matrix* $\partial \boldsymbol{x}_t / \partial \boldsymbol{\theta}|_0$ *such that* Z_{t+1} *is nonsingular for some* $T > 0$ *is that the linear system*

$$\xi_{t+1} = A_t \xi_t + C_t \omega_t \tag{3.3-3'}$$

be completely controllable (from time 0).

Proof Necessity is obvious.

For sufficiency, take

$$\left. \frac{\partial \boldsymbol{x}_t}{\partial \boldsymbol{\theta}} \right|_0 = C_t{}^{\mathrm{t}} \Phi_{T+1, t+1}{}^{\mathrm{t}} \tag{3.4-4}$$

where

$$\Phi_{t+1, \tau} = A_t \Phi_{t, \tau} \qquad \Phi_{t, t} = I \tag{3.4-5}$$

From (3.4-3)–(3.4-5)

$$\begin{aligned} Z_{T+1} &= \Phi_{T+1, 0} Z_0 + \sum_{\tau=0}^{T} \Phi_{T+1, \tau+1} C_\tau \left. \frac{\partial \boldsymbol{x}_\tau}{\partial \boldsymbol{\theta}} \right|_0 \\ &= \sum_{\tau=0}^{T} \Phi_{T+1, \tau+1} C_\tau C_\tau{}^{\mathrm{t}} \Phi_{T+1, \tau+1}{}^{\mathrm{t}} \end{aligned}$$

since $Z_0 = 0$.

By the complete controllability hypothesis for (3.4-3) (or (3.4-3')), there exists a T such that

$$\operatorname{rank}(C_T, \Phi_{T+1, T} C_{T-1}, \ldots, \Phi_{T+1, 1} C_0) = n$$

Then Z_{T+1} is nonsingular. By the implicit function theorem

$$\phi(T+1; y_0, \xi) = \boldsymbol{\eta}$$

has a solution $\boldsymbol{\eta} = G(\xi)$ for all $\boldsymbol{\eta}$ in an open neighborhood of $y_0 = 0$ in R^n.

3.5 CONTROLLABILITY OF BILINEAR SYSTEMS

When we restrict nonlinear dynamic systems† to the bilinear systems introduced in §2.5, we can say more about their controllability. This is the topic of this section.

One example of bilinear systems is the linearized approximation to (3.1-4′) of §3.1. Suppose $\{x_t\}$ and $\{s_t\}$ generate $\{\boldsymbol{k}_t\}$ according to (3.1-4′) of §3.1. Let the sequence of tax rates be changed to $\{x_t + v_t\}$. This change, in turn, induces a change in the saving rate from $\{s_t\}$ to $\{s_t + \sigma_t\}$. Consequently, the growth path of the capital stocks changes from $\{\boldsymbol{k}_t\}$ to $\{\boldsymbol{k}_t + \boldsymbol{\zeta}_t\}$. It is convenient to represent $\boldsymbol{\zeta}_t$ as $\boldsymbol{\xi}_t + \boldsymbol{\eta}_t$ where $\{\boldsymbol{\xi}_t\}$ is generated by

$$\boldsymbol{\xi}_{t+1} = \boldsymbol{g}(\boldsymbol{k}_t + \boldsymbol{\xi}_t, s_t + \bar{\sigma}_t) - \boldsymbol{g}(\boldsymbol{k}_t, s_t) + \{\boldsymbol{h}(\boldsymbol{k}_t + \boldsymbol{\xi}_t, s_t + \bar{\sigma}_t) - \boldsymbol{h}(\boldsymbol{k}_t, s_t)\}x_t$$
$$+ \boldsymbol{h}(\boldsymbol{k}_t + \boldsymbol{\xi}_t, s_t + \bar{\sigma}_t)v_t, \qquad \xi_0 = 0 \tag{3.5-1}$$

where

$$\bar{\sigma}_t = \frac{\partial s_t}{\partial x_t} v_t + \frac{\partial s_t}{\partial \boldsymbol{k}} \boldsymbol{\xi}_t$$

Then $\{\boldsymbol{\eta}_t\}$ is governed by

$$\begin{aligned}\boldsymbol{\eta}_{t+1} &= \boldsymbol{g}(\boldsymbol{k}_t + \boldsymbol{\xi}_t + \boldsymbol{\eta}_t, s_t + \sigma_t) - \boldsymbol{g}(\boldsymbol{k}_t + \boldsymbol{\xi}_t, s_t + \bar{\sigma}_t) \\ &\quad + \{\boldsymbol{h}(\boldsymbol{k}_t + \boldsymbol{\xi}_t + \boldsymbol{\eta}_t, s_t + \sigma_t) - \boldsymbol{h}(\boldsymbol{k}_t + \boldsymbol{\xi}_t, s_t + \bar{\sigma}_t)\}x_t \\ &\quad + \{\boldsymbol{h}(\boldsymbol{k}_t + \boldsymbol{\xi}_t + \boldsymbol{\eta}_t, s_t + \sigma_t) - \boldsymbol{h}(\boldsymbol{k}_t + \boldsymbol{\xi}_t, s_t + \bar{\sigma}_t)\}v_t, \\ \eta_0 &= \Delta \neq 0 \end{aligned} \tag{3.5-2}$$

where $\{\boldsymbol{\eta}_t\}$ is the perturbed growth path of the two-sector economy when there is a sudden redistribution of capital stocks between the sectors. The government uses the tax rate as the instruments.‡

To the first order of approximation, from (3.1-4′) of §3.1, (3.5-1) and (3.5-2), we have $\sigma_t - \bar{\sigma}_t = (\partial s_t / \partial \boldsymbol{k}_t)\boldsymbol{\eta}_t$. We see that $\{\boldsymbol{\eta}_t\}$ is generated by the bilinear dynamic equation

$$\boldsymbol{\eta}_{t+1} = A_t \boldsymbol{\eta}_t + v_t B_t \boldsymbol{\eta}_t, \qquad \eta_0 = \Delta \tag{3.5-3}$$

where $|\Delta|$ is a small perturbation parameter, where

$$(1 + \gamma)A_t = I\left\{1 + \frac{\partial s_t}{\partial \boldsymbol{k}_t}\begin{pmatrix}1\\0\end{pmatrix} f(\boldsymbol{k}_t + \boldsymbol{\xi}_t)\right\}$$
$$+ x_t\left\{\begin{pmatrix}-s_t\\1\end{pmatrix} f'(\boldsymbol{k}_t + \boldsymbol{\xi}_t) - \frac{\partial s_t}{\partial \boldsymbol{k}_t}\begin{pmatrix}1\\0\end{pmatrix} f(\boldsymbol{k}_t + \boldsymbol{\xi}_t)I\right\}$$

†This section is based on (Aoki, 1974a), (Goka, *et al.*, 1973) and (Tarn, *et al.*, 1973).
‡As in §3.1, we consider the simple case in which the consumption tax rate and the saving tax rate are the same. More detailed analysis using the two tax rates as two instruments is possible following the line of development of this section.

and

$$(1+\gamma)B_t = \left[\left(\begin{pmatrix}0\\1\end{pmatrix} - x_t \frac{\partial s_t}{\partial \boldsymbol{x}_t}\begin{pmatrix}1\\0\end{pmatrix}\right) f'(\boldsymbol{k}_t + \xi_t) - \frac{\partial s_t}{\partial \boldsymbol{k}_t}\begin{pmatrix}1\\0\end{pmatrix} f(\boldsymbol{k}_t + \xi_t) I\right] \tag{3.5-4}$$

Since $\{\xi_t\}$, $\{\boldsymbol{x}_t\}$ and $\{s_t\}$ are taken to be known functions of time as far as the analysis of $\boldsymbol{\eta}_t$ is concerned, (3.5-3) is a bilinear system as defined in §2.5.

3.5.1 Single Instrument with Rank $B_t = 1$

Write the sequence of $\boldsymbol{\eta}$'s (time path) generated by (3.5-3) as $\psi(t; \boldsymbol{\eta}_0, \boldsymbol{v}_{0,t})$ where $\boldsymbol{v}_{0,t}$ stands for the past history of the instruments $v_0, v_1, \ldots, v_{t-1}$ so that

$$\psi(0; \boldsymbol{\eta}_0, v_0) = \boldsymbol{\eta}_0$$

$$\psi(t+1; \boldsymbol{\eta}_0, \boldsymbol{v}_{0,t+1}) = (A_t + v_t B_t)\psi(t; \boldsymbol{\eta}_0, \boldsymbol{v}_{0,t})$$

We say the bilinear system (3.5-1) is controllable in a region $S \subset R^n$ if for every given pair of initial and terminal state $\boldsymbol{\eta}_0$ and $\boldsymbol{\eta}_F$ in S, there is a positive integer T and a finite sequence of instruments $\boldsymbol{v}_{0,T}$ such that $\boldsymbol{\eta}_F = \psi(T; \boldsymbol{\eta}_0, \boldsymbol{v}_{0,T})$.

First, consider the case in which

$$\text{rank } B_t = 1, \qquad \text{for all } t$$

In this case the saving rate is not responsive to the capital stocks, i.e. $s_p = 0$ and $s_g = 0$ in the example of §3.1. From (3.5-4) we see that B_t has rank one for all t. To remind ourselves of this assumption, we rewrite (3.5-3) as

$$\boldsymbol{\eta}_{t+1} = (A_t + v_t \boldsymbol{c}_t \boldsymbol{h}_t^{\mathrm{t}})\boldsymbol{\eta}_t \tag{3.5-3$'$}$$

where

$$B_t = \boldsymbol{c}_t \boldsymbol{h}_t^{\mathrm{t}}$$

is the matrix of rank 1. Since the development for the example of §3.1 where dim $\boldsymbol{\eta}_t = 2$ is the same as that for more general case dim $\boldsymbol{\eta}_t = n$ we take A_t and B_t to be $n \times n$ matrices in what follows. Equation (3.5-3′) can be rewritten further as

$$\boldsymbol{\eta}_{t+1} = A_t \boldsymbol{\eta}_t + \boldsymbol{c}_t g_t \tag{3.5-3a}$$

$$g_t = v_t \boldsymbol{h}_t^{\mathrm{t}} \boldsymbol{\eta}_t \tag{3.5-3b}$$

where g_t is the new scalar-valued instrument of the ordinary linear discrete-time control system (3.5-3a). Equation (3.5-3b) shows that the value of this newly introduced instrument is generated by the feedback of $\boldsymbol{\eta}_t$ through the time-varying feedback control gain $v_t \boldsymbol{h}_t^{\mathrm{t}}$. We have discussed controllability of systems such as (3.5-3a) in §3.2.1. If (3.5-3′) is controllable in R^n outside the origin, then there exists a finite T and a $\boldsymbol{v}_{0,T}$ and consequently $\boldsymbol{g}_{0,T}$ as defined by (3.5-3b) such that

$$\boldsymbol{\eta}_F - \phi_{T,0}\boldsymbol{\eta}_0 = \sum_{s=0}^{T-1} \phi_{T,s+1}\boldsymbol{c}_s g_s$$

has a solution. Namely, (3.5-3a) is a completely controllable dynamic system. Controllability of the bilinear system (3.5-3′) thus implies controllability of the linear system (3.5-3a).

Proposition 1 *The bilinear system* (3.5-3′) *with rank* $B_t = 1$ *for all* t *is controllable outside the origin in* R^n *only if* (3.5-3*a*) *is completely controllable.*

Suppose now that at some time t the n row vectors

$$\boldsymbol{h}_t^{\mathrm{t}},\ \boldsymbol{h}_{t+1}^{\mathrm{t}}\phi_{t+1,t},\ \ldots,\ \boldsymbol{h}_{t+n-1}^{\mathrm{t}}\phi_{t+n-1,t}$$

are not linearly independent, where $\phi_{\mathrm{t},\tau}$ is the transition matrix associated with (3.5-3a). Suppose that $\boldsymbol{\eta}_{\mathrm{t}}$ is such that

$$\Omega_{t,t-n+1}\boldsymbol{\eta}_t = 0$$

where

$$\Omega_{t,t-n+1} = \begin{bmatrix} \boldsymbol{h}_t^{\mathrm{t}} \\ \boldsymbol{h}_{t+1}^{\mathrm{t}}\phi_{t+1,t} \\ \vdots \\ \boldsymbol{h}_{t+n-1}^{\mathrm{t}}\phi_{t+n-1,t} \end{bmatrix}$$

Such $\boldsymbol{\eta}_t$ exists since $\boldsymbol{h}_t^{\mathrm{t}}, \ldots, \boldsymbol{h}_{t+n-1}^{\mathrm{t}}\phi_{t+n-1,t}$ are linearly dependent by assumption. Then

$$\begin{aligned} g_t &= 0 \\ g_{t+1} &= v_{t+1}\boldsymbol{h}_{t+1}^{\mathrm{t}}\boldsymbol{\eta}_{t+1} \\ &= v_{t+1}\boldsymbol{h}_{t+1}^{\mathrm{t}}(\phi_{t+1,t}\boldsymbol{\eta}_t + \boldsymbol{c}_t g_t) \\ &= v_{t+1}\boldsymbol{h}_{t+1}^{\mathrm{t}}\phi_{t+1,t}\boldsymbol{\eta}_t = 0 \\ &\ \ \vdots \end{aligned}$$

showing that no matter what instruments $v_t, v_{t+1}, \ldots, v_{t+n-1}$ are used, $0 = g_t = \ldots = g_{t+n-1}$ and (3.5-3a) is not controllable.

We have established then:

Proposition 2 *For* (3.5-3′) *to be controllable, it is necessary that rank* $\Omega_{t, t+n-1} = n$ *for all* t.

Thus far, we have obtained two necessary conditions for controllability of the bilinear dynamic system (3.5-3) with rank $B_t = 1$, for all t.

We next give a sufficient condition for controllability of bilinear systems with rank $B_t = 1$, for all t. Suppose that (3.5-3a) is completely controllable.

Then for any $\boldsymbol{\eta}_0$ and $\boldsymbol{\eta}_F$, a unique sequence of $\boldsymbol{g}_{t,T}$ exists that transfers $\boldsymbol{\eta}_0$ at time t to $\boldsymbol{\eta}_F$ at time T. Then from (3.3-3b), v_t is uniquely determined if $\boldsymbol{h}_\tau{}'\boldsymbol{\eta}_\tau \neq 0$, $\tau = t, t+1, \ldots, T-1$ where $\boldsymbol{\eta}$'s are the path generated by $\boldsymbol{g}_{t,T}$.

Since $\boldsymbol{\eta}$'s are functions of g's, we are done if we can show the existence of $\boldsymbol{g}_{t, t+n-1}$ for which $\boldsymbol{h}_\tau{}'\boldsymbol{\eta}_\tau \neq 0$, $\tau = t, \ldots, T-1$, while reaching η_F at the end. We have:

Theorem 1 *The bilinear system, with rank* $B_t = 1$ *for all* t, *is completely controllable if* (3.5-3*a*) *is completely controllable, rank* $\Omega_{t, t+n-1} = n$, $\boldsymbol{h}_t{}'\boldsymbol{c}_{t-1} \neq 0$ *for all* t, *and if* A_t *is nonsingular and* $\boldsymbol{h}_t{}'A_t{}^{-1}\boldsymbol{c}_t \neq 0$ *for all* t.

For proof, see (Goka *et al.*, 1973).

3.6 OBSERVABILITY (RECONSTRUCTIBILITY)

3.6.1 Linear Differential Systems

Loosely speaking, observability refers to the possibility of determining the current state from the future observation data, while reconstructibility refers to the possibility of determining the state at some past date from the observation data currently available.

The reconstructibility is defined formally as the ability to recover the initial condition uniquely from a set of output and control data currently available. Consider a dynamic system described by

$$\begin{aligned} \dot{z} &= \boldsymbol{f}(z, \boldsymbol{x}, t) \\ y &= \boldsymbol{g}(z, \boldsymbol{x}, t) \end{aligned} \tag{3.6-1}$$

Denote the output at time t when $z(t_0) = z_0$ by $y(t, t_0, z_0, \boldsymbol{x})$. We say the system (3.6-1) is completely reconstuctible if for all t_1, there exists t_0, $-\infty$

$< t_0 < t_1$, such that

$$y(t;\, t_0,\, z_0,\, x) = y(t;\, t_0,\, z_0',\, x), \qquad t_0 \leqslant t \leqslant t_1$$

for all $x(t)$, $t_0 \leqslant t \leqslant t_1$ implies $z_0 = z_0'$.

For a linear system

$$\begin{aligned} \dot{z} &= A(t)z + B(t)x(t) \\ y(t) &= C(t)z(t) + D(t)x(t) \end{aligned} \tag{3.6-2}$$

this definition is equivalent to:

Proposition *The system* (3.6-2) *is completely reconstructible if and only if for all* t_1, *there exists* $-\infty < t_0 < t_1$ *such that*

$$y(t;\, t_0,\, z_0,\, 0) = 0, \qquad t_0 \leqslant t \leqslant t_1 \tag{3.6-3}$$

implies $z_0 = 0$.

Proof If reconstructible, then (3.6-3) holds with $z_0 = 0$.

Since

$$y(t;\, t_0,\, z_0,\, x) = C(t)\left[\phi(t,\, t_0)z_0 + \int_{t_0}^{t} \phi(t,\, s)B(s)x(s)\, ds\right] + D(t)x(t),$$

$$y(t;\, t_0,\, z_0,\, x) = y(t;\, t_0,\, z_0',\, x)$$

implies

$$C(t)\phi(t,\, t_0)z_0 = C(t)\phi(t,\, t_0)z_0', \qquad t_0 \leqslant t \leqslant t_1$$

or

$$C(t)\phi(t,\, t_0)(z_0 - z_0') = 0$$

This is $y(t;\, t_0,\, z_0 - z_0',\, 0) = 0$.

Thus, if $y(t;\, t_0,\, z_0 - z_0',\, 0) = 0$ implies $z_0 - z_0' = 0$ then $z_0 = z_0'$, i.e., (3.6-2) is reconstructible.

The concept of observability is related to that of reconstructibility. We say the system (3.6-1) is uniformly completely observable if for all t_0, there exists t_1, $t_0 < t_1 < \infty$ such that $y(t;\, t_0,\, z_0,\, x) = y(t;\, t_0,\, z_0',\, x)$, $t_0 \leqslant t \leqslant t_1$ for all $x(t)$, $t_0 \leqslant t \leqslant t_1$ implies $z_0 = z_0'$.

Thus, for linear time-invariant system, these two concepts coincide.

Theorem 1 *A linear time-invariant system is completely reconstructible if and only if it is completely observable.*

We recall that controllability has to do with the range of the map $x \mapsto z(t; t_0, z_0, x)$. Observability is about the inverse map

$$\left\{\begin{array}{l}\text{the set of outputs}\\ y(t; t_0, z_0, x),\\ t_0 \leqslant t \leqslant t_1\end{array}\right\} \mapsto z_0$$

For a linear time-invariant linear system, we have a convenient test for reconstructibility when $D = 0$, namely

$$\mathfrak{N}(R) = \mathfrak{N}(C) \cap \mathfrak{N}(CA) \cap \ldots \cap \mathfrak{N}(CA^{n-1}) = \{0\} \quad (3.6\text{-}4)$$

where $\mathfrak{N}(\cdot)$ is the null space† of the observability matrix defined by

$$R = \begin{bmatrix} C \\ CA \\ \vdots \\ CA^{n-1} \end{bmatrix}$$

Theorem 2 *The system*

$$\dot{x} = Az + Bx$$

$$y = Cz$$

is completely reconstructible if and only if $\mathfrak{N}(R) = \{0\}$.

We can state the condition (3.6-4) equivalently as "the row vectors of R span R^n."

Proof Equation (3.6-3) becomes after denoting $y(t) - C\int_0^t e^{A(t-s)}Bx(s)\, ds$ by $w(t)$

$$w(t) = Ce^{A(t-t_0)}z_0 = 0, \qquad t_0 \leqslant t \leqslant t_1$$

by repeated differentiation, we get

$$w(t_0) = Cz_0$$

$$\dot{w}(t_0) = CAz_0$$

$$\vdots$$

$$w^{(n-1)}(t_0) = CA^{n-1}z_0$$

† $\mathfrak{N}(R) = \{z | Rz = 0\}$.

or

$$Rz_0 = 0$$

Then

$$z_0 = 0 \qquad \text{if } \mathcal{N}(R) = 0$$

Suppose

$$Ce^{A(t-t_0)}z_0 = 0 \text{ implies } z_0 = 0, \text{ for } t_0 \leqslant t \leqslant t_1.$$

Then

$$\mathcal{N}(Ce^{A(t-t_0)}) = \{0\} \text{ for } t_0 \leqslant t \leqslant t_1.$$

By the Cayley–Hamilton Theorem

$$Ce^{A(t-t_0)} = C\left[\alpha_0(t)I + \alpha_1(t)A + \cdots + \alpha_{n-1}(t)A^{n-1}\right]$$

or

$$\mathcal{N}(R) = \{0\}.$$

When this condition is met, we say (A, C) is a completely observable pair. Since

$$R^{t} = \begin{pmatrix} C^{t} & A^{t}C^{t} & \dots & (A^{t})^{n-1}C^{t} \end{pmatrix}$$

The condition for complete observability is that (A^{t}, C^{t}) *is a completely controllable pair. In this sense, these two concepts are dual.*

3.6.2 Linear Difference System

For systems described by linear difference equations we can define the concepts of reconstructibility and observability analogously. We illustrate the derivation of a criterion for observability for time-invariant linear discrete-time systems.

Suppose that at time t we know past values of the target vectors and the instrument vectors of a dynamic system

$$z_{\tau+1} = Az_\tau + Bx_\tau$$

$$y_\tau = Cz_\tau + Dx_\tau, \qquad \tau = 0, 1, \dots, t$$

Here the initial time is taken, without loss of generality, to be 0.

From (2.2-21),

$$y_\tau = C\phi_{\tau,0}z_0 + \sum_{s=0}^{\tau-1} C\phi_{\tau,s+1}Bx_s + Dx_\tau, \qquad \tau = 0, 1, \dots, t$$

Write this into the vector matrix form

$$\begin{bmatrix} w_0 \\ w_1 \\ \vdots \\ w_t \end{bmatrix} = R_t z_0 \qquad (3.6\text{-}5)$$

where

$$w_\tau = y_\tau - \sum_{s=0}^{\tau-1} C\phi_{\tau, s+1} B x_s - D x_\tau, \qquad \tau = 0, 1, \ldots, t$$

and

$$R_t = \begin{bmatrix} C \\ C\phi_{1,0} \\ C\phi_{2,0} \\ \vdots \\ C\phi_{t,0} \end{bmatrix} : m(t+1) \times n \text{ matrix}$$

Since w's are all known and z_0 is the only unknown vector by assumption, the observability question reduces to that of the existence of a unique solution to

$$\eta = R_t z_0$$

where η is a known vector.

The unique solution exists if and only if $\mathcal{N}(R_t) = 0$. This is exactly the same criterion of observability (3.6-4) obtained above.

3.6.3 Stochastic Controllability and Observability

We have discussed the concepts of controllability and observability for deterministic dynamic systems in state-space form. We can extend the scope of our discussion to stochastic systems in a natural way. Especially, the concept of observability finds a natural counterpart in statistics since observability is a condition on behavior of estimation error of some parameter or state vector as the size of observation data grows.

This is clearly seen from (3.6-5). A stacked set of 'observation' vectors is related to the unknown state vector z_0 by the algebraic equation (3.6-5). Consider $t < n$. Then from (3.6-5)

$$z_0 = R_t^{+} w_{0,t} + \rho, \qquad \rho \in \mathfrak{N}(R_t)$$

where $w_{0,t}$ is the stacked vector of $w_0, w_1, \ldots, w_t$, and ρ is some vector in the null space of R_t. The estimated or reconstructed z_0 from $w_0, \ldots, w_t$ is $R_t^{+} w_{0,t}$, where R_t^{+} denotes the Penrose pseudo-inverse.† The null space of R_t is nonincreasing with t and reduces to $\{0\}$ if and only if the system is observable.

When we have exogenous random noises, then estimated or reconstructed vectors do not converge to the true unknown vector when the observation data is finite. Stochastic observability may be defined then as a condition for estimates to converge in some probability sense. See, for example, Aoki (1967), Fitts (1972).

Stochastic controllability can similarly be defined by replacing deterministic descriptions by some probabilistic ones. See Connors (1967).

What is surprising is that deterministic concepts of controllability and observability are still useful in systems with random disturbances.

If a system is observable, then we could uniquely determine an initial state from a finite data. Observability is thus a special case of a broader problem of constructing (asymptotically) consistent estimators or of statistical hypothesis testing‡, such as determining which alternate models are consistent with the observed data set. If a system is observable, then the set of initial states that is consistent with the observed data eventually reduces to a singleton. See (Aoki and Li, 1973) on this topic.§ See Basman (1965) for an example of economic problems of determining alternate economic models consistent with observation.

Stochastic Stability Literature on stochastic controllability and observability is small compared with that on stochastic stability. As a matter of fact, a vast amount of papers, technical reports, and several books are available on the subject of stochastic stability of both continuous-time and

†See for example Appendix II in (Aoki, 1967).

‡For example, economic agents can statistically distinguish local disturbances from global ones, only if some observability conditions are satisfied.

§Decentralization is not the issue. The techniques of (Aoki and Li, 1973) can be applied to centralized systems as well, to construct a set of state vectors consistent with observation data.

discrete-time dynamic systems. For discrete-time systems, see for example Kushner (1974) or Aoki (1967) on several definitions of stochastic (Lyapunov) stability, and Beutler (1973) of discussion on them. See Curtain (1972) on stability of continuous-time stochastic systems.

3.6.4. Examples

In this section we present two examples of applications of observability. These examples show how interesting and perhaps unexpected implications may be drawn by examining suitably defined "observability" conditions. The first example shows a microeconomic model in which unobservability implies nonunique equilibria. The second example describes a monetary model in which nonexistence of a subset of markets being cleared is established by observability arguments.

Example 1 (Uniqueness of Equilibrium) We consider an industry composed of n firms. These firms all produce a homogeneous perishable good. Production takes one day so that the firms must make production decisions one day in advance. Because of the nonstorable nature of the goods put on the market by these firms, a unique market clearing price is established each day by

$$p_t = \gamma - \delta Q_t$$

where

$$Q_t = \Sigma_{j=1}^n q_{jt}$$

and where q_{jt} is the output put on the market by firm j on the tth day. Suppose for simplicity that firms adjust the outputs in response to changes in the market clearing price by

$$\Delta \boldsymbol{q}_{t+1} = F \Delta q_t + \boldsymbol{g} \Delta p_t$$

where F is an $n \times n$ matrix and $\boldsymbol{g}$ is an n-vector,

$$\Delta \boldsymbol{q}_t = \boldsymbol{q}_{t+1} - \boldsymbol{q}_t$$

and where

$$\boldsymbol{q}_t = (q_{1t}, \ldots, q_{nt})^{\mathrm{t}}$$

Suppose further that firms do not form coalitions so that they do not know other firms' production decisions. The market clearing price is the only signal observed by firms. If there are no changes in p_t over a period of n consecutive days, would a firm be justified to believe that every firm has reached an equilibrium?

From $\Delta p_t = \Delta p_{t+1} = \cdots = \Delta p_{t+n-1} = 0$ we obtain

$$0 = \begin{bmatrix} e^{t}\Delta q_t \\ e^{t}\Delta q_{t+1} \\ \vdots \\ e^{t}\Delta q_{t+n-1} \end{bmatrix} = \begin{bmatrix} e^{t} \\ e^{t}\phi \\ \vdots \\ e^{t}\phi^{n-1} \end{bmatrix} \Delta q_t$$

where

$$\phi = F - \delta g e^{t}$$

Thus, if (F, e) is an observable pair, i.e., if rank $(e, F^{t}e, \ldots, (F^{t})^{n-1}e) = n$, then we conclude that $\Delta q_t = 0$, i.e., every form is in equilibrium. If the rank is less than n, then the market clearing price can remain the same on n consecutive days while not all firms' outputs are at their equilibrium values.

Example 2 (Nonexistence of Partial Clearings of Markets) We describe an interesting and unexpected implication of lack of observability for a monetary disequilibrium model in which trading takes place out of equilibrium. Since this model is described in detail in §7.2, we give only brief descriptions of the model and take the dynamic equation as given. See also §7.1 for another example.

Suppose there are n traders indexed by i, $i = 1, \ldots, n$ and who receive constant flows of n goods, indexed by j as endowments. The money stock held by trader i at time t is $M_i(t)$, and his flow endowment of commodity j is denoted by w_j^{i}. For simplicity, assume the total money stock is constant,

$$\sum_{i=1}^{n} M_i(t) = M = \text{constant}$$

Trader i has a notional demand for commodity j, $d_j^{i}(t)$. Hence his excess demand for good j is

$$x_j^{i} = d_j^{i} - w_j^{i}, \qquad i, j = 1, \ldots, n$$

and the market excess demand for good j is

$$x_j = \sum_{i=1}^{n} x_j^i, \qquad j = 1, \ldots, n$$

Money is used in transactions to buy and sell goods. Assume that the price for good j, p_j, is changing according to

$$\begin{aligned} \dot{p}_j &= \beta_j p_j x_j \\ &= \beta_j z_j \end{aligned}$$

where we denote the nominal excess demand for good j by

$$z_j = p_j x_j, \qquad j = 1, \ldots, n$$

Because trading takes place out of equilibrium some traders may be rationed depending on the signs of x_j^i and x_j. We assume that the trader i is rationed proportionally when he is on the "long" side of the market, i.e., if $x_j^i x_j > 0$. His money stock changes according to

$$\dot{M}_i = -\sum_{j=1}^{n} p_j x_j^i k_j^i$$

where k_j^i is the proportionality due to rationing, i.e., $k_j^i = 1$ if trader is on the short side and $k_j^i < 1$ otherwise. We are not interested in the details of how it is determined since it is irrelevant to what we wish to discuss here.

Goods can be classified at any time t according to whether there exists a negative excess demand, positive excess demand or zero excess demand. The markets for those goods with zero excess demand clear at time t. Thus, the nominal excess demand vector z with components $z_1, \ldots, z_n$ can be partitioned as

$$z = \begin{bmatrix} z_+ \\ z_- \\ z_0 \end{bmatrix}$$

where actually z_0 is the null vector, with $\dim z_0 = n - \dim z_+ - \dim z_- = n_0$. Goods which go into z_+, z_- and z_0 change with time, in general.

It can be shown that an equilibrium state exists at $z = 0$ (all markets clear) and the total money stock is distributed in a unique way.

The linearized equation (rationing introduces nonlinearity into the price equation and the money stock equation) about the equilibrium can be derived as explained in §2.6 and can be written as

$$\dot{\zeta} = F\zeta \text{ and } \dot{z}_0 = S\zeta$$

where

$$\zeta = \begin{bmatrix} z_+ \\ z_- \\ \delta \boldsymbol{M} \end{bmatrix}$$

and where $\delta \boldsymbol{M}$ is the money stock vector for the n traders measured from the equilibrium money stock vector. The dimension of ζ is $2n - n_0$ and F and S are $(2n - n_0) \times (2n - n_0)$ and $n_0 \times (2n - n_0)$ matrices: Since $z_0 = 0$, it does not appear in the dynamic equation on the right-hand side of the above equations. Let $\dot{z}_0(t)$ be the change in the excess demand for the subset of goods which happen to clear at the time t.

We ask now if some markets stay cleared for some time in this model, i.e., does there exist a subset of goods the markets for which remain cleared? In other words, if the markets for the goods $i_1, i_2, \ldots, i_{n_0}$ clear at time t, will the markets for these goods remain cleared during the time interval $[t, t + \tau]$ for some $\tau > 0$?

Phrased in this way, we see we can apply the techniques introduced in proving controllability and observability criteria to answer the question.

From the condition $z_0(s) \equiv 0$ for $t \leqslant s \leqslant t + \tau$, we have

$$\dot{z}_0(t) = S\zeta = 0$$

$$z_0^{(j)}(t) = S\zeta^{(j-1)} = SF^{j-1}\zeta = 0, \qquad j = 2, \ldots, 2n - n_0$$

Putting them together in a vector-matrix form we have

$$\mathfrak{L}\zeta = 0 \tag{3.6-6}$$

where

$$\mathfrak{L} = \begin{bmatrix} S \\ SF \\ \vdots \\ SF^{2n-n_0-1} \end{bmatrix}$$

If $\mathcal{N}(\mathcal{L}) = \{0\}$, then $\zeta = 0$ is the only vector that satisfies (3.6-6). But then $\zeta = 0$ means that all markets clear and the system is at equilibrium. Therefore, if (F, S) is a completely observable pair, then there do not exist partial equilibria in which markets for some goods remain cleared over a positive time interval.

3.6.4 Nonlinear Systems

Interestingly, the question of observability for nonlinear dynamic systems has been investigated by Gale and Nikaido as the question of global univalence in the context of factor price equalization and the uniqueness of competitive equilibrium. Later their conditions were relaxed by Fitts and Eliott *et al.*, see (Gilbert, 1969), (Kou *et al.*, 1973) and (Nikaido, 1968) for further details.

4

Stability of Dynamic Systems

4.1 CONTINUOUS TIME SYSTEMS

Given a differential equation

$$\dot{y} = f(y) \tag{4.1-1}$$

any point y^e such that $f(y^e) = 0$ is called an equilibrium point. Make it the origin by a change of the coordinate system, if necessary.

Our definition of stability is the following: An equilibrium state $y = 0$ of (4.1-1) is said to be stable if for every $\epsilon > 0$, there is a $\delta = \delta(\epsilon)$ such that all solutions $y(t)$ of (4.1-1) satisfying $\|y(t_0)\| \leqslant \delta$ satisfy $\|y(t)\| \leqslant \epsilon$ for all $t \geqslant t_0$.

This concept of stability is also called stability in the sense of Lyapunov (Desoer, p. 142, 1970).

If *in addition* $y(t) \to 0$ as $t \to \infty$, then we say $y = 0$ is asymptotically stable.

4.1.1 Linear Dynamic Systems

We have already discussed in §2.2.1 the stability of linear dynamic systems

$$\dot{y}(t) = Ay(t)$$

where the $n \times n$ matrix A has n linearly independent eigenvectors. When A does not possess n linearly independent eigenvectors, our earlier development must be modified.

Since Jordan canonical forms are complicated to explain, we do not use them. Instead we follow Bellman (1953) and work with triangular matrices.

Theorem 1 (Bellman) *Given an $n \times n$ matrix A, there exists a nonsingular matrix T such that $T^{-1}AT$ is (upper) triangular.*

Proof The proof is by induction on n (see p. 21 of Bellman, 1953).

Corollary *The matrix T may be chosen so that $\Sigma_{i,j}|b_{ij}| < \epsilon$ for any $\epsilon > 0$ where*

$$T^{-1}AT = \begin{bmatrix} \lambda_1 & \cdots & \cdots & \cdots \\ 0 & \cdots & b_{ij} & \cdots \\ \cdots & \cdots & \cdots & \cdots \\ 0 & 0 & \cdots & \lambda_n \end{bmatrix}$$

Proof See p. 23 of Bellman (1953).

Given

$$\dot{z} = Az$$

where A is an $n \times n$ constant matrix, perform a change of variable

$$y = Tz$$

so that $T^{-1}AT$ is upper triangular. Then

$$\dot{y} = By$$

with

$$B = \begin{bmatrix} \lambda_1 & \cdots & \cdots & \cdots \\ 0 & \cdots & b_{ij} & \cdots \\ \cdots & \cdots & \cdots & \cdots \\ 0 & 0 & \cdots & \lambda_n \end{bmatrix}$$

In terms of the components, the differential equations are

$$\begin{aligned} \dot{y}_1 &= \lambda_1 y_1 + b_{12}y_2 + \cdots + b_{1n}y_n \\ \dot{y}_2 &= \qquad\qquad \lambda_2 y_2 + \cdots + b_{2n}y_n \\ &\vdots \\ \dot{y}_n &= \qquad\qquad\qquad\qquad\quad +\lambda_n y_n \end{aligned}$$

Solve this system of differential equations from y_n backwards to obtain $y_{n-1}, y_{n-2}, \ldots, y_1$.

First of all we have $y_n = y_n(0)e^{\lambda_n t}$. Substituting it into

$$\dot{y}_{n-1} = \lambda_{n-1}y_{n-1} + b_{n-1,n}y_n$$

solve this equation for y_{n-1}.

If $\lambda_{n-1} \neq \lambda_n$, then

$$y_{n-1} = y_{n-1}(0)e^{\lambda_{n-1}t} + \left(\int_0^t e^{\lambda_{n-1}(t-s)}e^{\lambda_n s}\,ds\right)c$$

$$= y_{n-1}(0)e^{\lambda_{n-1}t} + c(e^{\lambda_n t} - e^{\lambda_{n-1}t}) \Big/ (\lambda_n - \lambda_{n-1})$$

with $c = b_{n-1,n}y_n(0)$.

If $\lambda_{n-1} = \lambda_n$, then from the first line of the above expression,

$$y_{n-1} = (y_{n-1}(0) + ct)e^{\lambda_{n-1}t}$$

We now repeat the process by solving for y_{n-2} with the known functions y_{n-1} and y_n as the inhomogeneous terms.

We see that if λ's are all distinct, y_i, $i = 1, \ldots, n$ are linear combinations of exponentials. If some eigenvalue λ has multiplicity k, then the components of y will have the form $P_{k-1}(t)e^{\lambda t}$ where $P_{k-1}(t)$ is a polynomial of degree $k - 1$ at most.

Thus we have the following theorem.

Theorem 2 *A necessary and sufficient condition that all solutions of*

$$\dot{y} = Ay$$

tend to zero as $t \to \infty$ is that $\mathrm{Re}(\lambda_i) < 0$ *for all $i = 1, \ldots, n$ where λ_i is an eigenvalue of A. Namely, if* $\mathrm{Re}(\lambda_i) < 0$ *for all i, then the system is asymptotically stable.*

4.1.2 Nonlinear Dynamic Systems

In this section we extend the stability analysis of the previous section to nonlinear systems.

First we consider systems that are only slightly nonlinear in the sense that the nonlinear term in the differential equation is "smaller" that the linear term†

$$\dot{z} = Az + f(z) \tag{4.1-2}$$

†We know that the unique solution exists in a small neighborhood of $t = 0$, since $Az + f(z)$ satisfies the Lipschitz condition, when $f(z) = O(\|z\|)$.

where

$$f(0) = 0 \tag{4.1-3}$$

and $f(z)$ is small compared with z.

This smallness requirement may be expressed more precisely by

$$\|f(z)\|/\|z\| \to 0 \text{ as } \|z\| \to 0$$

This is often written as

$$f(z) = o(\|z\|)$$

From (4.1-3), $z(t) \equiv 0$ is clearly a solution of (4.1-2), or the origin is an equilibrium solution (or state). The solution $z = 0$ is also said to be a null solution of (4.1-2).

We establish that a solution exists on the entire positive axis for (4.1-2) if:

(a) all eigenvalues of A have negative real parts
(b) f is continuous in some neighborhood of $z = 0$
(c) $f(z) = o(\|z\|)$

by showing that (i) $\|z\|$ is uniformly bounded and that (ii) the bound thus obtained lies within R, the region in which f is continuous.

From (4.1-2)

$$z(t) = e^{At}c + \int_0^t e^{A(t-s)} f(z(s))\, ds$$

Assumption (a) implies the existence of a positive constant α such that $\|e^{At}c\| \leqslant \alpha\|c\|$ for all $t \geqslant 0$.

Fact For all $\|c\|$ sufficiently small, $\|z(t)\| < 2\alpha\|c\|$ for all $t \geqslant 0$.

The proof is by contradiction (see p. 80, Bellman 1953).

We state a local asymptotic stability theorem for (4.1-2).

Theorem *Under assumptions (a) ~ (c), $z(t) \to 0$ as $t \to \infty$ if $\|z(0)\|$ is sufficiently small.*

For a proof, see p. 80 of (Bellman, 1953).

Earlier, we discussed the existence of a nonsingular matrix T transforming A into an upper triangular form with $\Sigma|b_{ij}| < \epsilon$ for any $\epsilon > 0$. Let T be such a matrix and let $y = T^{-1}z$.

Then

$$\dot{y} = Uy + h(y)$$

with

$$U = T^{-1}AT \qquad \text{upper triangular.}$$

$$\boldsymbol{h}(y) = T^{-1}\boldsymbol{f}(Ty)$$

Note that:

(i) $\boldsymbol{h}(0) = 0$
(ii) $\boldsymbol{h}$ is continuous
(iii) $\boldsymbol{h}(y) = \boldsymbol{o}(\|y\|)$

Let A be such that all its eigenvalues λ satisfy

$$\mathrm{Re}(\lambda) \leqslant \mu < 0$$

Then, using $\overline{}$ to denote the complex conjugate

$$\frac{1}{2}\frac{d}{dt}\bar{y}^{\mathrm{t}}y = \bar{y}^{\mathrm{t}}(U + \overline{U})y + \tfrac{1}{2}\left(\bar{y}^{\mathrm{t}}\boldsymbol{h}(y) + \overline{\boldsymbol{h}(y)}^{\mathrm{t}}y\right)$$

$$< (\mu + \epsilon)\bar{y}^{\mathrm{t}}y + \|y\| \cdot \|\boldsymbol{h}(y)\| \tag{4.1-4}$$

where the Cauchy–Schwarz inequality is used.

Since $\mu < 0$, with $y(0)$ sufficiently small, we have

$$\left(\frac{d}{dt}\bar{y}^{\mathrm{t}}y\right)_{t=0} < 0$$

Actually we have

$$\bar{y}^{\mathrm{t}}(t)y(t) \leqslant \left[\bar{y}^{\mathrm{t}}(0)y(0)\right]e^{-\beta t}$$

for some $\beta > 0$, hence $y(t) \to 0$ as $t \to \infty$, i.e., the origin of the system (4.1-4) is asymptotically stable.

4.2 DISCRETE TIME SYSTEMS

Stability of the discrete-time time-invariant system

$$z_{t+1} = Az_t \tag{4.2-1}$$

where A is a constant $n \times n$ matrix, is immediate when A has n distinct eigenvalues. Then we know that there exists a nonsingular matrix T such that

$$T^{-1}AT = \mathrm{diag}(\lambda_1 \quad \ldots \quad \lambda_n)$$

The solution of (4.2-1) is

$$z_t = A^t z_0 \qquad t = 0, 1, \ldots$$

where

$$T^{-1}A^tT = \mathrm{diag}(\lambda_1{}^t \quad \ldots \quad \lambda_n{}^t)$$

Change the variable from z_t to w_t by

$$z_t = Tw_t$$

Then

$$w_t = T^{-1}ATw_{t-1}$$

Thus

$$w_t = \begin{bmatrix} \lambda_1{}^t & 0 & 0 & 0 \\ 0 & \cdots & 0 & 0 \\ & \vdots & & \\ 0 & 0 & \cdots & 0 \\ 0 & 0 & 0 & \lambda_n{}^t \end{bmatrix} w_0$$

and

$$z_t = T \begin{bmatrix} \lambda_1{}^t & 0 & 0 & 0 \\ 0 & \cdots & 0 & 0 \\ & \vdots & & \\ 0 & 0 & \cdots & 0 \\ 0 & 0 & 0 & \lambda_n{}^t \end{bmatrix} T^{-1}z_0$$

we see that:

Proposition 1 *The time-invariant discrete-time system* (4.2-1) *is stable if the eigenvalues of* A *are all distinct and if* $|\lambda_i| \leqslant 1$, $i = 1, \ldots, n$.

The system (4.2-1) *is asymptotically stable if* A *has* n *distinct eigenvalues and if* $|\lambda_i| < 1$, $i = 1, \ldots, n$.

When A has some repeated eigenvalues, we can repeat the analysis carried out for continuous-time systems. From the Corollary of Theorem 1, we can pick any $0 < c < 1/n$ and T such that $T^{-1}AT$ is upper triangular with the eigenvalues of A lying along the main diagonal line and the elements b_{ij} above the main diagonal line such that $|b_{ij}| < c < 1/n$.

Let $S = T^{-1}AT$.

Then S^2 is upper triangular, i.e., ${S_{ij}}^2 = 0$ for $i > j$, ${S_{ii}}^2 = \lambda_i^2$ and

$$
\begin{aligned}
{S_{ij}}^2 &= \sum_{k=i}^{j} S_{ik} S_{kj} \qquad \text{for } i < j \\
&= b_{ij}^{(2)},
\end{aligned}
$$

where

$$b_{ij}^{(2)} = \sum_{k=i+1} b_{ik} b_{kj} < nc^2 < c$$

By mathematical induction we see that S^t has ${\lambda_1}^t, \ldots, {\lambda_n}^t$ along the main diagonal line and the nonzero off-diagonal elements are less than c in modulus.

Changing the variable as before from z_t to $z_t = Tw_t$ we see that:

Proposition 2 *The solution of* (4.2-1) *is uniformly stable if the eigenvalues of A all have modulus not greater than* 1.

Proposition 2 states that the solution of (4.2-1) is uniformly bounded. Let

$$z_t = e^{\lambda t} y_t$$

where λ is a negative real number to be specified presently. Then from (4.2-1)

$$y_{t+1} = (Ae^{-\lambda}) y_t \tag{4.2-1'}$$

From

$$0 = |\mu I - Ae^{-\lambda}| = |\mu e^{\lambda} I - A| e^{-\lambda}$$

we see that we can choose $|\lambda|$ small enough so that the eigenvalues of $Ae^{-\lambda}$ all have modulus not greater than 1 if the eigenvalues of A all have modulus less than one.

From Proposition 2 the solution to (4.2-1′) is uniformly bounded, hence $z_t \to 0$ as $t \to \infty$.

We have established:

Proposition 3 *If the eigenvalues of A in* (4.2-1) *all have modulus less than one, then the origin is asymptotically stable.*

See Section 1.3 of Miller, 1968 for stability arguments for time-varying discrete time systems.

Example Let us examine the macroeconomic model introduced in Example 5 of §2.2 for stability. The matrix F of (2.2-E7) has eigenvalues f_{33} and 0 with multiplicity 2, where

$$
\begin{aligned}
f_{33} &= d_1 - \boldsymbol{\lambda}^t(a_1\boldsymbol{\gamma} + b_1\boldsymbol{\delta})/\Delta \\
&= d_1 - \{a_{31}(a_1a_{22} - b_1a_{12}) + a_{32}(-a_1a_{21} + b_1)\}/\Delta \\
&= d_1 - \{a_1(a_{22}a_{31} - a_{21}a_{32}) + b_1(a_{32} - a_{31}a_{12})\}/\Delta
\end{aligned}
$$

The system is therefore stable if $|f_{33}| \leqslant 1$. From their defining relations (b) and (2.2-E7), we have

$$
\begin{aligned}
a_{22}a_{31} - a_{21}a_{32} &= -d_2/c_2 \\
a_{32} - a_{31}a_{12} &= d_2(1 - a_2c_1)/c_2 \\
\Delta &= b_2(1 - a_2c_1)/c_2 - a_2
\end{aligned}
$$

Rewrite f_{33} using the above as

$$
f_{33} = d_1 + \frac{d_2}{c_2}\,\frac{a_1 - b_1(1 - a_2c_1)}{(1 - a_2c_1)b_2/c_2 - a_2}
$$

For stability, it is necessary that the coefficients in (2.2-E1)–(2.2-E4) satisfy

$$
-1 - d_1 \leqslant \frac{d_2}{c_2}\,\frac{a_1 - b_1(1 - a_2c_1)}{(1 - a_2c_1)b_2/c_2 - a_2} < 1 - d_1
$$

If the inflation rate is estimated by (2.2-E8), then the stability of the model is determined by the eigenvalues of Φ in (2.2-E9). They may be computed as follows.

$$
\begin{aligned}
0 = |\lambda I - \Phi| &= \begin{vmatrix} \lambda I - (F + \boldsymbol{\alpha f}^t) & -a_0\boldsymbol{\alpha} \\ -\boldsymbol{f}^t & \lambda - a_0 \end{vmatrix} \\
&= |\lambda - a_0|\,|\lambda I - F - \boldsymbol{\alpha f}^t - a_0\boldsymbol{\alpha f}^t/(\lambda - a_0)| \\
&= (\lambda - a_0)\left|\lambda I - F - \frac{\lambda}{\lambda - a_0}\boldsymbol{\alpha f}^t\right|
\end{aligned}
$$

where

$$\left|\lambda I - F - \frac{\lambda}{\lambda - a_0}\boldsymbol{\alpha f}^{\mathrm{t}}\right| = \det\begin{bmatrix} \lambda & -\dfrac{\lambda}{\lambda - a_0}\alpha_1 b_0 & -f_{13} \\ 0 & \lambda - \dfrac{\lambda}{\lambda - a_0}\alpha_2 b_0 & -f_{23} \\ 0 & -\dfrac{\lambda}{\lambda - a_0}\alpha_3 b_0 & \lambda - f_{33} \end{bmatrix}$$

$$= \lambda\left[(\lambda - f_{33})\left(\lambda - \frac{\lambda}{\lambda - a_0}\alpha_2 b_0\right) + f_{23}\left(-\frac{\lambda}{\lambda - a_0}\alpha_3 b_0\right)\right]$$

$$= \frac{\lambda^2}{\lambda - a_0}\left[(\lambda - f_{33})(\lambda - a_0 - \alpha_2 b_0) - f_{23}\alpha_3 b_0\right]$$

The eigenvalues of Φ are, then, two zeros and two roots of the quadratic polynomial in the above braces, where $(\alpha_1, \alpha_2, \alpha_3)^{\mathrm{t}} = \alpha$ as defined in (2.2-E7) of Example 5, §2.2.

We note that in both the systems we have examined above F and Φ, there is at least one zero eigenvalue. This means that without active intervention by the authority, it is not possible to reduce to zero deviations in some endogenous variables.

4.3 TIME-VARYING SYSTEMS

So far, we have discussed stability of time-invariant linear dynamic systems, that is, those whose characteristics do not change with time. Characteristics of economic systems usually change with time, however. The changes may occur gradually or may happen quickly, relative to some basic time scales in economic models. In the former, we may be able to model them by time-invariant systems in the short run. When changes in some important characteristics of the models may not be ignored, we model them as time-varying dynamic systems. We then have to decide on stability of time-varying systems. We know some sufficient conditions for asymptotic stability of time varying systems. A typical sufficient condition is as follows.

Theorem *Let* $\{z_t\}$ *be generated by* $z_{t+1} = (\Phi + \psi_t)z_t$ *where* $\|\Phi\| < 1$ *and* $\Sigma_t\|\psi_t\| < \infty$. *Then* $z_t \to 0$ *as* $t \to \infty$.

Proof There exists a number r such that $\|\Phi\| \leqslant r < 1$, by assumption. From the solution of the difference equation

$$z_t = \Phi^t z_0 + \sum_{j=0}^{t-1} \Phi^{t-1-j} \psi_j z_j$$

we have the inequality

$$\|z_t\| \leqslant r^t \|z_0\| + \sum_{j=0}^{t-1} r^{t-1-j} \|\psi_j\| \, \|z_j\|$$

Denote $\|z_t\| r^{-t}$ by $\|w_t\|$.

Then the above inequality is rewritten as

$$\|w_t\| \leqslant \|z_0\| + \frac{1}{r} \sum_{j=0}^{t-1} \|\psi_j\| \, \|w_j\|$$

From the difference equation version of the Bellman–Gronwall Lemma (See Miller, Sec. 1.3, 1968), we see that

$$\|w_t\| \leqslant c \exp\left(\frac{1}{r} \sum_{j=0}^{t-1} \|\psi_j\| \right)$$

From the assumption, we thus establish the boundedness of $\|w_t\|$ for all t, hence $\|z_t\| \to 0$ as $t \to \infty$.

4.4 DISCUSSION

We have shown how eigenvalues of the matrices associated with (linearized) dynamic equations determine stability behavior of the variables governed by the dynamic equations. When eigenvalues are easily determinable or can be shown to lie in some regions of stability,† eigenvalues are quite useful since they tell us a lot more than mere stability.

Unfortunately, in most cases, eigenvalues are rather difficult to obtain explicitly. We have some techniques to determine stability without explicit derivation of eigenvalues for a limited class of matrices such as matrices with all nonnegative matrices or Metzler matrices (matrices in which diagonal elements are of the same sign which is opposite to those of

†Gerschgorin's circle theorem is the most useful one of many results of this nature. See Householder, 1964.

nonnegative off-diagonal matrices. These matrices arise naturally in activity analysis, input-output models and in markets with the gross substitutability assumption. For positive or nonnegative matrices we have theorems due to Perron and Frobenius (Gantmacher 1959). See Bellman (1960), Karlin (1959), Newman (1959), Nikaido (1968), or Negishi (1962) for results on Metzler matrices.

Lyapanov's second method is a body of stability analysis technique that does not require explicit eigenvalue calculation and is applicable to a wider class of systems. Elementary exposition of the method is available in many places. See, for example, Bellman (1960), Newman (1961), LeSalle and Lefschetz (1961).

Lyapanov's method has been extended to cover stochastic systems. See for example (Kushner, 1967). For results on systems with uncertain parameters see (Peng, 1972).

As an illustration, we cite one theorem which is easily proved by Lyapunov's method. (Use a Lyapunov function $V(\boldsymbol{x}) = \Sigma_{i=1}^{n} d_i|x_i|$). We need the concept of diagonal dominance to state the theorem. We say that a matrix A has a dominant diagonal if $|a_{jj}| > \Sigma_{j \neq j}|a_{ij}|$, for all i. We say that A has a quasidominant diagonal if DA has a dominant diagonal for some diagonal matrix D with positive elements. The following result is due to Siljak (1973).

Let

$$\dot{\boldsymbol{x}} = A(t, \boldsymbol{x})\boldsymbol{x}$$

where

$$a_{ij}(t, \boldsymbol{x}) = \begin{cases} -\phi_i(t, \boldsymbol{x}) + e_i(t)\phi_{ii}(t, \boldsymbol{x}), & i = j \\ e_{ij}(t)\phi_{ij}(t, \boldsymbol{x}), & i \neq j \end{cases}$$

We assume that

$$\left.\begin{array}{l} \phi_i(t, \boldsymbol{x}) \geqslant \alpha_i > 0 \\ |\phi_{ij}(t, \boldsymbol{x})x_j| \leqslant \alpha_{ij}|x_j| \end{array}\right\} \quad \text{for all } t \geqslant 0 \text{ and } \boldsymbol{x},$$

with $\alpha_{ij} \geqslant 0$ such that $\alpha_i > \alpha_{ii}$.

Define the connectivity matrix corresponding to $e_j(t)$ above by

$$\bar{E} = (\bar{e}_{ij})$$

where $\bar{e}_{ij} = 1$ if variable x_j influences x_i and zero if x_j does not influence x_i.

Define a constant matrix $\bar{A}$ by

$$\bar{A} = (\bar{a}_{ij})$$

$$\bar{a}_{ij} = \begin{cases} -\alpha_i + \bar{e}_{ii}\alpha_{ii}, & i = j \\ \bar{e}_{ij}\,\alpha_{ij}, & i \neq j \end{cases}$$

By our assumption of $\alpha_i > \alpha_{ii}$, $\bar{A}$ is a Metzler matrix.

Theorem (Siljak, 1973) *The equilibrium $\boldsymbol{x} = 0$ of the system is asymptotically stable from any initial conditions if $\bar{A}$ is a quasidiagonal dominant matrix.*

One difficulty with the Lyapunov method has been and still is that we do not possess a general method of constructing Lyapunov functions. Recent research in vector Lyapunov functions (Matrosov, 1972) has alleviated this difficulty somewhat. In this method a number of Lyapunov functions is employed in stability analysis rather than a single Lyapunov function. A standard senario using this approach is to construct a vector valued Lyapunov function $\boldsymbol{v}$ and compare its time derivative expression with another function $\boldsymbol{w}$ with a known stability property such as $\boldsymbol{w} \to 0$ as $t \to \infty$, and establish $0 \leqslant \boldsymbol{v} \leqslant \boldsymbol{w}$. This inequality is usually established by appealing to some known results in differential inequality (Lakshmikantham, 1969). This is known as the comparison principle (Matrosov, 1968). See also Michel (1974).

PART II
DETERMINISTIC DYNAMIC ECONOMIC SYSTEMS

5
Stabilization Policies of Economic Systems

We consider several questions related to guiding economic activities in a stable manner. They are generally called stabilization problems. Policy questions related to stabilization of national and international economies are obviously of the greatest importance not only to policymakers but also to the welfare of everyone concerned.

Frisch and Tinbergen contributed greatly to the development of the theory of economic policy, although they worked primarily with static, i.e., one-period problems. We want to extend these pioneers' work to dynamic macroeconomic models and discuss intertemporal decision-making problems. In a dynamic framework, it is no longer sufficient to choose instrument values on a period to period basis since current policy decisions affect not only the current period, but also over a number of future periods. Target vectors may also be changing with time.

Very simply put, the key question is how to choose a sequence of control instruments over a period of time so as to affect in a desired way the values of macroeconomic variables that are not directly controlled (Tinbergen's noncontrolled variables). To do this, we assume an objective function of the policymaker is expressed by a quadratic function, and a linear macroeconomic (econometric) model is known.† This approach has been developed primarily by H. Theil, 1965, C. C. Holt, 1962, and C. C. Holt *et al.*, 1960. Under this restrictive set of assumptions, we will provide answers to many questions such as the question of the existence of stabilization policies, the instrument instability question and so on.

Arguments against intervention such as countercyclical policies, by policymakers exist because of the inevitable misspecification in the models used by policymakers (for example, see Baumol Chapter 18). Later we will

†Problems that arise when the model is not completely specified will be discussed later.

examine some of the effects of misspecification. We note that a countercyclical policy is only a very special example among feedback policies available to policymakers, and the mere fact that a particular countercyclical policy is destabilizing in some instances does not imply that all forms of stabilization policies are to be condemned.

It is necessary to determine what range of parameter values should be allowed in such a policy and what values should be excluded, but the structure of the policy should not be abandoned. The problem of finding stabilizing control laws, or of specifying stabilizing policy instrument generation schemes, is at the heart of macroeconomic policy problems. The models considered here are suitably generalized by including stochastic disturbances, errors in data and the delays in gathering data, misspecified or uncertain system parameters and so on.†

After giving some known results on linear models, we proceed to discuss the so-called instrument instability question before considering the use of cost functions or policymaker's (or societies') welfare functions to rank alternate stabilizing policies.

5.1 STABILIZABILITY

First, let us examine whether a given dynamic model can be stabilized by some control law (policy), before we introduce the concept of objective functions and extend our discussions to choosing the "best" control laws among those that result in stable dynamic systems. Given a dynamic system

$$\dot{z} = \boldsymbol{f}(z, \boldsymbol{x}, t)$$

we say the linear feedback control law

$$\boldsymbol{x} = \boldsymbol{F}(t)z + v \tag{5.1-1}$$

is a stabilizing (asymptotically stable) control law if the resultant dynamic system

$$\dot{z} = \boldsymbol{f}(z, Fz + v, t)$$

is stable (asymptotically stable).

The control law (policy) (5.1-1) means that the instrument value $\boldsymbol{x}$ is

†See Chapters 9, 10 and 11 for more extensive coverage.

generated partly automatically by a fixed rule as the time-varying constant matrix $F(t)$ times the state vector (feedback part) and partly by the newly redefined instrument $\boldsymbol{v}$. For example, the growth rate of the government expenditure on goods and services may be tied in part to a fixed formula (which may be varying with time) depending on the state of the economy and in part may be varied at the discretion of policymakers. The vector $\boldsymbol{v}$ could, of course, be absent in some policy generation schemes.

In the case of a linear control system

$$\dot{z} = A(t)z + B(t)\boldsymbol{x} \tag{5.1-2}$$

the control law (5.1-1) transforms it into

$$\dot{z} = [A(t) + B(t)F(t)]z + B(t)\boldsymbol{v}$$

If A and B are constant matrices, then the system is stable (asymptotically stable) if the real parts of all the eigenvalues of $A + BF$ are nonpositive (negative).

For time-invariant linear system where

$$\dot{z} = Az + B\boldsymbol{x} \tag{5.1-3}$$

where A and B are constant matrices, we have the following theorem.

Theorem *The eigenvalues of $A + BF$ can be assigned arbitrarily by a suitable choice of F, subject only to the complex conjugacy condition, if* (5.1-3) *is completely controllable. Thus there exists an asymptotically stable feedback control law with a constant matrix F if* (5.1-3) *is completely controllable.*

Proof We first prove this for the special case where B is an n-vector $\boldsymbol{b}$, and x is a scalar.

If (5.1-3) is completely controllable, there exists a nonsingular matrix T such that for $\boldsymbol{w}(t) = T^{-1}z(t)$,

$$\dot{\boldsymbol{w}} = G\boldsymbol{w} + \boldsymbol{f}x$$

where G is in the phase canonical form with the last row vector $(-\alpha_0 \quad \ldots \quad -\alpha_{n-1})$ and $\boldsymbol{f}^t = (0 \quad 0 \quad \ldots \quad 0 \quad 1)$ (see §3.2.4). Let

$$\boldsymbol{F} = (\phi_1, \ldots, \phi_n) : n\text{-row vector}$$

and

$$x = \boldsymbol{F}\boldsymbol{w} + v$$

Then

$$G + \boldsymbol{fF} = \begin{bmatrix} 010 & \dots & 0 \\ 001 & \dots & 0 \\ & \vdots & \\ 0 & \dots & 1 \\ g_1 & \dots & g_n \end{bmatrix}$$

where

$$g_i = -\alpha_{i-1} + \phi_i, \qquad i = 1, \dots, n$$

The characteristic polynomial of $G + \boldsymbol{fF}$ has coefficients g_i, $i = 1, \dots, n$. By suitable choices of ϕ_j's, any eigenvalue can be generated, subject only to the complex conjugacy conditions of eigenvalues.

To prove the case where B is a matrix, we appeal to the following Lemma.

Lemma (Heymann) *When (A, B) is a controllable pair, there exists a matrix F such that*

$$\langle A|B\rangle = \langle A + BF|\boldsymbol{b}\rangle$$

for any $\boldsymbol{b}$ vector that is in the range of B, where $\langle A|B\rangle$ = rank $(B \dots A^{n-1}B)$.

For example, choose the first column vector $\boldsymbol{b}_1$ from B. Then, with $\boldsymbol{v}^t = (v \quad 0 \quad \dots \quad 0)$ in the control rule $\boldsymbol{x} = Fz + v$, (5.1-3) becomes

$$\dot{z} = (A + BF)z + \boldsymbol{b}_1 v$$

Since $(A + BF, \boldsymbol{b}_1)$ is a controllable pair, after transforming this system into a phase canonical form there exists a feedback row vector $(\phi_1, \dots, \phi_n)$ such that

$$v = (\phi_1, \dots, \phi_n)z + v'$$

can assign any eigenvalues by the previous arguments.

For the proof of this Lemma, see (Heymann, 1968).

The complete controllability is therefore a sufficient condition for stability. This sufficient condition can be weakened to that of stabilizability (Wonham, 1967).

5.2 INSTRUMENT INSTABILITY

When economists switched from static models to dynamic models in their policymaking studies, they realized that because current policy decisions have effects in future periods, current policy decisions must not only offset current period exogenous disturbances but also cumulative effects of past policy decisions and past disturbances. Economists worry about the possibility of these cumulative effects getting ever larger and requiring ever greater changes in the instrument values to cancel them out. This is the so-called instrument instability, studied recently by Gramlich (1971), Holbrook (1972) and Chow (1973). Some endogenous variables being affected by instruments could be getting ever larger or exhibiting violent oscillations as well. Thus we see, the instrument instability is only one aspect of the overall dynamic model stabilization problem. As Chow realized correctly, it is to the existence of stabilization policies we should address ourselves.

For expositional simplicity we consider only deterministic linear models, even though the problem to be discussed here is obviously more serious and difficult to solve when models are nonlinear, have stochastic exogenous disturbances, have delays in data acquisition and in implementation of instruments and/or contain imprecisely known parameters in structural or reduced form equations. We will come back to these difficult cases later after we cover the topic for linear deterministic models.

We postpone to the next section questions related to choice of cost functions or welfare functions that may be used in tradeoffs of various competing targets or in tradeoffs of instability of instruments with that of some endogenous variables.† We demonstrate here and in the next section that systems properties discussed in §3.2 and 3.6 related to controllability and observability ensure the existence of stabilization policies in appropriate dynamic setting and give sufficient conditions (and necessary conditions when possible) for removing instability for a class of equations for generating instruments. If a set of instruments available to the policymakers is found to be unable to remove instability, these conditions will suggest ways to remedy the matter, e.g., by introducing additional instruments to make the model completely controllable, or by revising welfare functions to include all relevant variables.

Holbrook bases his development on the equation of the form

$$y_t = \xi_t + \sum_{i=0}^{N} w_i x_{t-i} \tag{5.2-1}$$

†In stochastic models we should compare variances of these variables.

where y_t is the target (goal) variable, ξ_t includes all current influences on y_t other than the current and lagged values of the scalar instruments x_{t-i}, $i = 0, \ldots, N$, and where N is the "maximum" period over which effects of current policy decisions are assumed to be felt. Holbrook then proceeds with the discussion of appropriate choice of the weights w's so that $\Sigma_i w_i \Delta x_{t-i} = 0$ is a stable equation.

This type of representation is not desirable because (1) N could be very large, making the model unwieldy, (we can usually use state-space representation of smaller dimensions to represent the same amount of information in (5.2-1) more compactly) and (2) discussions can be carried out better if we recognize explicitly that two sets of equations are involved in the stabilization questions; one for the natural or inherent dynamics of the model and the other for instrument generation. These two equations combined determine the dynamics governing the target-instrument pairs such as (5.2-1), so that the w's are only determined in part by policies. The rest depends on the dynamics as we shall show shortly.

Let us make our points clear. We assume a single instrument for ease of illustration. Suppose the model dynamics is given by

$$z_{t+1} = Az_t + \boldsymbol{b}x_t + \boldsymbol{d}_t, \qquad t = 0, 1, \ldots \tag{5.2-2}$$

where z_t is an n-dimensional vector at time t, called the state vector, x_t is a scalar instrument variable, and $\boldsymbol{d}_t$ is a vector of some exogenous variables. The matrix A is therefore $n \times n$ and $\boldsymbol{b}$ is an n-vector.

The target variable y_t is related to the state vector and the instrument variable by

$$y_t = \boldsymbol{c}^{\mathrm{t}} z_t + ex_t, \qquad t = 0, 1, \ldots \tag{5.2-3}$$

where $\boldsymbol{c}^{\mathrm{t}}$ is an n-dimensional row vector and e is a scalar.†

Equations (5.2-2) and (5.2-3) constitute an internal (state-space) description of the model.

Combining (5.2-2) and (5.2-3), we obtain

$$y_t = \boldsymbol{c}^{\mathrm{t}} A^t z_0 + \sum_{i=0}^{t-1} c^{\mathrm{t}} A^{t-1-i} (\boldsymbol{b}x_i + \boldsymbol{d}_i) + ex_t, \qquad t = 1, \ldots \tag{5.2-4}$$

†If the instrument has no impact effect on y_t, then $e = 0$. In general we are interested in the stability of z_t itself. Then $\boldsymbol{c}^{\mathrm{t}}$ is replaced by the identity matrix in (5.2-3).

This can be put into a form similar to (5.2-1) used by Holbrook,

$$y_t = \xi_t + \sum_{i=0}^{t} w_i x_{t-i} \tag{5.2-5}$$

where

$$\xi_t = \boldsymbol{c}^{\mathrm{t}} A^t z_0 + \sum_{i=0}^{t-1} \boldsymbol{c}^{\mathrm{t}} A^{t-1-i} \boldsymbol{d}_i$$

and where

$$\begin{aligned} w_i &= \boldsymbol{c}^{\mathrm{t}} A^{i-1} \boldsymbol{b}, \qquad i = 1, 2, \ldots, t \\ w_0 &= e \end{aligned} \tag{5.2-6}$$

For $t \geqslant N$, the summation is truncated at N in (5.2-1).

By the Cayley–Hamilton Theorem, we can express A^k, $k \geqslant n$, as a matrix polynomial involving $I, A, \ldots, A^{n-1}$. This means that the w_i's with $i > n$ are not independent of $w_0, \ldots, w_n$ in (5.2-5), and that even though $|w_t|$ may not be negligible for some $t > n$, it is not necessary to extend the summation in (5.2-5) from 0 to t but it suffices to sum from $i = 0$ to n, $t \geqslant n$, by appropriately redefining the w's. When this is done, we obtain a proper (i.e., minimal dimensional) input-output or reduced form representation of the dynamics of the model

$$y_t - \sum_{i=1}^{n} \alpha_i y_{t-i} = \sum_{i=0}^{n} w_i x_{t-i} \tag{5.2-7}$$

where

$$A^n = \sum_{1}^{n} \alpha_i A^{n-i}$$

and where

$$\begin{cases} w_0 = e \\ w_1 = -\alpha_1 e + \boldsymbol{c}^{\mathrm{t}} \boldsymbol{b} \\ \quad \vdots \\ w_n = -\alpha_n e - \alpha_{n-1} \boldsymbol{c}^{\mathrm{t}} \boldsymbol{b} - \cdots - \alpha_1 \boldsymbol{c}^{\mathrm{t}} A^{n-2} \boldsymbol{b} + \boldsymbol{c}^{\mathrm{t}} A^{n-1} \boldsymbol{b} \end{cases}$$

Note in particular that the w's are not independent but are related. They are linear combinations of e, $\mathbf{c}^t\mathbf{b}, \ldots, \mathbf{c}^tA^{n-1}\mathbf{b}$. Under the assumption that the value of the target variable is to be fixed at a constant value over n periods (5.2-7) is equivalent to

$$\Delta x_t = -\frac{1}{e}\sum_{0}^{n-1} w_s \Delta x_{t-s} \qquad (5.2\text{-}7')$$

This is exactly (9) of Holbrook (1972) if n is taken to be *equal* to the dimension of the economic dynamic system. If n is larger than the dimension, we have just shown how to reduce it. If n is taken to be less than the dimension, then equations such as (5.2-1) or (5.2-7′) must be regarded as a dynamic equation that generates new instrument values from the past ones, which at best reflect partially or approximately the dynamics of the system to be controlled. More general equations for instrument generation shall be introduced later.

Equation (5.2-7) points out one important fact not mentioned by Holbrook: If the dynamics of the economic system is modeled using the n-dimensional state vector as in (5.2-2), then the input–output relation such as (5.2-7) (or (5.2-5)) must have at least $n + 1$ weights $w_0, \ldots, w_n$ to include all the cumulative effects due to the current and past policy actions. Using fewer weights could mislead since some of the cumulative effects due to past policy actions are then improperly neglected. On the other hand, if the target-instrument relation is given by (5.2-7), then the state vector must have dimension n. If the input–output difference equation of order m is used to prescribe values for the current instrument variable (as is done by Holbrook), such as

$$x_t = -\left(y_t - \sum_{1}^{m} k_i x_{t-i}\right)\Big/ k_0, \qquad m < n \qquad (5.2\text{-}8)$$

then it is implicit in the procedure that either (1) the dynamics of the system are assumed to be at most m-dimensional, i.e. (5.2-8) replaces (5.2-7) or (2) we must consider two equations: (5.2-7) for the dynamics of the model and (5.2-8) as an equation for generating instrument variables.

The first alternative is usually not appropriate if m is too small, say one or two. The model for policy implication study may very well require 4 or 5th-dimensional state vector representation, for example.

We regard equations such as (5.2-8) as equations for instrument generation. Therefore, parameters such as k's or m are at the disposal of policymakers to some extent, k_i being the weights (gains) used in relating x_t to y_t and past y's or past x's. The form of the equations and the k's are choice variables for the policymakers.

We now formulate the stabilization problem of the combined system of equations i.e., when the dynamics is governed by (5.2-7) or equivalently by (5.2-2) or (5.2-3) and the system is driven by instruments generated by an equation such as (5.2-8). Although we continue to use the discrete-time (difference equation) formulation, parallel developments can be carried out using the continuous-time (differential equation) formulation.

It is perhaps appropriate to point out to the reader at this point that stabilization study of this section is preliminary in nature to that of §5.3. In this section, a set of target variables and a class of equations relating current policy instruments to past target values and current and/or past instrument values are assumed to be given. We then ask whether a set of parameter values exists (which singles out a particular instrument generation equation uniquely from the given class) that stabilizes the economy. We do not ask in this section whether policymakers prefer one set of time paths of the target variables over some others. We merely note here whether these variables can be stabilized within the chosen class of instrument generating equations.

Suppose we consider a class of equations for the instrument generation of the form

$$x_t = \sum_{i=0}^{m} \delta_i y_{t-i} + \sum_{j=1}^{m} l_j x_{t-j} \tag{5.2-9}$$

and apply it to the dynamic system described by (5.2-7). Equation (5.2-8) is a special case of (5.2-9). In other words, assume that policymakers have decided (perhaps arbitrarily) that they consider only equations of the type (5.2-9) in their stabilization considerations.

Suppose a particular choice of δ's and l's in (5.2-9) results in the system's instability. We want to examine how these parameters can be modified to stabilize the system, if this is possible. What are the conditions under which this is possible? To answer these questions, it is convenient to use the method of root-locus when the adjustable variables are one or two (see Appendix D for this approach). Otherwise it may be better to represent the problem in the state-space form first. We examine this possibility next.

5.2.1 State-Space Formulation of the Stabilization Problem

State-space representation of (5.2-9) is, by the method of §2.2,

$$\begin{aligned} \boldsymbol{s}_{t+1} &= E\boldsymbol{s}_t + \boldsymbol{f} y_t \\ x_t &= \boldsymbol{h}^{\mathrm{t}}\boldsymbol{s}_t + \gamma y_t \end{aligned} \tag{5.2-10}$$

where s_t is the m-dimensional vector that describes the "dynamic state" of the device generating the instrument values, E is an $m \times m$ matrix, $\boldsymbol{f}$ is some m-dimensional (column) vector, $\boldsymbol{h}^t$ is an m-dimensional (row) vector $(10 \quad \cdots \quad 0)$, and γ is a scalar constant.

Instrument generation equations such as

$$x_{t+1} = gx_t + fy_t \tag{5.2-11}$$

or

$$x_{t+1} - x_t = \gamma y_t \tag{5.2-12}$$

are special cases of (5.2-10).

To see that (5.2-8) can be put into (5.2-10), introduce m-dimensional state vector $\boldsymbol{s}_t$ with components $s_t^1, \ldots, s_t^m$ by

$$s_t^1 = x_t + y_t/k_0$$

$$s_{t-1}^j = s_t^{j-1} - \frac{k_j}{k_0} x_{t-1}$$

$$= s_t^{j-1} - \frac{k_j}{k_0} s_{t-1}^1 + \frac{k_j}{k_0^2} y_{t-1}, \qquad j = 2, \ldots, m-1$$

$$s_t^m = \frac{k_m}{k_0} s_{t-1}^1 + \frac{k_m}{k_0^2} y_{t-1}$$

with

$$E = \begin{bmatrix} \lambda_1 & 1 & 0 & 0 & \cdots & 0 \\ \lambda_2 & 0 & 1 & 0 & \cdots & 0 \\ & & & & \vdots & \\ \lambda_{m-1} & 0 & 0 & 0 & \cdots & 1 \\ \lambda_m & 0 & 0 & 0 & \cdots & 0 \end{bmatrix}, \quad \boldsymbol{f} = \frac{1}{k_0} \begin{bmatrix} \lambda_1 \\ \lambda_2 \\ \vdots \\ \lambda_{m-1} \\ \lambda_m \end{bmatrix},$$

$$\lambda_i = \frac{k_i}{k_0}, \; i = 1, \ldots, m, \qquad \gamma = -\frac{1}{k_0}$$

and where

$$\boldsymbol{h}^t = (1 \quad 0 \quad \ldots \quad 0)$$

We must remember that (5.2-10) is an equation for x_t in terms of y_t and past x's and y's so that x is the output and y is the input. This is the exact opposite of the situation in §2.4 since we were concerned with evaluations of the effects of x on y there. Here we are interested in generating x as a function of y. We can now state the combined stabilization problem. Given

$$\begin{aligned} &\text{system dynamics (5.2-2):} \quad && z_{t+1} = Az_t + \boldsymbol{b}x_t + \boldsymbol{d}_t \\ &\text{target (5.2-3):} && y_t = \boldsymbol{c}^{\mathrm{t}}z_t + ex_t \end{aligned} \tag{5.2-13}$$

Find instrument generation equation (5.2-10):

$$\begin{aligned} s_{t+1} &= Es_t + \boldsymbol{f}y_t \\ x_t &= \boldsymbol{h}^{\mathrm{t}}s_t + \gamma y_t \end{aligned} \tag{5.2-14}$$

such that the combined system is stable.

Solve x_t and y_t in terms of $\boldsymbol{c}^{\mathrm{t}}z_t$ and $\boldsymbol{h}^{\mathrm{t}}s_t$ as

$$y_t = \frac{1}{1 - e\gamma}\boldsymbol{c}^{\mathrm{t}}z_t + \frac{e}{1 - e\gamma}\boldsymbol{h}^{\mathrm{t}}s_t$$

and

$$x_t = \frac{\gamma}{1 - e\gamma}\boldsymbol{c}^{\mathrm{t}}z_t + \frac{1}{1 - e\gamma}\boldsymbol{h}^{\mathrm{t}}s_t$$

provided $e\gamma \neq 1$. Then the dynamics for the combined system are governed by

$$\begin{pmatrix} z_{t+1} \\ s_{t+1} \end{pmatrix} = F\begin{pmatrix} z_t \\ s_t \end{pmatrix} + \begin{pmatrix} \boldsymbol{d}_t \\ 0 \end{pmatrix} \tag{5.2-15}$$

where

$$F = \begin{bmatrix} A + \dfrac{\gamma}{1 - e\gamma}\boldsymbol{b}\boldsymbol{c}^{\mathrm{t}} & \dfrac{\boldsymbol{b}\boldsymbol{h}^{\mathrm{t}}}{1 - e\gamma} \\ \dfrac{\boldsymbol{f}\boldsymbol{c}^{\mathrm{t}}}{1 - e\gamma} & E + \dfrac{e}{1 - e\gamma}\boldsymbol{f}\boldsymbol{h}^{\mathrm{t}} \end{bmatrix} \tag{5.2-16}$$

Here x_t and y_t are both scalars. There is an obvious extension when they are vectors. The combined systems are therefore stable if and only if F is a stable matrix.

Let us now treat a special case of this that serves to clarify the point.

The instrument x_t could be the stock of money, for example. Then $x_{t+1} - x_t$ is the rate of increase in money stock, y_t could be the deviation of real output from the desired output by measuring the target from its desired or equilibrium level.

Suppose (5.2-12) is used to generate the instrument. The combined system dynamics are now described by (5.2-15) with F given by

$$F = \begin{pmatrix} A & \boldsymbol{b} \\ \gamma \boldsymbol{c}^{\mathrm{t}} & 1 + \gamma e \end{pmatrix} \tag{5.2-17}$$

Having chosen (5.2-12) as the equation for generating instrument values, γ is the only parameter at the disposal of policymakers to stabilize F of (5.2-17).

The total system is stable if the eigenvalues of F are all contained in the unit circle in the complex plane. Then we say F is an asymptotically stable matrix.

We first assume that A is an asymptotically stable matrix, i.e. the model of the economy (5.2-13) is taken to be inherently stable.†

The stability question of the whole system can therefore be posed as: Does there exist a real number γ in (5.2-17) to make F a stable matrix? The answer is contained in the next proposition.

Proposition 1 *With an asymptotically stable matrix A, the matrix F is stable for small $|\gamma|$, if and only if*

$$\gamma(e\mu(1) + \nu(1)) < 0 \tag{5.2-18}$$

where

$$\mu(\lambda) = |\lambda I - A| \text{ and } \nu(\lambda) = \boldsymbol{c}^{\mathrm{t}} \operatorname{adj}(\lambda I - A)\boldsymbol{b}$$

where adj$(\cdot)$ denotes the adjoint matrix. The condition (5.2-18) *can be met (by changing e slightly if necessary) if*

$$\operatorname{rank}(\boldsymbol{b}, A\boldsymbol{b}, \ldots, A^{n-1}\boldsymbol{b}) = n$$

The proof of this proposition may be carried out by arguments provided in Appendix D or Corfmat and Morse (1973). The reader may be

†Otherwise we can stabilize the system in two stages if $(A, \boldsymbol{b})$ is a stabilizable pair and if z_t is available to policymakers. First choose $x_t = \boldsymbol{k}^{\mathrm{t}} z_t$ to make $A + \boldsymbol{b}\boldsymbol{k}^{\mathrm{t}}$ a state matrix. Then apply the discussion here with A replaced by $A + \boldsymbol{b}\boldsymbol{k}^{\mathrm{t}}$ (see the last part of this section).

interested in the meaning of the condition before he turns to the proof. Next we offer an interpretation of (5.2-18).

Briefly, the interpretation is: when $e = 0$, i.e. when no impact effects of the instruments appear in the target variable, then $\nu(1) \neq 0$ if the economy is controllable. When $e \neq 0$, it could happen that $e\mu(1) + \nu(1) = 0$ since $\boldsymbol{b}$ and $\boldsymbol{c}$ are parts of the given structure of the model.

The next section spells these conditions out in detail.

From (5.2-17), we see that with $\gamma = 0$, i.e., with a constant instrument value, one eigenvalue of F is located exactly at one. This eigenvalue makes F stable but not asymptotically stable. The condition for this eigenvalue to move inside the unit circle on the complex plane for $\gamma \neq 0$ is that of (5.2-18).

We sketch the proof when A can be diagonalized, i.e., when there exists nonsingular transformation T such that $T^{-1}AT$ is diagonal. The idea of the proof remains essentially the same when T is such that $T^{-1}AT$ is upper triangular with small off-diagonal elements (see Corollary to Theorem 1 of §4.1).

Thus, we suppose that T is such that

$$\begin{pmatrix} T^{-1} & 0 \\ 0 & 1 \end{pmatrix} F \begin{pmatrix} T & 0 \\ 0 & 1 \end{pmatrix} = \begin{pmatrix} D & T^{-1}\boldsymbol{b} \\ \gamma \boldsymbol{c}^{\mathrm{t}} T & 1 + \gamma e \end{pmatrix}$$

where we write D for $T^{-1}AT$.

The eigenvalues of F are the roots of the characteristic equation

$$0 = \begin{vmatrix} \lambda I - D & -T^{-1}\boldsymbol{b} \\ -\gamma \boldsymbol{c}^{\mathrm{t}} T & \lambda - 1 - \gamma e \end{vmatrix}$$

$$= \prod_{i=1}^{n} (\lambda - \lambda_i) \cdot \left\{ \lambda - 1 - \gamma e - \gamma \boldsymbol{c}^{\mathrm{t}} T (\lambda I - D)^{-1} T^{-1} \boldsymbol{b} \right\}$$

See Lemmas 1 and 2 in Appendix B.

We note that $T(\lambda I - D)^{-1}T^{-1} = (\lambda I - A)^{-1} = \operatorname{adj}(\lambda I - A)/|\lambda I - A| = \nu(\lambda)/\mu(\lambda)$. Therefore, if

$$e\mu(1) + \nu(1) \neq 0 \tag{5.2-19}$$

then for small $|\gamma|$ such that (5.2-18) is satisfied, the root of the characteristic equation at $\lambda = 1$ can be moved inside the unit circle (also see Appendix C). All the roots of $|\lambda I - A| = 0$ are located inside the circle by assumption. Thus, small displacement of these roots by $|\gamma| \neq 0$ still places them all inside the circle.

From the Cayley–Hamilton Theorem, we know that

$$\mu(\lambda) = |\lambda I - A| = \lambda^n - \sum_1^n \alpha_i \lambda^{n-i}$$

and as we discussed in §2.3, we can write

$$\operatorname{adj}(\lambda I - A) = \lambda^{n-1}B_0 + \lambda^{n-2}B_1 + \cdots + \lambda B_{n-2} + B_{n-1}$$

where

$$B_0 = I$$

$$B_k = B_{k-1}A - \alpha_k I, \qquad k = 1, \ldots, n-1$$

We know that

$$\mu(1) = 1 - \alpha_1 - \alpha_2 - \cdots - \alpha_n > 0$$

(since $\mu(\lambda)$ is a stable polynomial), and

$$\begin{aligned}\nu(1) &= \boldsymbol{c}^{\mathrm{t}}\boldsymbol{b} + \boldsymbol{c}^{\mathrm{t}}B_1\boldsymbol{b} + \cdots + \boldsymbol{c}^{\mathrm{t}}B_{n-1}\boldsymbol{b} \\ &= (1 - \alpha_1 - \cdots - \alpha_{n-1})\boldsymbol{c}^{\mathrm{t}}\boldsymbol{b} \\ &\quad + (1 - \alpha_1 - \cdots - \alpha_{n-2})\boldsymbol{c}^{\mathrm{t}}A\boldsymbol{b} \\ &\quad + \cdots + \boldsymbol{c}^{\mathrm{t}}A^{n-1}\boldsymbol{b} - \alpha_n e\end{aligned}$$

The condition (5.2-18) will be satisfied if

$$\begin{aligned}0 &\neq [e\mu(1) + \nu(1)] \\ &= [(1 - \alpha_1 - \cdots - \alpha_{n-1})(e + \boldsymbol{c}^{\mathrm{t}}\boldsymbol{b}) - \alpha_n e \\ &\quad + (1 - \alpha_1 - \cdots - \alpha_{n-2})\boldsymbol{c}^{\mathrm{t}}A\boldsymbol{b} + \cdots + \boldsymbol{c}^{\mathrm{t}}A^{n-1}\boldsymbol{b}]\end{aligned} \qquad (5.2\text{-}20)$$

Since $\boldsymbol{b}$, $\boldsymbol{c}$ and e are parts of the structure of the model, it could happen that the right hand side of (5.2-20) vanishes.

Then (5.2-17) is not stable, even when the subsystems of (5.2-17) taken separately are stable, i.e. even when A is stable and $1 + \gamma e < 0$.

If at least one of $\boldsymbol{c}^{\mathrm{t}}\boldsymbol{b}, \boldsymbol{c}^{\mathrm{t}}A\boldsymbol{b}, \ldots, \boldsymbol{c}^{\mathrm{t}}A^{n-1}\boldsymbol{b}$ is nonzero, or if the system is output controllable, then

$$\begin{aligned}&(1 - \alpha_1 - \cdots - \alpha_{n-1})\boldsymbol{c}^{\mathrm{t}}\boldsymbol{b} \\ &\quad + (1 - \alpha_1 - \cdots - \alpha_{n-2})\boldsymbol{c}^{\mathrm{t}}A\boldsymbol{b} + \cdots + \boldsymbol{c}^{\mathrm{t}}A^{n-1}\boldsymbol{b} \neq 0\end{aligned}$$

Suppose the model is output-controllable. Then by continuity of the right-hand side of (5.2-20) with respect to e, we see that (5.2-20) is satisfied for some $e \neq 0$.

5.2.2 Examples

We will use some very simple deterministic macroeconomic models to illustrate the above comments about the state-space formulation of the stabilization problem.

Example 1 Let ΔY_t be the change in real output at the tth period (say the quarter) and let ΔM_t be the change in the money stock at the tth quarter. The change in the interest rate is Δr_t at the tth quarter.

Suppose we have the dynamics described by

$$\Delta Y_t = a_1 \Delta Y_{t-1} + a_2 \Delta Y_{t-2} + a_3 \Delta M_{t-1}$$

The dynamic equation can be put into the state-space form

$$z_{t+1} = A z_t + \boldsymbol{b} x_t$$

with

$$z_t = \begin{pmatrix} \Delta Y_t \\ \Delta Y_{t-1} \end{pmatrix} \quad \text{and} \quad x_t = \Delta M_t$$

$$A = \begin{pmatrix} a_1 & a_2 \\ 1 & 0 \end{pmatrix} \qquad \boldsymbol{b} = \begin{pmatrix} a_3 \\ 0 \end{pmatrix}$$

Take the target variable to be Δr_t that is assumed to be given by

$$\Delta r_t = b_1 \Delta Y_t + b_2 \Delta M_t$$

Namely, the target variable is given by

$$y_t = \boldsymbol{c}^{\mathrm{t}} z_t + e x_t$$

with

$$\boldsymbol{c}^{\mathrm{t}} = (b_1, 0), \qquad e = b_2$$

Since

$$\operatorname{rank}(\boldsymbol{b}, A\boldsymbol{b}) = \operatorname{rank}\left(\begin{pmatrix} 1 \\ 0 \end{pmatrix}, \begin{pmatrix} a_1 \\ 1 \end{pmatrix}\right) = 2$$

the dynamic system is controllable.

Suppose we adopt the policy, $x_t = \gamma y_t$ or $\Delta M_t = \gamma \Delta r_t$, of tying the change of the money stock to that of the interest rate. This policy modifies the economy's dynamics to

$$z_{t+1} = \{A + \gamma \boldsymbol{b}\boldsymbol{c}^{\mathrm{t}} / (1 - \gamma e)\} z_t$$

The stabilization question is then what choice of γ will make the matrix $A + \gamma \boldsymbol{b}\boldsymbol{c}^{\mathrm{t}}/(1 - \gamma e)$ asymptotically stable. We see that if a_2 is negative then any γ such that $a_1 + \gamma a_3 b_1$ is negative is stabilizing.

If the policy is $x_{t+1} - x_t = \gamma y_t$ or $\Delta M_{t+1} - \Delta M_t = \gamma \Delta r_t$, then (5.2-18) gives the necessary and sufficient condition for stabilization. Now,

$$\mu(\lambda) = |\lambda I - A| = \lambda^2 - a_1 \lambda - a_2$$

We assume that $a_1 < 0$ and $a_2 < 0$ so that A is stable. We have

$$\nu(\lambda) = \boldsymbol{c}^{\mathrm{t}} \operatorname{adj} (\lambda I - A) \boldsymbol{b}$$

If

$$e\mu(1) + \nu(1) = b_2(1 - a_1 - a_2) + a_3 b_1 \neq 0$$

then the equation for instrument generation

$$\Delta M_{t+1} - \Delta M_t = \gamma \Delta r_t$$

stabilizes the total system if γ is chosen to be such that $|\gamma|$ is small with

$$\operatorname{sgn} \gamma = -\operatorname{sgn} [a_3 b_1 + b_2(1 - a_1 - a_2)]$$

Example 2 This example is a slight modification of Example 2 found in Aoki (1973a). We consider a small open country with the demand for output

$$Y' = G + X - M + C + I \tag{5.2-21a}$$

where G is the government expenditure (the instrument), X is exports, M is imports, C is consumption and I is investment.

The balance of cumulative trade deficit D_t is given by

$$D_{t+1} - D_t = M_t - X_t \tag{5.2-21b}$$

We assume the relations

$$C_t = (1 - s) Y_t \tag{5.2-21c}$$

$$X_t = x Y_t \tag{5.2-21d}$$

where Y_t is the actual output, and

$$M_t = mY_t - \tau D_t \tag{5.2-21e}$$

$$\kappa_1(Y_{t+1} - Y_t) = (I_{t+1} - I_t + \kappa_2 I_t)/v \tag{5.2-21f}$$

where $\kappa_1, \kappa_2 > 0$, and v is the capital output ratio, and

$$Y_{t+1} - Y_t = v\sigma(Y_t' - Y_t) \tag{5.2.21g}$$

Define z_t by

$$z_t = (Y_t, I_t, D_t)^{\mathrm{t}}$$

Then

$$z_{t+1} = Az_t + \boldsymbol{b}G_t \tag{5.2-22}$$

We are interested in a situation where (5.2-22) is uncontrollable. We assume therefore that

$$m = x \tag{5.2-23}$$

For this case, A becomes

$$A = \begin{pmatrix} A_{11} & & A_{12} \\ 0 & 0 & 1 - \tau \end{pmatrix}, \qquad A_{11} = \begin{pmatrix} \alpha & \beta \\ \delta & \epsilon \end{pmatrix}$$

where $\alpha = v\sigma(1 - s)$, $\beta = v\sigma$, $\delta = (v\sigma)(v\kappa_1)(1 - s)$, $\epsilon = (v\sigma)(v\kappa_1) + (1 - \kappa_2)$

$$A_{12} = v\sigma\tau\begin{pmatrix} 1 \\ v\kappa_1 \end{pmatrix}$$

and

$$\boldsymbol{b} = v\sigma\begin{bmatrix} 1 \\ v\kappa_1 \\ 0 \end{bmatrix}$$

The dynamic system is not controllable since rank$(\boldsymbol{b}, A\boldsymbol{b}, A^2\boldsymbol{b}) = 2$. We assume that A is stable as before by assuming

$$|1 - \tau| < 1$$

$$\alpha + \epsilon < 0$$

$$\Delta = \alpha\epsilon - \beta\delta > 0$$

Take the target variable to be

$$D_t = (0 \quad 0 \quad 1)z_t$$

i.e. $e = 0$ in this case.

Consider the instrument generation equation

$$G_{t+1} - G_t = \gamma D_t \tag{5.2-24}$$

With the above formulation the characteristic equation of A is

$$\begin{aligned}\mu_3(\lambda) &= |\lambda I_3 - A| \\ &= \mu_2(\lambda)[\lambda - (1 - \tau)]\end{aligned}$$

where

$$\mu_2(\lambda) = \lambda^2 - (\alpha + \epsilon)\lambda + \Delta$$

Since

$$\begin{aligned}(\lambda I_3 - A)^{-1} &= \begin{pmatrix} (\lambda I_2 - A_{11})^{-1} & (\lambda I_2 - A_{11})^{-1} A_{12}/[\lambda - (1 - \tau)] \\ 0 & 1/[\lambda - (1 - \tau)] \end{pmatrix} \\ &= \begin{pmatrix} \operatorname{adj}(\lambda I_2 - A_{11})/\mu_2(\lambda) & \operatorname{adj}(\lambda I_2 - A_{11}) A_{12}/\mu_3(\lambda) \\ 0 & 1/[\lambda - (1 - \tau)] \end{pmatrix}\end{aligned}$$

we have

$$\begin{aligned}\nu(\lambda) &= \boldsymbol{c}^{\mathrm{t}} \operatorname{adj}(\lambda I_3 - A)\boldsymbol{b} \\ &= 0\end{aligned}$$

Then, since $e = 0$, the left hand side of (5.2-19) becomes

$$e\mu(1) + \nu(1) = \nu(1) = 0$$

Thus, no γ exists that stabilizes (5.2-22), even though A_{11} is asymptotically stable and (5.2-24) is not unstable, i.e., (5.2-24) has $|\lambda| < 1$.

If the target variable is taken to be the government deficit that is

$$y_t = G_t$$

since there is no tax in this model, then $e = 1$ in this case. Equation (5.2-19) becomes

$$e\mu_3(1) + \nu(1) = 1 - (\alpha + \epsilon) + \Delta \neq 0$$

thus, γ exists to stabilize the whole system.

5.2.3 Stabilization Analysis with Transforms

Let $\hat{z}(\zeta)$ denote the ζ-transform (lag-transform) of $\{z_t\}$ introduced in §2.4,

$$\hat{z}(\zeta) = \sum_{t=0}^{\infty} z_t \zeta^{-t}$$

Assume $\hat{z}(\zeta)$ is defined in an appropriate domain so that convergence is not a problem.

Given (5.2-7), the dynamic equation of the economy in the input-output framework, its ζ-transform is

$$\hat{y}(\zeta) - \sum_{i=1}^{n} \alpha_i \big(\hat{y}(\zeta)\zeta^{-i} + y_{-1}\zeta^{-i+1} + \cdots + y_{-i}\big)$$
$$= \sum_{i=0}^{n} w_i \big(\hat{x}(\zeta)\zeta^{-i} + x_{-1}\zeta^{-i+1} + \cdots + x_{-i}\big) \qquad (5.2\text{-}25)$$

or

$$M_D(\zeta)\hat{y}(\zeta) = N_D(\zeta)\hat{x}(\zeta) + \sum_{i=1}^{n} \alpha_i \big(y_{-1}\zeta^{-i+1} + \cdots + y_{-i}\big) + \sum_{i=0}^{n} w_i \big(x_{-1}\zeta^{-i+1} + \cdots + x_{-i}\big)$$

where

$$M_D(\zeta) = 1 - \sum_{i=1}^{n} \alpha_i \zeta^{-i} \quad \text{and} \quad N_D(\zeta) = \sum_{i=0}^{n} w_i \zeta^{-i}$$

Suppose we choose an equation for the instrument generation to control the dynamic system

$$x_t = \sum_{i=0}^{m} k_i y_{t-i} + \sum_{j=1}^{m} l_j x_{t-j} \qquad (5.2\text{-}26)$$

and apply it to the dynamic system described by (5.2-7). The ζ-transform of (5.2-26) is

$$N_1(\zeta)\hat{x}(\zeta) = M_1(\zeta)\hat{y}(\zeta) + (\text{initial condition terms}) \qquad (5.2\text{-}27)$$

where

$$N_1(\zeta) = 1 - \sum_{j=1}^{m} l_j \zeta^{-j}$$

and

$$M_1(\zeta) = \sum_{0}^{m} k_i \zeta^{-i}$$

Eliminating $\hat{x}(\zeta)$ from (5.2-25) and (5.2-27), we obtain

$$\left(M_D(\zeta) - \frac{N_D(\zeta) M_1(\zeta)}{N_1(\zeta)} \right) \hat{y}(\zeta) = \hat{w}(\zeta)$$

where $\hat{w}(\zeta)$ is due to the initial conditions in (5.2-25) and (5.2-27). These equations are the transforms of the difference equations governing the combined equations (5.2-7) and (5.2-26). If the roots of $M_D(\lambda)N_1(\lambda) - N_D(\lambda)M_1(\lambda) = 0$ all have moduli less than one, the sequence $\{y_t\}$ is asymptotically stable, as discussed in §4.2.

From (5.2-27)

$$\hat{x}(\zeta) = \frac{M_1(\zeta)}{N_1(\zeta)} \hat{y}(\zeta)$$

with $\hat{y}(\zeta)$ given by (5.2-25). This equation shows whether the instrument instability exists or not, i.e., again the roots of

$$M_D(\lambda)N_1(\lambda) - N_D(\lambda)M_1(\lambda) = 0$$

govern the stability of the instrument generation equation.

5.2.4 Perfect Output Controllability

In addition to insuring stability, instrument generating equations ought to be capable of making the target variable follow a desired or specified time path (e.g., maintaining a specified level).

As our discussions in §3.2 make clear, this property, called output reproduciability or perfect controllability, is not automatic and is more difficult to obtain than the stabilizability discussed above. The question is: Is a proposed instrument generating equation, such as (5.2-8), capable of achieving the desired target level or time path when y_t^* (desired value) is used in (5.2-8) instead of y_t? In this section we give a brief discussion on this point. We represent the dynamic equation

$$z_{t+1} = Az_t + bx_t$$

using the ζ-transforms as

$$(\zeta I - A)\hat{z}(\zeta) = b\hat{x}(\zeta) + \text{(initial condition terms)}$$

or

$$\hat{z}(\zeta) = (\zeta I - A)^{-1}b\hat{x}(\zeta) + \hat{w}(\zeta)$$

where $\hat{z}(\zeta)$ and $\hat{x}(\zeta)$ are the ζ-transforms (discrete-time analog of Laplace transform) of $\{z_t\}$ and $\{x_t\}$ and $w(\zeta)$ is due to the initial condition. (See §2.3 for more on the ζ-transforms.)

Since A is assumed to be stable, we can ignore $w(\zeta)$. From the target equation, the ζ-transform of the output is

$$\begin{aligned}\hat{y}(\zeta) &= c^t\hat{z}(\zeta) + e\hat{x}(\zeta)\\ &= \left[c^t(\zeta I - A)^{-1}b + e\right]\hat{x}(\zeta) + c^t\hat{w}(\zeta)\end{aligned}$$

This equation is in effect the transform of the input-output relation (5.2-7).

Briefly put, the condition of perfect controllability is the condition for the invertibility of $[c^t(\zeta I - A)^{-1}b + e]$. If $c^t(\zeta I - A)^{-1}b + e$ is not identically zero, then one can solve the above for $\hat{x}(\zeta)$ as

$$\hat{x}(\zeta) = \left[c^t(\zeta I - A)^{-1}b + e\right]^{-1}\left[\hat{y}(\zeta) - c^t w(\zeta)\right]$$

Then denoting the transform of the desired time trajectory of the target by $\hat{y}(\zeta)^*$,

$$\hat{x}(\zeta) = \left[c^t(\zeta I - A)^{-1}b + e\right]^{-1}\hat{y}(\zeta)^*$$

is the equation for instrument generation to be used for the economy described by (5.2.2). Then the target variable is related to the desired time path by

$$\hat{y}(\zeta) = \hat{y}(\zeta)^* + c^t\hat{w}(\zeta)$$

When the instrument generation equation is taken to be

$$\hat{x}(\zeta) = M(\zeta)\hat{y}(\zeta)^* \tag{5.2-28}$$

where $M(\zeta)$ is some polynomial lag operator, then its effect on the target can be evaluated by

$$\hat{y}(\zeta) = \left[\boldsymbol{c}^{\mathrm{t}}(\zeta I - A)^{-1}\boldsymbol{b} + e\right]M(\zeta)\hat{y}(\zeta)^* + \boldsymbol{c}^{\mathrm{t}}\hat{\boldsymbol{w}}(\zeta)$$

We can use the final-value theorem to see if $y_t - y_t^* \to 0$ as $t \to \infty$ or not.

Fact

$$\lim_{t\to\infty} y_t = \lim_{\zeta\to 1} (\zeta - 1)\hat{y}(\zeta)$$

We can state this fact as a proposition.

Proposition *The instrument generation equation* (5.2-28) *is such that* $y_t - y_t^* \to 0$ *if and only if*

$$\lim_{\zeta\to 1} (\zeta - 1)\left\{\left[\boldsymbol{c}^{\mathrm{t}}(\zeta I - A)^{-1}\boldsymbol{b} + e\right]M(\zeta) - I\right\}\hat{y}(\zeta)^* = 0$$

So far, we have emphasized the necessity of specifying two dynamic equations; one for the economy and the other for instrument generation. Some parameter of the equation for the instrument has then been chosen to make the combined system stable. We have not discussed how the equation for the instrument is chosen in the first place.

The question has been considered in the literature by adopting some cost or welfare functions and deriving the instruments using (intertemporal) optimization of the chosen functions. For example, a quadratic or piecewise quadratic function of $\boldsymbol{y}_t$ and $\boldsymbol{x}_t$ is often used as the objective function. For example, see Aoki (1967) or Benjamin Friedman (1972) and the references quoted.

In the case of quadratic objective functions in which $\boldsymbol{x}_t$ appears with positive weight, a large body of results is available. This is the subject of §5.3.

We conclude this section by discussing the stabilization when A is not stable.

Suppose A is not stable in

$$\begin{cases} z_{t+1} = Az_t + \boldsymbol{b}x_t \\ y_t = \boldsymbol{c}^{\mathrm{t}}z_t + ex_t \end{cases}$$

and suppose $\mathrm{rank}(\boldsymbol{b}, A\boldsymbol{b}, \ldots, A^{n-1}\boldsymbol{b}) = n$.

Stabilization of A may be discussed under various assumptions such as:

(i) Data on z_t is available to the policymakers or

(ii) only z_{t-1} is available at time t to policymakers due to delay in data gathering and transmission, or
(iii) only $y_t, y_{t-1}, \ldots,$ are available to policymakers.

Under (i), we have shown in §5.1 that some n-dimensional vector $\boldsymbol{k}^{\mathrm{t}}$ exist such that

$$x_t = \boldsymbol{k}^{\mathrm{t}} z_t$$

stabilizes the dynamics, i.e. $A + \boldsymbol{b}\boldsymbol{k}^{\mathrm{t}}$ is a stable matrix.

Under (ii), with the instrument generation equation

$$x_t = \boldsymbol{k}^{\mathrm{t}} z_{t-1}$$

the dynamic equation may be rewritten as

$$\begin{pmatrix} z_{t+1} \\ z_t \end{pmatrix} = \begin{pmatrix} A & \boldsymbol{b}\boldsymbol{k}^{\mathrm{t}} \\ I & 0 \end{pmatrix} \begin{pmatrix} z_t \\ z_{t-1} \end{pmatrix} \tag{5.2-26}$$

The system is stabilizable if $\boldsymbol{k}$ can be chosen to make the matrix in (5.2-26) stable. (See Appendix 3 for further discussion on this point.)

Under (iii), we discuss as an example

$$x_t = \gamma y_t$$

Substituting it into the above, we see that

$$x_t = \frac{\gamma}{1 - \gamma e} \boldsymbol{c}^{\mathrm{t}} z_t$$

and the use of this instrument generation equation implies that the dynamics are modified to

$$z_{t+1} = (A + \tau \boldsymbol{b}\boldsymbol{c}^{\mathrm{t}}) z_t$$

where

$$\tau = \gamma / (1 - \gamma e)$$

A necessary and sufficient condition for stabilization for the scalar target variable is that the system is output controllable, i.e.,

$$\operatorname{rank}(\boldsymbol{c}^{\mathrm{t}}\boldsymbol{b}, \boldsymbol{c}^{\mathrm{t}}A\boldsymbol{b}, \ldots, \boldsymbol{c}^{\mathrm{t}}A^{n-1}\boldsymbol{b}) = 1$$

or there exists a nonnegative integer d such that†

$$\boldsymbol{c}^{\mathrm{t}} A^d b \neq 0.$$

†For a proof see Bhattacharyya *et al.*, (1972).

When A is diagonalizable, then the eigenvalues of $A + \tau \boldsymbol{b}\boldsymbol{c}^{\mathrm{t}}$ are given by the roots of

$$
\begin{aligned}
0 &= |\lambda I - A - \tau \boldsymbol{b}\boldsymbol{c}^{\mathrm{t}}| \\
&= |\lambda I - A| \cdot \left\{1 - \tau \boldsymbol{c}^{\mathrm{t}}(\lambda I - A)^{-1}\boldsymbol{b}\right\}
\end{aligned}
$$

where Corollary 2 to Lemma 2 of Appendix B is used. The condition, therefore, is seen to be essentially the same as (5.2-18) with γ replaced by τ.

5.3 QUADRATIC WELFARE FUNCTION

Now we assume that a quadratic function† in the target vector $\boldsymbol{y}$ and the instrument vector $\boldsymbol{x}$ is chosen as the appropriate function expressing social welfare in general or tradeoffs of mutually incompatible objectives in particular society and the policymakers. The target vectors may contain such things as GNP, rate of inflation, rate of unemployment, and so on.

As we indicated at the beginning of this chapter, we assume that a deterministic linear dynamic equation is given to describe the time behavior of a macroeconomic model. Also given is a quadratic objective function to be minimized. The policymakers have desired time paths for various macroeconomic variables (control instruments and uncontrolled variables). By minimizing the quadratic objective function they minimize the deviations from these desired paths.

We discuss discrete-time systems using dynamic programming. Continuous-time systems are discussed in §5.4 using calculus of variations.

A discrete-time version of the same problem is to minimize

$$J = z_T{}^{\mathrm{t}} P z_T + \sum_{\tau=0}^{T-1} W_\tau \tag{5.3-1}$$

where

$$W_\tau = z_\tau{}^{\mathrm{t}} Q_\tau z_\tau + \boldsymbol{x}_\tau{}^{\mathrm{t}} R_\tau \boldsymbol{x}_\tau \tag{5.3-2}$$

subject to

$$z_{\tau+1} = A_\tau z_\tau + B_\tau \boldsymbol{x}_\tau, \qquad \tau = 0, 1, \ldots, T \tag{5.3-3}$$

†If a quadratic function is objectionable since it implies underachieving a desired level or path, and overachieving them is penalized equally, we may wish to use piecewise quadratic functions and assign different penalties to deviations on either side (see B. Friedman 1972, 1974). This is, however, a detail of technicalities. Basic analytical tools will be the same.

where†

$$Q_\tau^{\mathrm{t}} = Q_\tau \geqslant 0, \qquad R_\tau^{\mathrm{t}} = R_\tau > 0, \qquad P^{\mathrm{t}} = P \geqslant 0$$

The first term in (5.3-1) may be regarded as the terminal stock valuation or penalty against deviations of certain target variables from their targets at the end of the planning horizon. We will discuss (5.3-4) in detail, since the analyses are basically the same for other quadratic functions.

In some cases, we may want to include cross product terms in J, so that

$$W_\tau = (z_\tau^{\mathrm{t}} \quad x_\tau^{\mathrm{t}})\begin{pmatrix} Q_\tau & S_\tau \\ S_\tau^{\mathrm{t}} & R_\tau \end{pmatrix}\begin{pmatrix} z_\tau \\ x_\tau \end{pmatrix}, \qquad \text{where}\begin{pmatrix} Q_\tau & S_\tau \\ S_\tau^{\mathrm{t}} & R_\tau \end{pmatrix} \geqslant 0 \qquad R_\tau > 0. \tag{5.3-4a}$$

In some cases, we may want to use y's instead of z's.

$$W_\tau = (y_\tau^{\mathrm{t}} \quad x_\tau^{\mathrm{t}})\begin{pmatrix} N_\tau & L_\tau \\ L_\tau^{\mathrm{t}} & M_\tau \end{pmatrix}\begin{pmatrix} y_\tau \\ x_\tau \end{pmatrix}, \qquad N_\tau \geqslant 0,\ M_\tau > 0 \tag{5.3-4b}$$

All of these cases can be treated uniformly by introducing an artificial observation equation

$$\bar{y}_\tau = C_\tau z_\tau + D_\tau x_\tau \tag{5.3-5}$$

and rewriting (5.3-1) as

$$J = z_T^{\mathrm{t}} P z_T + \sum_{\tau=0}^{T-1} \bar{y}_\tau^{\mathrm{t}} \bar{y}_\tau \tag{5.3-6}$$

In case of (5.3-2),

$$C_\tau = \begin{pmatrix} Q_\tau^{1/2} \\ 0 \end{pmatrix} \quad \text{and} \quad D_\tau = \begin{pmatrix} 0 \\ R_\tau^{1/2} \end{pmatrix}$$

will put it in the form of (5.3-5) and (5.3-6).

When the cross product terms are present in W_τ as in (5.3-4a),

$$C_\tau = \begin{bmatrix} [Q_\tau - S_\tau R_\tau^{-1} S_\tau^{\mathrm{t}}]^{1/2} \\ R_\tau^{-1/2} S_\tau^{\mathrm{t}} \end{bmatrix} \quad \text{and} \quad D_\tau = \begin{pmatrix} 0 \\ R_\tau^{1/2} \end{pmatrix}$$

†The assumption $R_t > 0$ may be replaced with $R_\tau \geqslant 0$ by using pseudo-inverses instead of the inverses.

will accomplish the same purpose. Note that by assumption

$$\begin{pmatrix} Q_\tau & S_\tau \\ S_\tau^{t} & R_\tau \end{pmatrix} \geqslant 0,$$

hence the square root of $Q_\tau - S_\tau R_\tau^{-1} S_\tau^{t}$ is well-defined. Finally, if W_τ is given by (5.3-4b) and if the target vector is related to z and x by

$$y_\tau = H_\tau z_\tau + G_\tau x_\tau$$

then

$$W_\tau = (z_\tau^{t} \quad x_\tau^{t}) \begin{pmatrix} H_\tau^{t} \\ G_\tau^{t} \end{pmatrix} \begin{pmatrix} N_\tau & L_\tau \\ L_\tau^{t} & M_\tau \end{pmatrix} (H_\tau \quad G_\tau) \begin{pmatrix} z_\tau \\ x_\tau \end{pmatrix}$$

All we need to do to reduce it to the form in (5.3-6), is to define

$$\begin{pmatrix} Q_\tau & S_\tau \\ S_\tau^{t} & R_\tau \end{pmatrix} = \begin{pmatrix} H_\tau^{t} \\ G_\tau^{t} \end{pmatrix} \begin{pmatrix} N_\tau & L_\tau \\ L_\tau^{t} & M_\tau \end{pmatrix} (H_\tau \quad G_\tau)$$

and construct C_τ and D_τ accordingly. Without loss of generality, then, we may say that the problem is to minimize (5.3-6) subject to (5.3-3) and (5.3-5) (we drop the overbar from the y's):
Minimize

$$z_T^{t} P z_T + \sum_{\tau=0}^{T-1} W_\tau$$

subject to

$$z_{\tau+1} = A_\tau z_\tau + B_\tau x_\tau$$

$$y_\tau = C_\tau z_\tau + D_\tau x_\tau$$

where

$$W_\tau = y_\tau^{t} y_\tau$$

Optimal Control Law *Denote by $J_{t,T}(z_t, P; \{x_{t,T}\})$ the value of*

$$z_T^{t} P z_T + \sum_{\tau=t}^{T-1} W_\tau$$

where $\{x_{t,T}\}$ stands for $x_t, x_{t+1}, \ldots, x_{T-1}$.

Let

$$J_{t,T}^0\left(z_t, P; \{x_{t,T}^0\}\right) = \min_{\{x_{t,T}\}} J_{t,T}(z_t, P; \{x_{t,T}\})$$

Then by the principle of optimality of Bellman, $J_{t,T}^0$ satisfies the functional equation

$$J_{t,T}^0\left(z_t, P; \{x_{t,T}^0\}\right) = \operatorname*{Min}_{x_t}\left[\|y_t\|^2 + J_{t+1,T}^0\left(z_{t+1}, P; \{x_{t+1,T}^0\}\right)\right] \tag{5.3-7}$$

A standard dynamic programming backward induction argument gives us the sequence of optimal instruments $\{x_{t,T}^0\}$.

Consider time $T-1$ so that x_{T-1} is the only instrument yet to be determined. Since

$$\begin{aligned} J_{T-1,T}(z_{T-1}; P; x_{T-1}) &= z_T{}^t P z_T + y_{T-1}{}^t y_{T-1} \\ &= z_T{}^t P z_T + \|C_{T-1} z_{T-1} + D_{T-1} x_{T-1}\|^2 \end{aligned}$$

we obtain

$$x_{T-1}{}^0 = -(B_{T-1}{}^t P B_{T-1} + D_{T-1}{}^t D_{T-1})^+ (B_{T-1}{}^t P A_{T-1} + D_{T-1}{}^t C_{T-1}) z_{T-1}$$

and

$$J_{T-1,T}{}^0\left(z_{T-1}, P; x_{T-1}{}^0\right) = \|z_{T-1}\|^2_{\Pi_{T-1,T}(P)}$$

where

$$\begin{aligned} \Pi_{T-1,T}(P) = {} & (A_{T-1}{}^t P A_{T-1} + C_{T-1}{}^t C_{T-1}) \\ & - (B_{T-1}{}^t P A_{T-1} + D_{T-1}{}^t C_{T-1})^t (B_{T-1}{}^t P B_{T-1} + D_{T-1}{}^t D_{T-1})^+ \\ & \cdot (B_{T-1}{}^t P A_{T-1} + D_{T-1}{}^t C_{T-1}) \end{aligned}$$

where $(\ )^+$ is the Penrose pseudo-inverse.† See Aoki (1967) for the pseudo-inverse.

---

†If the column vectors of D_{T-1} are linearly independent, then the inverse and the pseudo-inverse are the same. Readers not familiar with the concept of pseudo-inverse may proceed by replacing the pseudo-inverse with the inverse.

By mathematical induction, we can easily show that the solution to (5.3-7) is given by

$$J_{t,T}^{0}\left(z_t, P; \left\{x_{t,T}^{0}\right\}\right) = \|z_t\|^2_{\Pi_{t,T}(P)} \tag{5.3-8}$$

where $\Pi_{t,T}(P)$ satisfies what is called the algebraic Riccati equation

$$\begin{aligned}\Pi_{t,T}(P) = {} & A_t^{\mathrm{t}}\Pi_{t+1,T}(P)A_t + C_t^{\mathrm{t}}C_t \\ & - (B_t^{\mathrm{t}}\Pi_{t+1,T}(P)A_t + D_t^{\mathrm{t}}C_t)^{\mathrm{t}}(B_t^{\mathrm{t}}\Pi_{t+1,T}(P)B_t + D_t^{\mathrm{t}}D_t)^{+} \\ & \cdot(B_t^{\mathrm{t}}\Pi_{t+1,T}(P)A_t + D_t^{\mathrm{t}}C_t)\end{aligned} \tag{5.3-9}$$

and

$$x_t^{0} = -(B_t^{\mathrm{t}}\Pi_{t+1,T}(P)B_t + D_t^{\mathrm{t}}D_t)^{+}\,(B_t^{\mathrm{t}}\Pi_{t+1,T}(P)A_t + D_t^{\mathrm{t}}C_t)z_t \tag{5.3-10}$$

Note that the optimal instrument is linear in z_t with time-varying weights assigned to the components of it. The computation of these optimal combinations of the components of z_t requires that the values of the matrices A, B, C and D are known. This concludes the derivation of the sequence of optimal instruments for the discrete-time dynamic systems.

5.4 LINEAR CONTINUOUS TIME DYNAMIC SYSTEMS AND THE RICCATI EQUATION

5.4.1 Derivation of Optimal Control Policies

In this section, we derive an optimal control law for a linear system

$$\dot{z} = A(t)z + B(t)x$$

$$z(t_0) = z_0$$

By optimal, we mean in the sense that the cost function

$$J = z^{\mathrm{t}}(T)Pz(T) + \int_{t_0}^{T}\left[z^{\mathrm{t}}(t)Q(t)z(t) + x^{\mathrm{t}}(t)R(t)x(t)\right]dt$$

is minimized, where $P = P^{\mathrm{t}} \geqslant 0$, $Q(t) = Q^{\mathrm{t}}(t) > 0$ and $R^{\mathrm{t}}(t) = R(t) > 0$,

$t_0 \leqslant t \leqslant T$.† We use the method of calculus of variation, rather than repeating the dynamic programming argument we used for discrete systems. We assume functions of time such as $A(t)$ and $B(t)$ are all continuous and bounded.

Suppose an optimal control $\boldsymbol{x}^*(t)$, $t_0 \leqslant t \leqslant T$, exists. We consider a perturbed control

$$\boldsymbol{x}(t) = \boldsymbol{x}^*(t) + \epsilon\tilde{\boldsymbol{x}}(t), \qquad t_0 \leqslant t \leqslant T$$

where ϵ is a small positive number and $\tilde{\boldsymbol{x}}(t)$ is an *arbitrary* function of t.

As the result, we have

$$z(t) = z^*(t) + \epsilon\tilde{z}(t)$$

where the time path $z^*(\cdot)$ corresponds to $\boldsymbol{x}^*(\cdot)$, namely

$$\dot{z}^*(t) = A(t)z^*(t) + B(t)\boldsymbol{x}^*(t), \qquad z^*(t_0) = z_0 \tag{5.4-1}$$

and

$$\dot{\tilde{z}}(t) = A(t)\tilde{z}(t) + B(t)\tilde{\boldsymbol{x}}(t), \qquad \tilde{z}(t_0) = 0 \tag{5.4-2}$$

From (5.4-2), $\tilde{z}$ can be related to $\tilde{\boldsymbol{x}}$ by

$$\tilde{z}(t) = \int_{t_0}^{t} \phi(t, s)B(s)\tilde{\boldsymbol{x}}(s)\, ds, \qquad t_0 \leqslant t \leqslant T \tag{5.4-3}$$

where

$$\dot{\phi}(t, s) = A(t)\phi(t, s), \ \phi(s, s) = I$$

The criterion function now becomes a function of ϵ, $J(\epsilon)$,

$$J(\epsilon) = J^* + 2\epsilon K + (\text{quadratic terms in } \epsilon)$$

where

$$J^* = \langle z^*(T), Pz^*(T)\rangle + \int_{t_0}^{T} [\langle z^*(t), Q(t)z^*(t)\rangle + \langle \boldsymbol{x}^*(t), R(t)\boldsymbol{x}^*(t)\rangle]\, dt$$

†In case $R(t) \geqslant 0$, use pseudo-inverses instead of the inverses.

and

$$K = \int_{t_0}^{T} \{ \tilde{z}^{t}(t) Q(t) z^{*}(t) + \tilde{x}^{t}(t) R(t) x^{*}(t) \} \, dt + \langle \tilde{z}(T), P z^{*}(T) \rangle \tag{5.4-4}$$

where $\langle \ , \ \rangle$ is a notation for the inner product.

Since x^* is optimal, $J(\epsilon)$ is minimal at $\epsilon = 0$. It is, therefore, necessary and sufficient for optimality that $K = 0$, assuming the quadratic term is positive.

Substitute (5.4-3) into (5.4-4) to express K as

$$0 = K = \int_{t_0}^{T} \langle \tilde{x}(t), B^{t}(t) \int_{t}^{T} \phi^{t}(s, t) Q(s) z^{*}(s) \, ds + R(t) x^{*}(t) + B^{t}(t) \phi^{t}(T, t) P z^{*}(T) \rangle \, dt$$

The above expression suggests that we introduce a new variable $p(t)$, called the adjoint vector, by

$$p(t) = \int_{t}^{T} \phi^{t}(s, t) Q(s) z^{*}(s) \, ds + \phi^{t}(T, t) P z^{*}(T) \tag{5.4-5}$$

Then the condition $K = 0$ is expressible as

$$0 = K = \int_{t_0}^{T} \langle \tilde{x}(t), B^{t}(t) p(t) + R(t) x^{*}(t) \rangle \, dt$$

Since $\tilde{x}(t)$ is arbitrary, $K = 0$ if and only if

$$B^{t}(t) p(t) + R(t) x^{*}(t) = 0, \qquad t_0 \leqslant t \leqslant T$$

or

$$x^{*}(t) = -R^{-1}(t) B^{t}(t) p(t), \qquad t_0 \leqslant t \leqslant T \tag{5.4-6}$$

Equation (5.4-6) expresses the optimal control in terms of the adjoint variable.

From (5.4-5), at $t = T$,

$$p(T) = P z^{*}(T)$$

By differentiating (5.4-5), we see that $p(t)$ is governed by the differential equation

$$\dot{p}(t) = -Q(t)z^*(t) - A^t(t)p(t) \tag{5.4-7}$$

Combining (5.4-1) and (5.4-7), we obtain the set of differential equations governing the optimal time path of the system and of the instruments

$$\frac{d}{dt}\begin{pmatrix} z^*(t) \\ p(t) \end{pmatrix} = \begin{pmatrix} A(t) & -B(t)R^{-1}(t)B^t(t) \\ -Q(t) & -A^t(t) \end{pmatrix}\begin{pmatrix} z^*(t) \\ p(t) \end{pmatrix} \tag{5.4-8}$$

with a set of boundary conditions given at t_0 and T

$$\begin{cases} z^*(t_0) = z_0 \\ p(T) = Pz^*(T) \end{cases} \tag{5.4-9}†$$

Let $\Phi(t, t_0)$ be the fundamental solution of (5.4-8), and partition it as

$$\Phi(t, t_0) = \begin{pmatrix} \phi_{11}(t, t_0) & \phi_{12}(t, t_0) \\ \phi_{21}(t, t_0) & \phi_{22}(t, t_0) \end{pmatrix} \tag{5.4-10}$$

Using (5.4-10), we can express $z^*(t)$ and $p(t)$ in terms of $z^*(T)$ and of $p(T)$,

$$\begin{aligned} z^*(t) &= \phi_{11}(t, T)z^*(T) + \phi_{12}(t, T)p(T) \\ &= [\phi_{11}(t, T) + \phi_{12}(t, T)P]z^*(T) \\ p(t) &= \phi_{21}(t, T)z^*(T) + \phi_{22}(t, T)P(T) \\ &= [\phi_{21}(t, T) + \phi_{22}(t, T)P]z^*(T) \end{aligned}$$

Eliminating $z^*(T)$ from the above two equations, we can relate the adjoint vector $p(t)$ to $z^*(t)$

$$p(t) = S(t)z^*(t)$$

where

$$S(t) = [\phi_{21}(t, T) + \phi_{22}(t, T)P][\phi_{11}(t, T) + \phi_{12}(t, T)P]^{-1} \tag{5.4-11}$$

†Note that the boundary conditions are split into n conditions at $t = t_0$ and another n condition at $t = T$, i.e., at two points on the time axis rather than all at one point. This type of problem is known as the two-point boundary value problem.

From (5.4-6) and (5.4-11), the optimal control is expressed as

$$x^*(t) = -R^{-1}(t)B^{t}(t)S(t)z^*(t) \tag{5.4-12}$$

This is the solution of the optimal regulation problem. The optimal control is in the form of the feedback. The initial condition z_0 is related to $z^*(T)$ by

$$z_0 = [\phi_{11}(t_0, T) + \phi_{12}(t_0, T)P]z^*(T)$$

or

$$z^*(t) = [\phi_{11}(t, T) + \phi_{12}(t, T)P][\phi_{11}(t_0, T) + \phi_{12}(t_0, T)P]^{-1}z_0$$

The matrix $S(t)$ in (5.4-11) satisfies the following matrix differential equation, known as the Riccati differential equation.

$$\begin{aligned} -\dot{S}(t) &= Q(t) - S(t)B(t)R(t)^{-1}B^{t}(t)S(t) \\ &\quad + S(t)A(t) + A^{t}(t)S(t), \quad S(T) = P \end{aligned} \tag{5.4-13}$$

This equation is obtained by direct differentiation of $S(t)$. The fundamental solution Φ is the unique solution of the linear differential equation (5.4-8). Thus we can state the following theorem.

Theorem 1 *The solution to the Riccati equation is unique. When* (5.4-12) *is substituted into* (5.4-1), *the dynamic equation for the optimal trajectory is obtained as*

$$\dot{z}^*(t) = (A(t) - B(t)F(t))z^*(t), \qquad z^*(t_0) = z_0 \tag{5.4-14}$$

with

$$F(t) = R^{-1}(t)B^{t}(t)S(t)$$

(5.4-13) *may also be written as*

$$\begin{aligned} -\dot{S}(t) &= Q(t) + S(t)B(t)R(t)^{-1}B^{t}(t)S(t) \\ &\quad + S(t)[A(t) - B(t)F(t)] \\ &\quad + [A(t) - B(t)F(t)]^{t}S(t), \qquad S(T) = P \end{aligned} \tag{5.4-15}$$

We next establish the following theorem.

Theorem 2

$$\operatorname{Min} J = z_0{}^{\mathrm{t}} S(t_0) z_0 \quad \textit{and} \quad S(t) \geqslant 0, \qquad t_0 \leqslant t \leqslant T$$

Proof Since the optimal control is given by (5.4-12), the cost function can be rewritten as

$$\operatorname{Min} J = z^{\mathrm{t}}(T) P z(T) + \int_{t_0}^{T} z^{\mathrm{t}}(t) T(t) z(t)\, dt$$

where

$$T(t) = Q(t) + S(t) B(t) R^{-1}(t) B^{\mathrm{t}}(t) S(t)$$

where we drop the * from z for simplicity. Denote by $\psi(t, t_0)$ the fundamental solution of (5.4-14). Then

$$z(t) = \psi(t, t_0) z_0$$

Thus

$$\operatorname{Min} J = z_0{}^{\mathrm{t}} \left[\psi^{\mathrm{t}}(T, t_0) P \psi(T, t_0) + \int_{t_0}^{T} \psi^{\mathrm{t}}(t, t_0) T(t) \psi(t, t_0)\, dt \right] z_0$$

Define

$$V(t) = \psi^{\mathrm{t}}(T, t) P \psi(T, t) + \int_{t}^{T} \psi^{\mathrm{t}}(s, t) T(s) \psi(s, t)\, ds$$

We have

$$V(T) = P$$

Thus

$$\operatorname{Min} J = z_0{}^{\mathrm{t}} V(t_0) z_0$$

We next show that $V(t)$ satisfies the same Riccati equation that $S(t)$ does, with the same boundary condition. Since

$$\frac{\partial \psi(s, t)}{\partial t} = -\psi(s, t)\{A(t) - B(t) F(t)\}$$

and hence

$$\frac{\partial \psi^{t}(s, t)}{\partial t} = -\{A(t) - B(t)F(t)\}^{t}\psi^{t}(s, t)$$

we obtain

$$-\dot{V}(t) = T(t) + G^{t}(t)V(t) + V(t)G(t)$$

where $V(T) = P$, $G(t) = A(t) - B(t)F(t)$. This equation is exactly the same as (5.4-15). The terminal condition is also the same. From the uniqueness of the solution of the Riccati equation, we conclude that $V(t) = S(t)$, $t_0 \leqslant t \leqslant T$. See Silverman and Payne (1971).

The non-negative definiteness of $S(t)$ comes from the non-negativeness of the cost function value over (t, T) for any z_0, $t_0 \leqslant t \leqslant T$.

As the last item of discussion of this section we show that under a set of conditions to be stated later

$$S(t) \to S^* \geqslant 0, \qquad t \to \infty$$

Treating t as a variable, denote the optimal criterion function value over $[t, T]$ by $z^{t}(t)S(t, T)z(t)$.

Claim 1 *If the system* (5.4-1) *is completely controllable, then* $z^{t}(t)S(t, T)$ $z(t)$ *is bounded above for all* $t \leqslant T$.

Proof By the complete controllability, there exists an input that transfers $z(t)$ to the origin at some time $T' > t$. Then,

$$z^{t}(t)S(t, T)z(t) \leqslant \int_{t}^{T'} [z^{t}(s)Q(s)z(s) + x^{t}(s)R(s)x(s)]\, ds < \infty$$

Claim 2 *Suppose* $P = 0$, *then* $z^{t}(t)S(t, T)z(t)$ *is monotonically nondecreasing in* T.

Proof Suppose this statement is false, then there exist $T' < T''$ such that

$$z^{t}(t)S(t, T')z(t) > z^{t}(t)S(t, T'')z(t)$$

Apply the control optimal over $[t, T'']$ to the interval $[t, T']$. Since the contribution of the integral over (T', T'') is non-negative

$$z^{t}(t)S(t, T')z(t) \leqslant z^{t}(t)S(t, T'')z(t)$$

Hence, a contradiction exists and thus the statement must true.

Combining the two Claims 1 and 2, we have the following theorem.

Theorem 3 *If the system* (5.4-1) *is completely controllable, then there exists* $S^*(t) \geqslant 0$, *for the problem with* $P = 0$, *such that*

$$z^t(t)S(t, T)z(t) \rightarrow z^t(t)S^*(t)z(t) \quad \text{as} \quad T \rightarrow \infty$$

for all $z(t)$, *i.e.*, $S(t, T) \rightarrow S^*(t)$ *as* $T \rightarrow \infty$.

We have established in Theorem 3 that if the dynamic system is completely controllable, and if the terminal condition is zero then $S(t, T) \rightarrow S^*(t)$ as $T \rightarrow \infty$, for all t. This limit function satisfies (5.4-13) since the solutions of the Riccati equation are continuous with respect to the initial conditions. (Let us rewrite the solution of (5.4-13) as $\Pi(t; P, T)$, then

$$S^*(t) = \lim_{s \rightarrow \infty} \Pi(t; 0, s) = \lim_{s \rightarrow \infty} \Pi(t; \Pi(T; 0, s), T)$$

$$= \Pi\left(t; \lim_{s \rightarrow \infty} \Pi(T; 0, s), T\right) = \Pi(t; S^*(T), T)$$

is as shown.)

5.4.2 Regulation over the Infinite Horizon

Assume that A, B, Q and R matrices are all constant for the sake of simplicity, then S^* is also a constant matrix. It satisfies the algebraic Riccati equation, $0 = Q - S^*BR^{-1}B^tS^* + S^*A + A^tS^*$.

The optimal control law in this case is, of course, given by the time-invariant version of (5.4-12)

$$x(t) = -R^{-1}B^tS^*z(t)$$

This is the (long run) optimal control law that minimizes

$$J = \int_0^\infty (z^tQz + x^tRx)\, dt \tag{5.4-16}$$

With the optimal control law, the integrand in J is expressible as

$$z^tQz + x^tRx = z^tTz$$

where

$$T = Q + S^*BR^{-1}B^tS^*$$

and

$$z(t) = e^{Ft}z_0$$

with

$$F = A - BR^{-1}B^{t}S^{*}$$

Since T is symmetric and at least non-negative definite, we can write $T = C^{t}C$, where C is of maximum rank. Then

$$z^{t}Qz + x^{t}Rx = z^{t}C^{t}Cz = \|Ce^{Ft}z_0\|^2$$

and

$$z_0{}^{t}S^{*}z_0 = \int_0^{\infty}(z^{t}Qz + x^{t}Rx)\,dt = \int_0^{\infty}\|Ce^{Ft}z_0\|^2\,dt$$

If (C, F) is an observable pair, then

$$\int_0^{\infty}\|Ce^{Ft}z_0\|^2\,dt > 0, \qquad \text{for } z_0 \neq 0$$

since $\mathfrak{N}(Ce^{Ft}) = \{0\}$, where $\mathfrak{N}(\cdot)$ is the null space.

This means $S^{*} > 0$; hence $V(t) = z(t)^{t}S^{*}z(t)$ is a Lyapunov function (Desoer 1970) since

$$\begin{aligned}\dot{V}(t) &= z^{t}(t)(F^{t}S^{*} + S^{*}F)z(t)\\ &= -z^{t}(t)Tz(t)\\ &= -\|Ce^{Ft}z_0\|^2 < 0\end{aligned}$$

Hence

$$\dot{z} = Fz$$

is asymptotically stable.

Note that the observability involves C, which depends on Q and R.

Theorem 4 (Kalman 1960) *If (A, B) is completely controllable and if (C, A) is completely observable, then $S^{*} > 0$ and $A - BR^{-1}B^{t}S^{*}$ is asymptotically stable.*

Considerably more detail is known about (5.4-15) (for example, see Kučera 1972). The sufficiency condition in Theorem 4 can be weakened (Payne and Silverman 1973).

We next discuss two simple macroeconomic models. They serve to illustrate controllability and observability as sufficient conditions for the existence of optimal control laws.

5.5 EXAMPLES

Example 1 We illustrate† the above first by an example that augments the Phillips model used in Aoki (1973a).

Let Y be the supply, and Y' be the demand of output. The demand is composed of

$$Y' = G + X - M + C + I \tag{5.5-1}$$

where G is government expenditure; X is exports; M is imports; C is consumption; and I is investment.

All quantities are measured from the equilibrium values, so that they are the deviations from the equilibrium levels. Let v be the marginal capital-output ratio. Then, we have

$$\dot{Y} = I/v \tag{5.5-2}$$

Assume a behavioral rule for investment

$$I = \sigma(Y' - Y) \tag{5.5-3}$$

and consumption

$$C = (1 - s)Y \tag{5.5-4}$$

where σ and s are taken to be constants, $\sigma > 0$, $0 < s < 1$. Assume further that imports and exports are related to Y by

$$X = xY \tag{5.5-5}$$

and

$$M = mY + Y' - Y \tag{5.5-6}$$

where x and m are assumed to be constants, i.e., we assume that the imports are some fixed fraction of the actual output plus the excess demand. Combining (5.5-1)–(5.5-6), we obtain

$$\dot{Y} = \alpha Y + \beta G \tag{5.5-7}$$

where

$$\alpha = \sigma(x - m - s)/(2 - \sigma)v$$

†This example is an expanded account of the example given in (Aoki, 1973a).

and

$$\beta = \sigma / (2 - \sigma)v$$

Let D be the total foreign debt accumulated up to the present. It satisfies

$$\begin{aligned}\dot{D} &= M - X \\ &= (m - x)Y + Y' - Y \qquad (5.5\text{-}8)\end{aligned}$$

From (5.5-2), (5.5-3), (5.5-7), we rewrite (5.5-8) as

$$\dot{D} = \gamma Y + \delta G \qquad (5.5\text{-}9)$$

where

$$\begin{aligned}\gamma &= m - x + v\alpha/\sigma \\ &= [(m - x)(1 - \sigma) - s]/(2 - \sigma)\end{aligned}$$

and

$$\delta = v\beta/\sigma = 1/(2 - \sigma)$$

Thus, we have a state-space model, where the system dynamics is governed by

$$\dot{z} = Az + Bx \qquad (5.5\text{-}10)$$

where

$$A = \begin{pmatrix} \alpha & 0 \\ \gamma & 0 \end{pmatrix}, \quad B = \begin{pmatrix} \beta \\ \delta \end{pmatrix}, \quad z = \begin{pmatrix} Y \\ D \end{pmatrix}, \quad x = G$$

Note that β and δ are positive for $\sigma < 2$. We take the criterion function to be J [Equation (5.4-16)], with $R = I$.

Test for Controllability We test (5.5-10) for controllability. From $(B, AB) = \begin{pmatrix} \beta, & \alpha\beta \\ \delta, & \beta\gamma \end{pmatrix}$, the rank $(B \quad AB) = 2$ if and only if $\alpha\delta \neq \beta\gamma$. Since $\alpha\delta - \beta\gamma = \sigma(x - m)/(2 - \sigma)v$, the controllability criteria becomes the following: the system (5.5-10) is controllable if and only if

$$m \neq x \qquad (5.5\text{-}11)$$

To see that time paths of uncontrollable systems are confined to some subset of R^2, consider $m = x$. Then $\gamma = \alpha v/\sigma$ and $\delta = \beta v/\sigma$, i.e., $\dot{D} = \dot{Y}(v/\sigma)$.

Thus, if $Y(0) = Y_0$ and $D(0) = D_0$, the only values attainable by the system in (Y, D) space are those on a linear variety through (Y_0, D_0), i.e., Y and D values cannot be independently chosen. Under normal circumstances, we should construct models that are controllable, even though we may have no such freedom in some cases and must live with uncontrollable or "nearly uncontrollable" systems in the real world.

Test for Observability There are four possibilities here.

Case 1 First, consider the case where $Q = \mathrm{diag}(l_1^2, l_2^2)$, with $l_1, l_2 > 0$. This means that we are interested in the levels of both Y and D. Since the rank of $\mathrm{diag}(l_1, l_2) = 2$ for $l_1 \cdot l_2 \neq 0$, the observability condition is always met.

Case 2 Next, suppose $l_1 = 0$, $l_2 = l > 0$. Then, from $Q = H^{\mathrm{t}}H$, $H = (0, l)$. By definition (A, H) is an observable pair if and only if H^{t} and $A^{\mathrm{t}}H^{\mathrm{t}}$ are linearly independent. This is true if $\gamma \neq 0$ or if $(m - x)(1 - \sigma) \neq s$. Assuming the controllability condition $m \neq x$, (A, H) is an observable pair if

$$\sigma \neq 1 - s/(m - x) \qquad (5.5\text{-}12)$$

In this case, the level of Y does not appear in the welfare function. Still the observability condition is met provided (5.5-12) is true.

Case 3 Suppose $l_1 = l > 0$ and $l_2 = 0$. Then $H = (l, 0)$ and (A, H) is not an observable pair since $(l, 0)$ and $(\alpha l, 0)$ are linearly dependent. In this case, the level of D is not represented in the welfare function even though the level of Y is. In the next section, we show that the optimal policy is not stable for this case.

Case 4 Suppose that $Q = H^{\mathrm{t}}H$ with $H = (1, l)$. Then (A, H) is not an observable pair if $\alpha + \gamma l = 0$. This case is more subtle than the previous one since both Y and D are present in the objective function. We show in the next section that this case also leads to an unstable optimal policy.

Nature of Optimal Controls

Case 1 Case 1 requires no further comments.

Case 2 Under the assumption of (5.5-11) and (5.5-12), a symmetric positive definite solution to the algebraic Riccati equation (5.5-5) Π_∞ exists by Theorem 4 and all eigenvalues of $A - BB^{\mathrm{t}}\Pi_\infty$ have negative real parts and the optimal control policy results in a stable system.

Case 3 Under the assumption of (5.5-11), a symmetric positive semi-definite solution is given by

$$\Pi_\infty = \begin{pmatrix} \pi & 0 \\ 0 & 0 \end{pmatrix}$$

where

$$\pi = \left(\alpha + \sqrt{\alpha^2 + \beta^2 l^2}\right)/\beta^2 > 0$$

Then

$$A - BB^{\mathrm{t}}\Pi_\infty = \begin{pmatrix} \alpha - \beta^2\pi & 0 \\ \gamma - \beta\delta\pi & 0 \end{pmatrix}$$

showing that $A - BB^{\mathrm{t}}\Pi_\infty$ is *not* asymptotically stable. The reason for this is intuitively obvious. A relevant variable D was not present in the objective function thus reducing the model effectively from (5.5-10) to that of (5.5-7) alone. The optimal control is given by $x = -\beta\pi y$. The variable y is asymptotically stable since

$$\dot{Y} = (\alpha - \beta^2\pi)Y = -\sqrt{\alpha^2 + \beta^2 l^2}\; Y$$

but D becomes unstable since the exponent of the D becomes positive for the parameter combination that satisfies

$$\gamma - \beta\delta\pi = \gamma - \frac{\alpha\delta}{\beta} - \frac{\delta}{\beta}\sqrt{\alpha^2 + \beta^2 l^2} > 0$$

One might state that the unstable behavior of D is due to the failure to include D in the welfare function, i.e., the misspecification of the welfare function in this case.

Case 4 We now discuss that even when D is included in the cost function due care is needed to preserve the observability condition. To be specific, assume that

$$0 < \sigma < 1$$

$$x - m > 0$$

$$x - m - s < 0$$

Then, we have $\alpha < 0$, $\beta > 0$, $\gamma < 0$, $\delta > 0$ and

$$\Delta = \alpha\delta - \beta\gamma > 0 \tag{5.5-13}$$

The choice of l given by

$$l = -\alpha/\gamma < 0$$

makes (A, H) an unobservable pair.

Let

$$\prod_{\infty} = \begin{pmatrix} p & s \\ s & q \end{pmatrix}$$

be a positive semidefinite solution of the algebraic Riccati equation with $H = (1, l)$, $l = -\alpha/\gamma$.†

The optimal policy results in

$$A - BB'\prod_{\infty} = \begin{pmatrix} \alpha - \beta\xi & \beta\eta \\ \gamma - \delta\xi & \delta\eta \end{pmatrix}$$

where

$$\xi = \beta p + \delta s \quad \text{and} \quad \eta = -(\beta s + \delta q)$$

By solving the Riccati equation, it is shown that $\eta = l < 0$. This matrix has one unstable root since the characteristic equation is given by $\lambda^2 - (\alpha - \beta\xi - \delta\eta)\lambda + \eta\Delta = 0$ where $\Delta > 0$ but $\eta < 0$.

Example 2 A slightly more realistic example is obtained by

(1) modifying the import equation (5.5-6) to

$$M = mY + \theta(Y' - Y) \tag{5.5-6'}$$

where $0 < \theta < 1$ is a constant,

(2) and introducing the lag in the investment equation (5.5-2)

$$\kappa\dot{Y} = (\dot{I} + \kappa I)/v \tag{5.5-2'}$$

where $\kappa > 0$ is a constant.

Then (5.5-7) is replaced by

$$\dot{Y} = v\sigma(Y' - Y) \tag{5.5-7'}$$

†System parameters can be chosen to make $\Pi_{\infty} \geqslant 0$, i.e., to satisfy $p, q \geqslant 0$, $s^2 \leqslant pq$.

The state vector of this example now has three components

$$z = \begin{bmatrix} Y \\ I \\ D \end{bmatrix}$$

For this model, the system is uncontrollable only if

$$m = x \quad \text{and} \quad \theta = 0$$

a condition more unlikely to be realized in view of (5.5-6′), than (5.5-11). This example points out that the controllability condition will be affected very much by the model one chooses. One should choose a model the controllability of which is "robust" if at all possible. Similar sort of care must be exercised in choosing Q.

Two explicit sufficient conditions for optimal stabilization policy were pointed out in this section. Even though the controllability condition is rather natural and likely to be met for any reasonable models of economic systems, the observability condition requires more care. Inappropriate choice of $Q = H^{\mathrm{t}}H$ may result in (A, H) being an unobservable pair even though all variables of interest may be present in the criterion function. Inappropriate choice of the dynamic model and/or Q matrix is more likely to occur in models with greater dimension.

6

Stabilization of a Decentralized System: Example—Open Economy Model

6.1 INTRODUCTION

A decentralized dynamic system† is one in which two or more decision makers' actions will jointly affect the dynamic behavior of the system. Decision makers base their actions on partial and imperfect information on the various states of the dynamic system (and usually on each other's action).‡ In other words, in decentralized systems, the information necessary to make "optimal" decisions is decentralized, and thus is not available in any one place. Therefore, this assumption represents a radical departure from the Walrasian systems, in which all the necessary information is assumed to be available to the auctioneer or to the central coordinating (or planning) agency.

Stabilization of a decentralized dynamic system is more difficult than for a centralized dynamic system since all the information needed to make a "correct" decision is not available in any one place. A decentralized information pattern thus implies certain structural restrictions on stabilization policies. A degree of cooperation is necessary among decision makers so that their actions can be coordinated to work together to stabilize the dynamic system. The problem of stabilizing a decentralized system is therefore that of team decision-making, which is a special case of the theory of teams (Marschak and Radner 1971).

We do not discuss here a large body of literature on tâtonnement and nontâtonnement adjustment schemes that exist for static decentralized systems. We are interested in dynamic decentralized control laws, and, in particular, stabilization policies for dynamic decentralized systems (see, for

†This section is based in part on an unpublished paper by Aoki.

‡Control problems with imperfect information are discussed in Chapter 10.

example, Corfmat and Morse 1973, McFadden 1969, Patrick 1973 and Waud 1973).

The need to stabilize or, more generally, control economic activities in a decentralized manner arises most frequently in interdependent economic systems, as well as in closed economic systems. In interdependent economies, the so-called assignment problem has been discussed by Mundell (1962, 1968), Aoki (1974c) and others. The assignment problem is also considered in Chapter 8. Although the idea behind the assignment problem is intuitive and very appealing, Mundell did not formulate the assignment problem in any quantitative form and did not provide a criterion which can be used to test whether instruments can be assigned to targets on a one-to-one basis in a given model of the economy.

The purpose of this section is (1) to point out the complications that arise in the assignment problem in a general dynamic framework, and (2) to put on a quantitative basis some of the discussions surrounding the notion of comparative advantages in international trade.

Before we formulate a general case, let us briefly describe some existing models. McFadden (1969) formulated the assignment problem as follows. In the neighborhood of equilibrium the target variables are assumed to be influenced by the instruments by

$$y_{t+1} - y_t = Ax_t \tag{6.1-1}$$

where y_t is the vector of target variables (measured from some desired values) at time period t, where x_t is the vector of instrument variables at time period t. The instruments are assumed to be adjusted by

$$x_t = Sy_t \tag{6.1-2}$$

where S is some policy assignment matrix.

Restrictions are placed on the structure of S, e.g., that S be a (block) diagonal matrix, that would make the policy equation "decentralized." When S is restricted to be diagonal, for example, the assumption is that the ith instrument is under the control of the ith decision maker (a government agency, say) charged with the responsibility of regulating the ith target variable. Corfmat (1973) also considered this problem and gave the solution using the root-locus method (1973).†

Patrick (1973) considered decentralized policy assignment for a static model with a dynamic instrument generating equation. In his model,

†See Appendix C for the description of the root-locus method.

instruments are generated dynamically by

$$\dot{x} = Ky \tag{6.1-3}$$

where K is some policy assignment matrix.† In a decentralized mode of control, K is therefore constrained to have some specified structural form such as block diagonal.

Patrick assumed, however, a static relationship for targets and instruments

$$y = Ax \tag{6.1-4}$$

Substituting (6.1-4) into (6.1-3), Patrick then discusses stabilizability of

$$\dot{x} = KAx$$

where K is constrained by a decentralization requirement.

In this model, therefore, the dynamics are associated only with the instrument generation (6.1-3) and the target variables are assumed to attain values prescribed by (6.1-4) infinitely fast. Instead of this assumption on the system (6.1-4), it is desirable to consider the assignment problem in a truly dynamic setting where the system as well as the instrument generation are *both* dynamically modeled, recognizing explicitly that the adjustment speeds of targets for given instrument values are finite.

There is a basic issue involved here. Ignoring the finite adjustment speeds as is done in (6.1-4) allows one to concentrate on the instrument stability. As we have discussed in §5.2, the instrument stability is merely a necessary but not a sufficient condition for stable economic systems operation when in fact the adjustment speeds of the target variables are finite.

6.2 MODEL

We consider a model that is dynamic not only in its policy adjustment mechanism but also in the way target variables respond to changes in the instrument variables, and reexamine the assignment problem without the

†There may be cases in which imperfect knowledge about the basic structure of the economy, such as the values of the coefficients in the model or the nature of exogenous disturbances forces the decision maker to base his decisions only on limited information about the "state" of the economy (see, for example, Swoboda 1972). This type of limited information shows up as structural constraints on admissible policy combinations, constraints on S in (6.1-2) or K in (6.1-3) matrices, for example. Discussion of control problems with imperfect information is deferred to Chapter 10.

special assumption on adjustment made by McFadden (1969) and without the static system assumption made by Patrick (1973).†

Since we consider local behavior, i.e., behavior of variables near equilibrium values, we restrict ourselves to the linearized approximation of the systems.

As we have elaborated in §5.2, we will have the following set of equations in a general case

system dynamics: $\dot{z} = Az + Bx$

target: $y = Cz + Dx$

instrument generation: $x = Ks + Ly$

where $\dot{s} = Fs + Gy$

where all or some of the matrices A, B, C, D, F, G, K and L could be time-varying. Some of the matrices could be zero. In a decentralized system some structural constraints are imposed on some of these matrices.

The dynamic equations for s and z become, after relating x to s and z,

$$\frac{d}{dt}\begin{pmatrix} z \\ s \end{pmatrix} = \begin{pmatrix} \Phi_{11} & \Phi_{12} \\ \Phi_{21} & \Phi_{22} \end{pmatrix}\begin{pmatrix} z \\ s \end{pmatrix} \tag{6.2-1}$$

where

$$\Phi_{11} = A + B(I - LD)^{-1}LC, \qquad \Phi_{12} = B(I - LD)^{-1}K$$

$$\Phi_{21} = G\left[C + D(I - LD)^{-1}LC\right], \quad \Phi_{22} = F + GD(I - LD)^{-1}K$$

$$= G(I - DL)^{-1}C \qquad\qquad = F + G(I - DL)^{-1}DK$$

Here the policymakers may choose some or all of the matrices F, G, K and L subject to the information decentralization constraint to stabilize (6.2-1).

We do not discuss this most general case. See Aoki (1972) for some results along this line. We treat some special cases. Thus, we assume the economy is governed by

$$B\dot{y} = Cy + Dx \tag{6.2-3}$$

$$\dot{x} = Ky \tag{6.2-4}$$

†The use of differential rather than difference equations is not essential to our present purpose.

where (6.2-3) is the dynamic equation for y and (6.2-4) is the equation for instrument generation.† The x and y are to be interpreted as deviations from equilibrium values.

Assume the matrix B is nonsingular so that we can work with the reduced form instead of (6.2-3)

$$\dot{y} = Ey + Ax \tag{6.2-5}$$

where

$$E = B^{-1}C, \qquad A = B^{-1}D$$

Assume $m = \dim y = \dim x$, i.e., we assume the numbers of targets and instrument variables are the same. When we generalize economic models to include stochastic disturbances, we generally can do better by making $\dim x$ larger than $\dim y$, (Brainard 1967). We do not discuss this possibility here. The system is thus composed of two parts given by (6.2-4) and (6.2-5).

6.3 STABILIZATION OF A TWO-COUNTRY MODEL

Putting these two subsystems together as

$$\frac{d}{dt}\begin{pmatrix} y \\ x \end{pmatrix} = \begin{pmatrix} E & A \\ K & 0 \end{pmatrix}\begin{pmatrix} y \\ x \end{pmatrix} \tag{6.3-1}$$

we see that the stability of the instrument generation equation by proper choice of K is necessary, but may not be sufficient for the stability of the total system (6.3-1). This may be seen more explicitly by computing the eigenvalues associated with the dynamics of (6.3-1). We have

$$\begin{aligned} 0 &= \det\begin{pmatrix} \lambda I - E & -A \\ -K & \lambda I \end{pmatrix} \\ &= |\lambda I - E|\,|\lambda I - K(\lambda I - E)^{-1}A| \end{aligned} \tag{6.3-2}$$

where I is the $m \times m$ identity matrix. The second factor in (6.3-2)

$$|\lambda I - K(\lambda I - E)^{-1}A| \tag{6.3-3}$$

†Putting the dynamics as shown here implies that the vector y_t will generally contain not only the current target variables at t but also some lagged target variables.

clearly shows the special nature of the stability examined by Patrick, since the stability of his system is determined by the eigenvalues of KA,

$$0 = |\lambda I - KA| \tag{6.3-4}$$

The problem of decentralized stabilization of a dynamic economy may then be formulated as the problem of finding K with some prescribed structure such that the matrix

$$\begin{pmatrix} E & A \\ K & 0 \end{pmatrix}$$

is a stable matrix, i.e., has eigenvalues with negative real parts. For example, suppose that the first m_1 of the instruments and the targets, respectively, are under the control of Agency 1 and the rest under Agency 2. Then

$$K = \begin{pmatrix} K_1 & 0 \\ 0 & K_2 \end{pmatrix}$$

where

$$K_i : m_i \times m_i, \qquad i = 1, 2$$

$$m_1 + m_2 = m$$

Note that the presence of E, i.e., the dynamics of the target vector, complicates the stabilization problem substantially. We now give one explicit interpretation of (6.3-1) in terms of a model of a two-country open economy.

Suppose y is partitioned into three subvectors

$$y = \begin{bmatrix} y_1 \\ y_c \\ y_2 \end{bmatrix} : m \times 1$$

where y_c is the target subvector with components common to both countries, so that $\begin{pmatrix} y_1 \\ y_c \end{pmatrix}$ is the target vector for Country 1 and $\begin{pmatrix} y_c \\ y_2 \end{pmatrix}$ is target vector for Country 2. For example, y_i may contain the unemployment rate in Country i, the level of international reserves in Country i, $i = 1, 2$, and y_c may be the interest rate on bonds which are traded internationally. Then the interest rate will be common to both countries, and the components of the reserve level will satisfy a linear relation, if the total amount of the international reserves is assumed constant. Partition E, A and K conformably.

Assume that A is of the special structure

$$A = \begin{bmatrix} A_{11} & 0 \\ A_{21} & A_{22} \\ 0 & A_{32} \end{bmatrix} : n \times m$$

This assumption means that the instruments of Country 1 have no direct impact on $\dot{y}_2$, and x_2 has no direct impact on $\dot{y}_1$, but must work through various indirect transmission mechanisms.

We assume further that E is of the form

$$E = \begin{bmatrix} E_{11} & E_{12} & 0 \\ E_{21} & E_{22} & E_{23} \\ 0 & E_{32} & E_{33} \end{bmatrix} : m \times m$$

This assumption means that y_2 has no direct transmission mechanism to affect $\dot{y}_1$ and vice versa. We assume E is stable, both to keep the analysis reasonably simple for purposes of exposition and to consider a special case. Suppose the matrix K is constrained to be

$$K = \begin{pmatrix} k_1 \\ k_2 \end{pmatrix} : 2 \times m$$

This assumption implies that policymakers take into account only target variables in their own countries in generating new instrument values. Other possibilities, such as coordinating generation of instrument values by communicating target variables to each other, can be similarly formulated. Let the row vectors be

$$k_1 = g_1(l_1, l_2, 0)$$

and

$$k_2 = g_2(0, l_3, l_4)$$

where g_1 and g_2 are scalars. We assume therefore that in adjusting the rate of change in the instruments $\dot{x}_1$, policymakers assign relative weights l_1 and l_2 to y_1 and y_c. The variable g_1 is the "gain" factor that determines the actual adjustment rate of the instrument, row vectors l_3 and l_4 have similar interpretations.

The characteristic equation (6.3-2) becomes

$$\rho(\lambda) = \mu(\lambda)|\lambda I - K(\lambda I - E)^{-1}A| \tag{6.3-5}$$

where $\mu(\lambda) = |\lambda I - E|$. We have

$$(\lambda I - E)^{-1} = \frac{1}{|\lambda I - E|}\operatorname{adj}(\lambda I - E)$$

By the Cayley–Hamilton Theorem, $E^n = \sum_{i=1}^{n}\alpha_i E^{n-i}$. By the assumed stability of E, $\alpha_i < 0$, $i = 1, \ldots, n$. We know from §2.3 that

$$\operatorname{adj}(\lambda I - E) = \lambda^{n-1}I + \lambda^{n-2}F_1 + \cdots + \lambda F_{n-2} + F_{n-1}$$

where

$$F_k = E^k - \alpha_1 E^{k-1} - \cdots - \alpha_k I$$

We can write the 2×2 matrix $K(\lambda I - E)^{-1}A$ as

$$K(\lambda I - E)^{-1}A = \frac{K}{\mu(\lambda)}(\lambda^{n-1}I + \lambda^{n-2}F_1 + \cdots + \lambda F_{n-2} + F_{n-1})A \tag{6.3-6}$$

Let

$$G(\lambda) = (\lambda^{n-1}I + \lambda^{n-2}F_1 + \cdots + F_{n-1})$$

When $\dim y_1 = \dim y_c = \dim y_2 = 1$, E is a 3×3 matrix and

$$G(\lambda) = (\lambda^2 - \alpha_1\lambda - \alpha_2)I + (\lambda - \alpha_1)E + E^2 \tag{6.3-6$'$}$$

$$|\lambda I - KG(\lambda)A/\mu(\lambda)| = \lambda^2 - \gamma\lambda/\mu + \delta/\mu^2 \tag{6.3-7}$$

where

$$\gamma = \operatorname{tr} KG(\lambda)A$$

$$\delta = |KG(\lambda)A|$$

Since we imposed a structural constraint on the matrix K as shown here

$$K = \begin{pmatrix} g_1 l_1 & g_1 l_2 & 0 \\ 0 & g_2 l_3 & g_2 l_4 \end{pmatrix}$$

we can express γ and δ as

$$\gamma = g_1 \boldsymbol{h}^t G(\lambda)\boldsymbol{a}_1 + g_2 \boldsymbol{m}^t G(\lambda)\boldsymbol{a}_2$$

$$\delta = g_1 g_2 [\boldsymbol{h}^t G(\lambda)\boldsymbol{a}_1 \boldsymbol{m}^t G(\lambda)\boldsymbol{a}_2 - \boldsymbol{h}^t G(\lambda)\boldsymbol{a}_2 \boldsymbol{m}^t G(\lambda)\boldsymbol{a}_1]$$

where short-hand notations are used

$$\boldsymbol{h}^t = (l_1 \quad l_2 \quad 0), \boldsymbol{m}^t = (0 \quad l_3 \quad l_4) : 1 \times n$$

$$\boldsymbol{a}_1 = \begin{bmatrix} A_{11} \\ A_{21} \\ 0 \end{bmatrix} \quad \text{and} \quad \boldsymbol{a}_2 = \begin{bmatrix} 0 \\ A_{22} \\ A_{32} \end{bmatrix} : n \times 1$$

Let

$$\begin{aligned} \nu_1(\lambda) &= \boldsymbol{h}^t G(\lambda)\boldsymbol{a}_1, \quad & \nu_2(\lambda) &= \boldsymbol{m}^t G(\lambda)\boldsymbol{a}_2 \\ \nu_3(\lambda) &= \boldsymbol{h}^t G(\lambda)\boldsymbol{a}_2, \quad & \nu_4(\lambda) &= \boldsymbol{m}^t G(\lambda)\boldsymbol{a}_1 \end{aligned} \tag{6.3-8}$$

With these notations, the characteristic polynomial (6.3-5) becomes, after substituting (6.3-7) into (6.3-5)

$$\begin{aligned} \rho(\lambda) &= (\mu^2\lambda^2 - \mu\gamma\lambda + \delta)/\mu \\ &= \{\lambda\mu(\lambda)(\lambda\mu(\lambda) - g_2\nu_2(\lambda)) \\ &\quad - g_1[\nu_1(\lambda)(\lambda\mu(\lambda) - g_2\nu_2(\lambda)) + g_2\nu_3(\lambda)\nu_4(\lambda)]\}/\mu \end{aligned}$$

The roots of $\rho(\lambda)$ are equivalently given as those λ's that satisfy

$$1 = \frac{g_1}{\lambda\mu(\lambda)[\lambda\mu(\lambda) - g_2\nu_2(\lambda)]} \{\nu_1(\lambda)[\lambda\mu(\lambda) - g_2\nu_2(\lambda)] + g_2\nu_3(\lambda)\nu_4(\lambda)\} \tag{6.3-9}$$

For any pair of g_1 and g_2, there will be a λ satisfying (6.3-9). It depends continuously on the g's. We have then a two parameter family of possible root-loci (see Appendix on the root-locus method).

This equation clearly shows how policymakers in the two countries can or should cooperate and coordinate their stabilization policies. For example, Country 2 can choose g_2 to make $\lambda\mu(\lambda) - g_2\nu_2(\lambda)$ stable. Then Country 1 can choose g_1 to make λ satisfying (6.3-9) stable.

We now discuss the stabilization problem for the case $n = 3$, i.e., when y_1, y_c and y_2 are all scalars. Then $l_1 \sim l_4$ as well as A_{ij} all become scalars in addition to g_1 and g_2. We may discuss the following two subcases.

Case 1

$$0 \neq \nu_2(0) = \boldsymbol{m}^t G(0)\boldsymbol{a}_2 = \boldsymbol{m}^t(E^2 - \alpha_1 E - \alpha_2 I)\boldsymbol{a}_2$$

This will be the case if $(E, \boldsymbol{a}_2)$ is a controllable pair, then by definition $\boldsymbol{a}_2$, $E\boldsymbol{a}_2$ and $E^2\boldsymbol{a}_2$ are linearly independent. They cannot all lie in the null space of $\boldsymbol{m}^t$ since the dimension of the null space of $\boldsymbol{m}^t$ is at most 2.

Case 2 $0 = \nu_2(0)$ (This case happens only if $(E, \boldsymbol{a}_2)$ is not a controllable pair.)

We discuss Case 1 first.

Consider

$$\lambda\mu(\lambda) - g_2\nu_2(\lambda) = 0$$

or

$$-1 = (-g_2)\frac{\nu_2(\lambda)}{\lambda\mu(\lambda)} \tag{6.3-10}$$

By assumption $\mu(\lambda)$ is a stable polynomial. Consider the loci of λ that satisfy (6.3-10) parametrized by g_2, i.e., consider λ that satisfies (6.3-10) as a function of g_2.

Let θ be the angle of departure of the locus from $\lambda = 0$. For the system to be stable, θ must be π, so that the locus moves into the left half of the complex plane. From the right-hand side of (6.3-10), we have

$$\angle(-g_2) + \angle\nu_2(0) - \theta - \angle\mu(0) = \pi(\text{mod } 2\pi)$$

where $\angle$ stands for the argument of a (complex) number,

$$\angle\alpha = 0 \qquad \text{if } \alpha > 0$$

$$\angle\alpha = \pi \qquad \text{if } \alpha < 0$$

Thus

$$\theta = \pi$$

if

$$\angle(-g_2) + \angle\nu_2(0) - \angle\mu(0) = 2\pi(\text{mod } 2\pi)$$

Thus, by choosing g_2 such that

$$\angle(-g_2) = 2\pi + \angle\mu(0) - \angle\nu_2(0)(\text{mod } 2\pi)$$

the locus of (6.3-10) stays in the left half of the complex plane for $|g_2|$ small, i.e., the roots of (6.3-10) are stable.

Suppose Country 2 has stabilized (6.3-10) as above. Then following a similar argument, Country 1 can choose g_1 to stabilize (6.3-9). In (6.3-9), let

$$\rho(\lambda, g_2) = \lambda\mu(\lambda) - g_2\nu_2(\lambda)$$

$$-1 = \frac{-g_1}{\lambda\mu(\lambda)\rho(\lambda, g_2)}\{\nu_1(\lambda)\rho(\lambda, g_2) + g_2\nu_3(\lambda)\nu_4(\lambda)\}$$

In Case 1, we know that g_2 is chosen such that $\rho(0, g_2) \neq 0$.

Then from (6.3-8), $\nu_1(0) = \boldsymbol{h}^{\mathrm{t}}G(0), \boldsymbol{a}_1 \neq 0$, if $(E, \boldsymbol{a}_1)$ is also a controllable pair. Then $\nu_3(0)\nu_4(0) \neq 0$ from the assumption that $(E, \boldsymbol{a}_1)$ and $(E, \boldsymbol{a}_2)$ are controllable pairs.

Let θ be the angle of departure of the locus from $\lambda = 0$. We must have $\theta = \pi$ for stability in the following equation,

$$\angle(-g_2) + \angle\left[\nu(0)\rho(0, g_2) + g_2\nu_3(0)\nu_4(0)\right] - \theta - \angle\mu(0) - \angle\rho(0, g_2) = \pi(\mathrm{mod}\ 2\pi)$$

or $(-g_2)$ must be chosen as

$$\angle(-g_1) + \angle\left[\nu(0)\rho(0, g_2) + g_2\nu_3(0)\nu_4(0)\right] - \angle\mu(0) - \angle\rho(0, g_2) = 0(\mathrm{mod}\ 2\pi)$$

If $(E, \boldsymbol{a}_1)$ and $(E, \boldsymbol{a}_2)$ are not both controllable, so that $\nu_1(0)\rho(0, g_2) + g_2\nu_3(0)\nu_4(0) = 0$, then from the stability of $\mu(\lambda)\rho(\lambda, g_2)$, the existence of $(-g_2)$ for which all the loci are in the left half-complex plane is immediate.

In Case 2, $\nu_2(\lambda)$ becomes

$$\nu_2(\lambda) = \lambda\boldsymbol{m}^{\mathrm{t}}(\lambda I - \alpha_1 I + E)\boldsymbol{a}_2$$

g_2 must then be chosen to stabilize

$$\lambda[\mu(\lambda) - g_2\boldsymbol{m}^{\mathrm{t}}(\lambda I - \alpha_1 I + E)\boldsymbol{a}_2]$$

We have

$$-1 = (-g_2)\frac{(\boldsymbol{m}^{\mathrm{t}}\boldsymbol{a}_2)\lambda - \boldsymbol{m}^{\mathrm{t}}(\alpha_1 I - E)\boldsymbol{a}_2}{\mu(\lambda)}$$

instead of (6.3-9).

If $\boldsymbol{m}^{\mathrm{t}}\boldsymbol{a}_2 \neq 0$ and the signs of $(\boldsymbol{m}^{\mathrm{t}}\boldsymbol{a}_2)$ and $\boldsymbol{m}^{\mathrm{t}}(\alpha_1 I - E)\boldsymbol{a}_2$ are opposite, then roots of the numerator and denominator polynomials are both in the left half-complex plane, hence the root loci will be in the left half-plane for $g_2 \leqslant g_2 \leqslant 0$, for some $g_2 < 0$.

If the signs of $\boldsymbol{m}^{\mathrm{t}}\boldsymbol{a}_2$ and $\boldsymbol{m}^{\mathrm{t}}(\alpha_1 I - E)\boldsymbol{a}_2$ are the same, the numerator has one root in the right-hand side of the complex plane. Again the root loci remain in the left half-plane for some small $|g_2|$.

Finally suppose $\boldsymbol{m}^{\mathrm{t}}\boldsymbol{a}_2 = 0$. Again for some g_2, the system is stable.

Since E is stable by assumption, the roots of $\mu(\lambda) = |\lambda I - E|$ all lie in the left-half of the complex plane. There are only three possible configurations as drawn in Fig. 6.1.

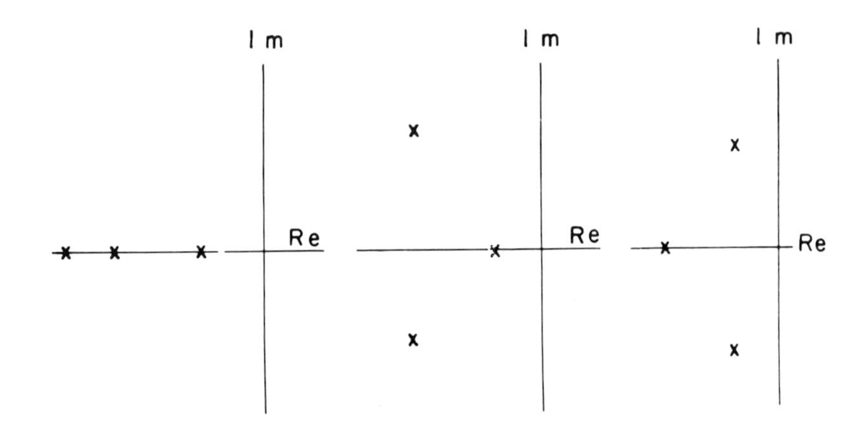

Fig. 6.1 Three possible pole configurations.

When one zero is located in the right half-plane, the root loci qualitatively become as drawn in Fig. 6.2. In all cases, there is a range of g_2 values for which all loci remain in the left-hand plane.

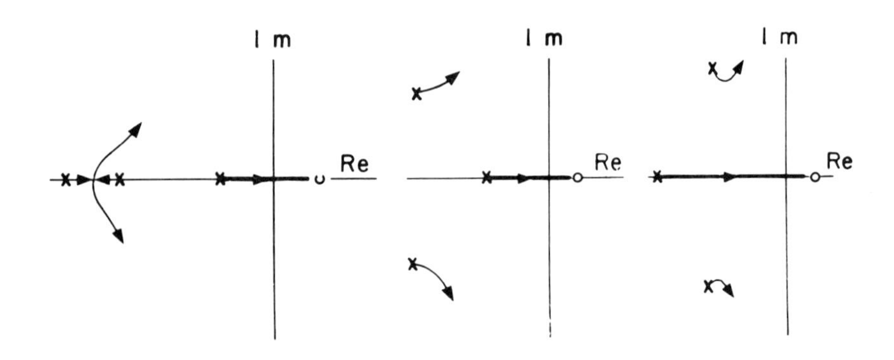

Fig. 6.2 Root-loci with one real zero.

Thus, regardless of the controllability condition of $(E, \boldsymbol{a}_2)$, provided E is stable, g_1 and g_2 can be so chosen that they cooperatively stabilize the dynamic system.

6.4 ASSIGNMENT OF INSTRUMENTS OR DECOUPLING PROBLEM IN DECENTRALIZED DYNAMIC SYSTEMS

We give a dynamic interpretation of definitions of comparative advantages as discussed by Patrick, (1973), and resolve some ambiguities in the meaning of assigning instruments to targets. His definition is "if one target is changed by adjustment of the instrument assigned to it, and then some or all other instruments are subsequently adjusted to eliminate the effects on their respective targets of the disturbances initiating from the change in the first instrument, the initial impact of the first instrument on its target is not completely offset."

We show that this implies $|A| \neq 0$ in his static model (6.1-4). We provide an alternate definition of comparative advantage in the dynamic model. A weaker definition of comparative advantage will be shown to be related to the so-called decoupling problem later.

In the static model of Patrick

$$y = Ax$$

where A is square, his definition is equivalent to A being nonsingular. To see this, it suffices to partition $\boldsymbol{x}$ and $\boldsymbol{y}$ into two subvectors and consider the case where we assume the instrument $\boldsymbol{x}_i$ is assigned to the target $\boldsymbol{y}_i$, $i = 1, 2$.

$$y = \begin{pmatrix} y_1 \\ y_2 \end{pmatrix}, \quad x = \begin{pmatrix} x_1 \\ x_2 \end{pmatrix} \quad \text{and} \quad A = \begin{pmatrix} A_{11} & A_{12} \\ A_{21} & A_{22} \end{pmatrix}$$

Suppose $\boldsymbol{x}_1$ is changed by some amount while $\boldsymbol{x}_2 = 0$. Then $\boldsymbol{y}_1 = A_{11}\boldsymbol{x}_1$ is the direct change. The change in $\boldsymbol{y}_2$, $\boldsymbol{y}_2 = A_{21}\boldsymbol{x}_1$, is eliminated by changing $\boldsymbol{x}_2$ to achieve

$$A_{21}\boldsymbol{x}_1 + A_{22}\boldsymbol{x}_2 = 0$$

or

$$\boldsymbol{x}_2 = -A_{22}{}^{-1}A_{21}\boldsymbol{x}_1$$

where $A_{22}{}^{-1}$ is assumed to exist.

Then due to this nonzero $\boldsymbol{x}_2$, $\boldsymbol{y}_1$ is modified to

$$\boldsymbol{y}_1 = (A_{11} - A_{12}A_{22}{}^{-1}A_{21})\boldsymbol{x}_1$$

Patrick's definition requires that

$$|A_{11} - A_{12}A_{22}{}^{-1}A_{21}| \neq 0$$

Similarly $\boldsymbol{x}_2$ is assigned to $\boldsymbol{y}_2$ yielding the condition

$$|A_{22} - A_{21}A_{11}{}^{-1}A_{12}| \neq 0$$

where $A_{11}{}^{-1}$ is assumed to exist. Since

$$\begin{aligned} |A| &= |A_{11}|\,|A_{22} - A_{21}A_{11}{}^{-1}A_{12}| \\ &= |A_{22}|\,|A_{11} - A_{12}A_{22}{}^{-1}A_{21}| \end{aligned}$$

his definition is seen to be equivalent to $|A| \neq 0$.

In the dynamic model, the condition becomes $|EA| \neq 0$, when A is square.

It is convenient to use Laplace transforms in order to interpret these considerations. Use $\hat{}$ to indicate the Laplace transforms with s as the Laplace transform variable as usual. From (6.2-5),

$$s\hat{y}(s) = E\hat{y}(s) + A\hat{x}(s) + y(0)$$

or

$$\hat{y}(s) = (sI - E)^{-1}A\hat{x}(s) + (sI - E)^{-1}y(0) \tag{6.4-1}$$

Suppose an instrument x_i is assigned to target i. Without loss of generality, we carry out the discussion when the first instrument x_1 is assigned to the first target variable y_1.

Partition y, x, E and A conformably,

$$\boldsymbol{y} = \begin{pmatrix} y_1 \\ \tilde{\boldsymbol{y}} \end{pmatrix}, \qquad \boldsymbol{x} = \begin{pmatrix} x_1 \\ \tilde{\boldsymbol{x}} \end{pmatrix}$$

$$E = \begin{pmatrix} e_{11} & e_{12} \\ e_{21} & \tilde{E} \end{pmatrix}, \qquad A = (\boldsymbol{a}^1, \tilde{A})$$

Thus,

$$\hat{y}_1(s) = (I \quad 0 \quad \ldots \quad 0)(sI - E)^{-1}\left(A\begin{pmatrix} \hat{x}_1(s) \\ 0 \end{pmatrix} + A\begin{pmatrix} 0 \\ \hat{\tilde{\boldsymbol{x}}}(s) \end{pmatrix}\right)$$

where the first term represents the effect on y_1 of x_1, while the second term represents the effect on y_1 of all the other instruments. The impact effect of a sudden change in x_1 (such as a step change) is obtained by the initial value theorem and the effects when all adjustments are completed is given by the final value theorem of §2.3. If a step change in x_1 is involved, then we can proceed as above by setting

$$\hat{x}_1(s) = \delta/s$$

where δ is the value of the change in x_1.

Thus the direct impact effect is given by

$$y_1(0+) = \lim_{s\to\infty}(1, 0, \ldots, 0)(sI - E)^{-1}A\begin{pmatrix}\delta\\0\end{pmatrix} = 0$$

the indirect impact effect is given by

$$y_1(0+) = \lim_{s\to\infty}(1, 0, \ldots, 0)(sI - E)^{-1}\delta A\begin{bmatrix}0\\1\\\vdots\\1\end{bmatrix} = 0$$

Namely, there is no impact effect for this model. Due to the step change in x_1, y_1 will be changed by

$$\begin{aligned} y_1(\infty) &= \lim_{s\to 0}(1, 0, \ldots, 0)(sI - E)^{-1}A\begin{pmatrix}\delta\\0\end{pmatrix} \\ &= -(1, \ldots, 0)E^{-1}A\begin{bmatrix}1\\\vdots\\0\end{bmatrix}\delta \\ &= -(e^{11}, e^{12})\boldsymbol{a}^1\delta \end{aligned}$$

where e^{11} and e^{12} are the (1, 1), and (1, 2) submatrices of E^{-1} respectively. This change in x_1 will change $\tilde{\boldsymbol{y}}$ by

$$\begin{aligned} \tilde{\boldsymbol{y}}(\infty) &= \lim_{s\to 0}(0, I)(sI - E)^{-1}A\begin{pmatrix}1\\0\end{pmatrix}\delta \\ &= -(0, I)E^{-1}\boldsymbol{a}^1\delta \\ &= -(e^{21}, \tilde{E}^{22})\boldsymbol{a}^1\delta \end{aligned}$$

where e^{21} is the (2, 1) element of E^{-1}, and $\tilde{E}^{22}$ is the (2, 2) element of E^{-1}. To offset this effect, $\tilde{\boldsymbol{x}}$ must be changed as a step function so that

$$(e^{21}, \tilde{E}^{22})(\boldsymbol{a}^1\delta + \tilde{A}\tilde{\boldsymbol{x}}) = 0$$

or

$$\tilde{\boldsymbol{x}} = -\left[(e^{21}, \tilde{E}^{22})\tilde{A}\right]^{-1}(e^{21}, \tilde{E}^{22})\boldsymbol{a}^1\delta$$

This $\tilde{\boldsymbol{x}}$ will modify $y_1(\infty)$ to

$$\begin{aligned} y_1(\infty) &= -(e^{11}, e^{12})(\boldsymbol{a}^1\delta + \tilde{A}\tilde{\boldsymbol{x}}) \\ &= -(e^{11}, e^{12})\left[I - \tilde{A}\left[(e^{21}, \tilde{E}^{22})\tilde{A}\right]^{-1}(e^{21}, \tilde{E}^{22})\right]\boldsymbol{a}^1\delta \end{aligned}$$

The condition that the initial change in x_1 is not to be nullified by the compensating change in $\tilde{\boldsymbol{x}}$ so that $\tilde{\boldsymbol{y}}$ is not affected by x_1 becomes:

$$(e^{11}, e^{12})\left[I - \tilde{A}\left[(e^{21}, \tilde{E}^{22})\tilde{A}\right]^{-1}(e^{21}, \tilde{E}^{22})\right]\boldsymbol{a}^1 \neq 0$$

or

$$(EA)_{11} - (EA)_{12}(EA)_{22}{}^{-1}(EA)_{21} \neq 0$$

The left-hand side may also be interpreted as the condition for "asymptotic decoupling," i.e. the effect of a step change in x_1 on $\tilde{\boldsymbol{y}}$ goes to zero as $t \to \infty$.

So far, we have treated the dynamic model assuming no instrument generation equation such as (6.1-3). There are $m!$ possible assignments, where $m = \dim \boldsymbol{y} = \dim \boldsymbol{x}$, which may be too large to be practicable when m is large, e.g. $4! = 24$, $5! = 120$. Thus $m = 4$ or 5 seems to be an upper limit for the method to remain manageable.

Then we may be interested in dividing $\boldsymbol{y}$ and $\boldsymbol{x}$ into two or more subsets and assigning a subset of $\boldsymbol{x}$ to a subset of $\boldsymbol{y}$ in a one-to-one way, and require that targets in one subset are not influenced by the instrument subvectors not assigned to it. This is a special case of decoupling problems to be discussed in Chapter 8.

Let $\boldsymbol{y}$ and $\boldsymbol{x}$ be conformably partitioned into K subvectors, and assume K to be block diagonal. We show in Chapter 8 that a necessary and sufficient condition for decoupling is that $(sI - E)^{-1}A$ be a block diagonal matrix.

With the instrument generation equation given by (6.1-3), we obtain from (6.1-3) and (6.4-1),

$$\hat{y}(s) = \left(I - \frac{(sI - E)^{-1}AK}{s}\right)^{-1}(sI - E)^{-1}y(0)$$

$$= \left(sI - E - \frac{AK}{s}\right)^{-1} y(0), \qquad \text{where } \boldsymbol{x}(0) = 0$$

If $[sI - E - (AK/s)]^{-1}$ is asymptotically stable and if it is block diagonal, then subsets of the targets are decoupled. However, since E is not generally block diagonal, it is not generally asymptotically decoupled. To achieve this, it is necessary to consider a more general instrument generation equation such as $\dot{\boldsymbol{x}} = G\boldsymbol{x} + K\boldsymbol{y}$. This topic will be taken up later in Chapter 8.

7

Stability Analysis of DynamicAdjustment Models

In this chapter, we consider two deterministic disequilibrium adjustment models. One is a Marshallian quantity adjustment model for a number of firms producing a nonstorable homogeneous good. Each firm adjusts its output of the good for the tth market day based on the market clearing price on the $(t - 1)$th day and its own supply price schedule. This model is therefore primarily a quantity adjustment model. No learning is considered here. The other is a model of a pure exchange economy with money in which each trader has a stock of money and flow endowments of goods and exchanges goods for money and money for goods. In this model, planned and realized transactions are not usually the same, and the dynamics of the model become nonlinear due to rationing that takes place when ex-ante and ex-post quantities are different.

In these two models we discuss equilibrium states and stability of adjustment processes. We describe the first situation by a set of difference equations and the second by a set of differential equations, illustrating discrete-time and continuous-time analyses. Stochastic adjustment models are taken up later in Chapter 12.

7.1 A MARSHALLIAN ADJUSTMENT SCHEME: CASE OF PERISHABLE GOOD†

7.1.1 Aggregated Model

We analyze a quantity adjustment scheme for producers of a perishable good, recently considered by Axel Leijonhufvud (1974). In a model of what he refers to as the Marshallian homeostat, quantities produced by a

†Also see §12.1 for related topics.

representative firm are adjusted according to

$$p^d = d(q) \tag{M.1}$$

$$p^s = s(q) \tag{M.2}$$

$$\Delta q = h(s(q) - d(q)), \quad h < 0 \tag{M.3}$$†

where q is the actual output rate, p^d is the maximum price at which consumers are willing to absorb a given rate of output and p^s is the minimum price required to induce producers to continue a given rate of output; $d(\cdot)$ is the demand price schedule, $s(\cdot)$ is the supply price schedule, and h is a negative scalar.

In the above Marshallian quantity adjustment scheme, the demand price p^d must be known for the representative firm to adjust the output rate according to (M.3). Since p^d is not actually known,‡ Leijonhufvud replaces (M.3) with (M.3′), which is taken to be the adjustment scheme for the representative firm

$$\Delta q_t = h[s(q_t) - p_t^*] \tag{M.3′}$$

where

$$p_t^* = d(q_t)$$

is the market clearing price for the actual output rate q_t.§ In other words, the price adjusts "rapidly" during a market day according to an adjustment mechanism of the type

$$\Delta p_{i,t} = \lambda_w(D(p_{i,t}) - q_t), \qquad i = 0, 1, \ldots$$

where i is the iteration index, completing its adjustment within a single day. The price p_t^* thus obtained is the market clearing price at the end of the tth market day. The quantity adjustment equation (M.3′) takes the

†A more general adjustment mechanism is

$$\Delta q = f(s(q) - d(q)) \text{ with } f(0) = 0, f' < 0.$$

‡Another possibility is to discuss parameter-learning behavior of the representative firm. See §12.3 for techniques on analyzing parameter adaptive situations.

§In imperfect information models, one could use subjective estimates of the revenue or profit rate such as marginal revenue products instead. See §12.2 or Aoki (1974d, 1975d).

place of the more familiar cobweb type quantity adjustment given by $\Delta q_t = q_{t+1} - q_t = S(p_t) - S(p_{t-1})$.† We defer the discussion of adjustment mechanisms with exogenous disturbances until §12.1.

7.1.2 Disaggregated Model

Leijonhufvud next considered the disaggregated version of (M.3′) in an atomistic market of n firms producing a single homogeneous nonstorable good. The jth producer is assumed to adjust his output rate by

$$\Delta q_{j,t} = h_j[s_{j,t}(q_{j,t}) - p_t^*], \qquad j = 1, \ldots, n \tag{7.1-1}$$‡

and the market clearing price is given by the joint output rates of n producers

$$p_t^* = d\left(\sum_{j=1}^{n} q_{j,t}\right) \tag{7.1-2}$$

Each firm observes p_t^*. Equations (7.1-1) and (7.1-2) explicitly introduce the individual firms' behavior and their interactions at the market level. With these interactions explicitly spelled out, we can ask and answer some interesting questions. For example, can the market be in equilibrium in the sense that the total output on the market $\Sigma_{j=1}^{n} q_{j,t}$ remains the same on each market day and the market clearing price remains the same while individual firms are not necessarily in equilibrium? We now show how this question can be answered. (See §12.2 for the market behavior with exogenous random shocks.) Although we could treat $s(\cdot)$ and $d(\cdot)$ as general nonlinear functions, we analyze the adjustment mechanisms (7.1-1) and (7.1-2), assuming that the firms' supply price schedules are linear in q,

$$s_j(q) = \alpha_j + \beta_j q, \qquad j = 1, \ldots, n \tag{7.1-3}$$

where α's and β's are deterministic scalars.

We assume also that the market clearing price is linearly related to the total amount of the good put on the market by

$$p_t^* = \gamma - \delta Q_t$$

---

†We assume that production delay is one day in this section and Chapter 12.

‡A more general case where the adjustment step size is changing with time can be considered analogously, under some regularity conditions. See Aoki (1975d).

where

$$Q_t = \sum_{j=1}^{n} q_{j,t} \tag{7.1-4}$$

Let q_{jt}^* be such that $p_t^* = s_j(q_{jt}^*)$. Suppose that the supply price schedule is such that this output rate is uniquely determined. Then firm j's adjustment equation may equivalently be rewritten as

$$\begin{aligned} \Delta q_{jt} &= h_j\left[s_j(q_{jt}) - s_j(q_{jt}^*)\right] \\ &= h_j \frac{\partial s_j}{\partial q}(q_{jt}^*)(q_{jt} - q_{jt}^*) + \cdots \end{aligned}$$

provided either $s_j(\cdot)$ is affine or $q_{jt} - q_{jt}^*$ is sufficiently small if $s_j(\cdot)$ is not affine. We assume the former, i.e., $s_j(q) = \alpha_j + \beta_j q$. Then the quantity adjustment equation (7.1-1) may be rewritten as

$$\Delta q_{jt} = \theta_j(q_{jt} - q_{jt}^*) \tag{7.1-1$'$}$$

where $\theta_j = h_j\beta_j$ is the decision variable of firm j, and is the adjustment step size in closing the discrepancy between q_{jt} and q_{jt}^*. We see then that (7.1-1) and (7.1-1′) are equivalent. We will use (7.1-1) in our subsequent discussion.

Equation (7.1-1′) shows that the adjustment equation (7.1-1) is similar to the small step gradient method, also known as the hill-climbing method, in nonlinear programming. It differs from the hill-climbing method in one important respect, however. In nonlinear programming problems, the hill, the local maximum, does not change with decisions such as the size of the steps and the direction of the climb. In our problem q_{jt}^* is influenced by not only firm j's output decision but also by all other firms' decision. This is an example of information externality. The presence of information externality makes this problem very difficult to solve in general. We discuss some special cases in the next section and in §12.2. Also see Aoki (1975d).

7.1.3 Difference Equations for Output Adjustments

As before, denote the adjustment of the output rate by

$$\Delta q_{j,t} = q_{j,t+1} - q_{j,t}, \qquad j = 1, \ldots, n$$

From (7.1-1), using the assumed linear schedule (7.1-3) we see that the jth firm's output is adjusted according to

$$\Delta q_{j,\,t+1} - \Delta q_{j,\,t} = h_j \beta_j \Delta q_{j,\,t} - h_j \Delta p_t^*, \qquad j = 1, \ldots, n \tag{7.1-5}$$

Using the vector-matrix notation, we can write it conveniently as

$$\Delta \boldsymbol{q}_{t+1} = (I + HB)\Delta \boldsymbol{q}_t - \boldsymbol{h}\Delta p_t^* \tag{7.1-5'}$$

where

$$H = \mathrm{diag}(h_1 \quad \ldots \quad h_n)$$

$$\Delta \boldsymbol{q}_t = (\Delta q_{1t} \quad \ldots \quad \Delta q_{nt})^{\mathrm{t}}$$

$$B = \mathrm{diag}(\beta_1 \quad \ldots \quad \beta_n)$$

$$\boldsymbol{h} = (h_1 \quad \ldots \quad h_n)^{\mathrm{t}}$$

From (7.1-4), we note that the change in the market clearing price is related to the change in the amount of goods put on the market by

$$\Delta p_t^* = -\delta \Delta Q_t$$
$$= -\delta \boldsymbol{e}^{\mathrm{t}} \Delta \boldsymbol{q}_t$$

where

$$\boldsymbol{e} = (1 \ldots 1)^{\mathrm{t}}$$

Substitute this into (7.1-5′) to obtain the difference equation for output adjustment. Then $\Delta \boldsymbol{q}_t$ is governed by the linear difference equation

$$\Delta \boldsymbol{q}_{t+1} = \Phi \Delta \boldsymbol{q}_t$$

where

$$\Phi = (I + HB + \delta \boldsymbol{h}\boldsymbol{e}^{\mathrm{t}}) : n \times n \tag{7.1-6}$$

where from (7.1-1) and (7.1-3), the initial condition is

$$\Delta \boldsymbol{q}_0 = H[\boldsymbol{\alpha} - \gamma \boldsymbol{e} + (B + \delta \boldsymbol{e}\boldsymbol{e}^{\mathrm{t}})\boldsymbol{q}_0], \qquad \text{where } \boldsymbol{\alpha} = (\alpha_1 \ldots \alpha_n)^{\mathrm{t}}$$

This is the dynamic equation governing the time evolution of the output adjustment vector involving all the firms that are dynamically coupled together by the fact that they observe the same signal, the market clearing price. The total output produced by the ith firm on the $(t + 1)$th day is shown as the ith components of $\boldsymbol{q}_{t+1}$, the latter is given by

$$\begin{aligned}\boldsymbol{q}_{t+1} &= \boldsymbol{q}_0 + \Delta\boldsymbol{q}_0 + \cdots + \Delta\boldsymbol{q}_t \\ &= \boldsymbol{q}_0 + (I + \Phi + \cdots + \Phi^t)\Delta\boldsymbol{q}_0 \\ &= \boldsymbol{q}_0 + (I - \Phi)^{-1}(I - \Phi^{t+1})\Delta\boldsymbol{q}_0\end{aligned}$$

where we see that $(I - \Phi)^{-1}$ exists from (7.1-6) provided $h_i\beta_i \neq 0$, $i = 1, \ldots, n$ (see Corollary 1 of Lemma 2, Appendix B).

We show in the next section that $\Phi^t \to 0$ as $t \to \infty$. Then the firm-by-firm equilibrium output rate $\boldsymbol{q}_\infty$ exists and is given by

$$\boldsymbol{q}_\infty = \boldsymbol{q}_0 + (I - \Phi)^{-1}\Delta\boldsymbol{q}_0$$

From (7.1-6), this may be shown to be independent of $\boldsymbol{h}$ and is equal to

$$\boldsymbol{q}_\infty = \frac{\gamma + \delta\boldsymbol{e}^{\mathrm{t}}B^{-1}\boldsymbol{\alpha}}{1 + \delta\boldsymbol{e}^{\mathrm{t}}B^{-1}\boldsymbol{e}}B^{-1}\boldsymbol{e} - B^{-1}\boldsymbol{\alpha}$$

where Lemma 2 of Appendix B is used to obtain the inverse $(I - \Phi)^{-1}$. In other words, when every firm is in equilibrium, the ith firm's output rate is proportional to $1/\beta_i$ and is given by $(c - \alpha_i)/\beta_i$ where

$$c = \left(\gamma + \delta\sum_j \alpha_j/\beta_j\right)\bigg/\left(1 + \delta\sum_j 1/\beta_j\right)$$

7.1.4 Eigenvalue Computation

Next we compute the eigenvalues of Φ to show that the adjustment process is asymptotically stable as claimed above.

The eigenvalues are the zeroes of the characteristic polynomial:

$$\begin{aligned}g(\lambda) &= |\lambda I - \Phi| \\ &= |(\lambda - 1)I - HB - \delta\boldsymbol{h}\boldsymbol{e}^{\mathrm{t}}| \\ &= |(\lambda - 1)I - HB| \cdot \{1 - \delta\boldsymbol{e}^{\mathrm{t}}[(\lambda - 1)I - HB]^{-1}\boldsymbol{h}\} \\ &= \{\Pi_{j=1}{}^n(\lambda - 1 - \theta_j)\}\left(1 - \delta\Sigma_{j=1}{}^n\frac{h_j}{\lambda - 1 - \theta_j}\right) \qquad (7.1\text{-}7)\end{aligned}$$

where $\theta_j = h_j\beta_j$ (see Corollary 1 of Lemma 2, Appendix B).

For the purpose of the proof only, assume that

$$0 > \theta_1 \geqslant \theta_2 \geqslant \cdots \geqslant \theta_n$$

Denote the eigenvalues of Φ by $\lambda_1, \lambda_2, \ldots, \lambda_n$. We note that the sign of $g(\lambda)$ alternates at $\lambda = 1 + \theta_1, 1 + \theta_2, \ldots$. Therefore $1 > 1 + \theta_1 \geqslant \lambda_1 \geqslant 1 + \theta_2 \geqslant \lambda_2 \geqslant \cdots \geqslant 1 + \theta_n \geqslant \lambda_n > -1$ if $|h_j \beta_j| < 2$ for all j and if $\delta \Sigma_j |h_j| + |h_n| \beta_n < 2$.† Recalling our discussion in §4.2,

$$\Delta \boldsymbol{q}_t = \Phi^t \Delta \boldsymbol{q}_0 \to 0 \qquad \text{as } t \to \infty$$

This shows that as time passes, each firm will reach an individual equilibrium level $\boldsymbol{q}_\infty$ of the output rate.

The convergence of the adjustment process will be dominated by the eigenvalue with the largest magnitude. If $\theta_1 + \theta_n < -2$, then $|1 + \theta_n| > |1 + \theta_1|$. Then λ_n has the largest modulus lying between 1 and $1 + \theta_n$. This means that the speed with which the adjustment process approaches the equilibrium depends primarily on the slope of the supply price schedule of the nth firm.

7.1.5 Equilibrium for the Individual Firms and the Market

The quantity of the good available to the market on the tth day Q_t changes with time as‡

$$\Delta Q_t = \boldsymbol{e}^{\mathrm{t}} \Delta \boldsymbol{q}_t$$

An interesting question is: Can ΔQ_t be zero from some time on without $\Delta \boldsymbol{q}$ being zero? That is, can the market be in equilibrium when not all the individual firms being are in equilibrium? We examine this for a special case in which $h_j = h$ for all j.

Suppose

$$0 = \Delta Q_{t+\tau} = \boldsymbol{e}^{\mathrm{t}} \Delta \boldsymbol{q}_{t+\tau}, \qquad \tau = 0, 1, \ldots$$

for some t.

Then from (7.1-6) and the Cayley–Hamilton Theorem, this condition is

†This condition is a generalization of the well-known cob-web stability condition, §12.1.

‡We can derive the difference equation for ΔQ_t's by eliminating $\Delta \boldsymbol{q}$'s, if we wish.

equivalent to

$$0 = \begin{bmatrix} \boldsymbol{e}^{\mathrm{t}} \\ \boldsymbol{e}^{\mathrm{t}}\Phi \\ \boldsymbol{e}^{\mathrm{t}}\Phi^2 \\ \vdots \\ \boldsymbol{e}^{\mathrm{t}}\Phi^{n-1} \end{bmatrix} \Delta \boldsymbol{q}_t \tag{7.1-8}$$

Now from (7.1-6), we see that

$$\begin{aligned} \boldsymbol{e}^{\mathrm{t}}\Phi &= \boldsymbol{e}^{\mathrm{t}} + \boldsymbol{h}\boldsymbol{\beta}^{\mathrm{t}} + n\delta \boldsymbol{h}\boldsymbol{e}^{\mathrm{t}} \\ &= (1 + n\delta \boldsymbol{h})\boldsymbol{e}^{\mathrm{t}} + \boldsymbol{h}\boldsymbol{\beta}^{\mathrm{t}} \\ &= \boldsymbol{h}\boldsymbol{\beta}^{\mathrm{t}} \quad (\mathrm{mod}\ \boldsymbol{e}^{\mathrm{t}}) \end{aligned}$$

where

$$\boldsymbol{\beta} = (\beta_1 \quad \ldots \quad \beta_n)^{\mathrm{t}}$$

and

$$\begin{aligned} \boldsymbol{\beta}^{\mathrm{t}}\Phi &= \boldsymbol{\beta}^{\mathrm{t}} + \boldsymbol{h}(\beta^{(2)})^{\mathrm{t}} + \boldsymbol{h}\delta(\Sigma\beta_j)\boldsymbol{e}^{\mathrm{t}} \\ &= \boldsymbol{h}(\beta^{(2)})^{\mathrm{t}} \quad \mathrm{mod}(\boldsymbol{e}^{\mathrm{t}}, \boldsymbol{\beta}^{\mathrm{t}}) \end{aligned}$$

where $\boldsymbol{\beta}^{(2)} = (\beta_1^2 \quad \ldots \quad \beta_n^2)^{\mathrm{t}}$, and where "mod" means that the vectors on the right and left sides of the equality sign are congruent modulo the subspace spanned by the vectors indicated. We also see then that (7.1-8) is equivalent to

$$0 = \begin{bmatrix} \boldsymbol{e}^{\mathrm{t}} \\ \boldsymbol{\beta}^{\mathrm{t}} \\ \boldsymbol{\beta}^{(2)\mathrm{t}} \\ \vdots \\ \boldsymbol{\beta}^{(n-1)\mathrm{t}} \end{bmatrix} \Delta \boldsymbol{q}_t \tag{7.1-8'}$$

where

$$\boldsymbol{\beta}^{(i)} = (\beta_1^{\,i} \quad \ldots \quad \beta_n^{\,i})^{\mathrm{t}}$$

The matrix in (7.1-8′) is known as the Vandermonde matrix, the determinant of which is given by†

$$\begin{vmatrix} 1 & 1 & \cdots & 1 \\ \beta_1 & \beta_2 & \cdots & \beta_n \\ \beta_1^2 & \beta_2^2 & \cdots & \beta_n^2 \\ & & \vdots & \\ \beta_1^{n-1} & \beta_2^{n-1} & \cdots & \beta_n^{n-1} \end{vmatrix} = \prod_{i>j} (\beta_i - \beta_j)$$

Hence, if not all individual firms are in equilibrium the market can stay in equilibrium only if two or more firms have the same θ's, e.g., have the same marginal supply price schedules if h_j is the same for all j. This is because the firms with the same marginal supply price schedules are unable to distinguish the equilibrium condition from a nonequilibrium condition when h_j = const for all j.

This inability is a usual problem of observability. The matrix in (7.1-8) is precisely the stacked observability matrix of the dynamic equation for the $\Delta \boldsymbol{q}$'s since $\Delta p_t^* = -\delta \boldsymbol{e}^t \Delta \boldsymbol{q}_t$ is the observation made on $\Delta \boldsymbol{q}_t$. As a matter of fact, when the market clearing prices (and own outputs) are the only observed signals, the above analysis shows that subsets of firms with the same marginal supply price schedules may be aggregated with the same equation as (7.1-5). Thus, the total output of a class of firms with the same marginal supply price schedule must be zero in the market equilibrium, though the output of an individual firm of the class need not be zero. To see this more directly, consider the kernel or the null space‡ of the observability matrix in (7.1-8). We have seen that if some of β's are the same then the null space is a subspace of dimension one or more. To take the extreme case, suppose the β's are all the same and equal $\bar{\beta}$. Then Φ of (7.1-6) becomes $(1 + h\bar{\beta})I + h\delta \boldsymbol{e}\boldsymbol{e}^t$. Therefore any $\Delta \boldsymbol{q}_t$ such that $\boldsymbol{e}^t \Delta \boldsymbol{q}_t = 0$ implies that $\boldsymbol{e}^t \Phi^k \Delta \boldsymbol{q}_t = 0$ for $k = 1, 2, \ldots$. In this case, if $\Delta \boldsymbol{q}_t$ satisfies $\boldsymbol{e}^t \Delta \boldsymbol{q}_t = 0$, then the industry-wide output rate will remain the same and the same market clearing price prevails from the tth market day on even though $\Delta \boldsymbol{q}_t \neq 0$, i.e., at least one firm's output rate has not reached the equilibrium rate. Next, suppose $\beta_1 \neq \beta_2 = \cdots = \beta_n = \bar{\beta}$. Then

$$\Phi = \begin{bmatrix} 1 + h\beta_1 & 0 \\ 0 & (1 + h\bar{\beta})I_{n-1} \end{bmatrix} + h\delta \boldsymbol{e}\boldsymbol{e}^t$$

†With nonidentical adjustment step sizes, $\Pi_{i>j}(\theta_i - \theta_j)$, where $\theta_j = h_j \beta_j$.

‡The null space of an $n \times n$ matrix H is the subspace $\{\boldsymbol{q} | H\boldsymbol{q} = 0\}$. $\mathfrak{N}(H)$ consists of $\{0\}$ if and only if H is nonsingular.

$0 = \boldsymbol{e}^{\mathrm{t}}\Delta \boldsymbol{q}_t$ implies $\Delta q_{1t} + \Sigma_{j=2}^{n}\Delta q_{jt} = 0$. The condition $0 = \boldsymbol{e}^{\mathrm{t}}\Phi\Delta \boldsymbol{q}_t$ implies that $\Delta q_{1t} = 0$ and $\Sigma_{j=2}^{n}\Delta q_{jt} = 0$. Then $\boldsymbol{e}^{\mathrm{t}}\Phi^k\Delta \boldsymbol{q}_t = 0$ for $k = 2, 3, \ldots$. In other words, firm 1's output rate must be at its equilibrium rate. The output rates of firms 2 through n need not be zero individually as long as $\Sigma_{j=2}^{n}\Delta q_{jt} = 0$.

7.2 A MONETARY DISEQUILIBRIUM MODEL: STABILITY ANALYSIS OF A NONLINEAR ECONOMIC SYSTEM

7.2.1 Introduction

We next introduce a model in which money is used as a medium of exchange. The model we consider is a slight generalization of the one considered by Howitt (1973) and is related to that discussed by Grandmont and Younes (1972). We examine the existence of an equilibrium and carry out stability analysis of adjustments near the equilibrium.

In the process of the analysis, we hope to illustrate some analytical techniques which are applicable to many similar or more realistic models.

We consider a deterministic dynamic model of an exchange economy in which money is the only medium of exchange, i.e., goods are exchanged for money and only money can be used to acquire goods. For the purpose of exposition only, let us divide time into periods, called market days. The economy is composed of n traders who are immortal and who are endowed with initial stocks of money, and flows of n nonstorable goods which perish after one market day. Money has no direct utility but serves as the only means of storing purchasing power at no cost from one market day to the next. In addition it is the sole medium of exchange although this former property is de-emphasized in this model as we shall see shortly. We assume that trading takes place out of equilibrium in the following manner: Each trader is designated as the store-keeper for one specific good, so that goods are sold through their respective storekeepers. At the beginning of each market day, traders receive commodity bundles, and store-keepers post prices which they think will clear the market for that good.

Traders place purchase or sales orders for each good with its store-keeper. Store-keepers will accept purchase orders (i.e., notional demands) only if they are backed by sales orders for other goods or money holdings.

We incorporate in our model a feature that reflects an economic "fact" that traders' purchase plans may not be realized exactly, either because of rationing (by finding themselves on the long side of the market, for example), or because of the discrepancies between the money receipt they counted on from their sales plan and the actual money received. There are several ways such a model may be constructed. The basic idea behind such models is as discussed in (Grandmont and Younes, 1972). A trader cannot use, for various obvious reasons, all the proceeds from sales for that day before the market closes to finance his purchase for the same day. We choose the following simple model: We assume that trader i regards a dollar's worth of sales receipts in his hand as equivalent to γ_i dollar's worth of money in his purchase plan.† Furthermore, trader i uses as his budget constraint only δ_i of money plus γ_i times sales receipt on hand.‡ The case of $\gamma_i = 1$ and $\delta_i = 1$ is an idealized economy in which all plans are always realized and in which purchases and sales are perfectly synchronized.

After everybody places his order, the store-keepers carry out the orders exactly at the end of each market day if excess demands for their goods are zero; if not, the store-keepers ration order proportionately, i.e., fulfill a certain fixed percentage of all orders on the long side of the markets depending on the sizes of excess demands (or supplies). They then adjust prices as functions of these excess demands and post new prices at the beginning of the next market day.§

The model is simple enough to allow us to examine analytically the natures of adjustment behavior and their dependence on the system parameters such as the relative magnitudes of adjustment speeds and so on. Yet it retains enough features to make it economically interesting. For example, the only spillover effect in this model occurs through the cash balance of the traders and not directly between markets. No dynamic learning or expectation formation behavior due to past rationing is incorporated in the model, since they are not expected to alter the model behavior in any essential way. In spite of the simple structures, the model can give answers to some important and interesting questions about the dynamic behavior of market adjustments in disequilibrium with rationing and stock of money. For example, the so-called fix-price method of analysis and the Archibald-Lipsey type analysis will be shown to be the two extreme cases of the possible adjustments of the model. Using the

†An alternate interpretation will be provided later.

‡δ_i could presumably come from solving the usual dynamic programming functional equation of two period intertemporal optimization involving the current period consumption plus all future consumptions.

§For this price adjustment scheme to make sense, we assume that there are n types of agents and that there are many agents of each type. We then deal with a representative agent of each type.

technique and framework established in this section, we can also evaluate the effects of price controls imposed on some of the commodities on the prices of the rest of the commodities and on excess demands. This type of analysis seems to be a little more dynamic than what is available in the literature. See, for example, Gould and Henry (1967).

The proportional rationing rule (as a matter of fact, any nonprice rationing rules) will introduce nonlinearity in the model. We shall discuss a technique for dealing with the nonlinearity caused by rationing and show that a unique equilibrium state exists for the model, that all markets clear in equilibrium and that the adjustment processes for the prices of the goods and the individuals' money stocks are stable near the equilibrium under conditions to be stated in §7.2.5. We also show by a simple system theoretic argument that all excess demands for goods are different from zero throughout the adjustment process, i.e., no subset of markets remains cleared in reaching the equilibrium state. What makes the stability analysis of the model interesting and nonroutine is the fact that the model switches its dynamic behavior depending on which goods are in excess demand or in excess supply. They are determined in turn by the price and money stock dynamics.

We now give a brief summary of the section. In §7.2.2, the model is formally described. Its equilibrium state is described in §7.2.3 and its properties are shown to be economically reasonable. In preparation for stability analysis about the equilibrium state, a simple analytical device using a step function is employed to express nonlinearity due to rationing in §7.2.4. A rather detailed analysis of the model is performed, the nature of nonlinearity that enters in the differential equations governing exchanges is elucidated, and linearized differential equations for the prices and money stocks are derived. This analysis yields the asymptotic stability of the model in §7.2.5 where we compute the characteristic equation of the linearized system, the roots of which are the eigenvalues governing the adjustment speeds. We show that all the roots except one are negative and we show how they depend on the parameters of the system in three special cases. One eigenvalue is zero due to the constraint imposed by the assumption of a fixed total amount of money. All variables are shown to approach the equilibrium values exponentially in all three cases.

§7.2.6 is devoted to short-run and long-run adjustment behavior implied by the model.

Adjustment processes with price control are subsequently discussed.

Instead of the discrete-time descriptions used in previous sections for motivating the model, we consider the duration of a market day to be short and use a differential equation description of the model from now on.

7.2.2 Model

Let us now describe more precisely the model to be analyzed. There are n types of transactors indexed by i, $i = 1, \ldots, n$† and n pure flow endowments of perishable goods indexed by j, $j = 1, \ldots, n$. In addition, there is a stock commodity called money that is the only store of value (wealth asset) and the medium of exchange in the model. Let $M_i(t) > 0$ be the money stock held by trader i, $i = 1, \ldots, n$, at time t. Let $p_j(t)$ be the price of the commodity j, $j = 1, \ldots, n$, prevailing at time t. We assume for simplicity that the total money stock M in the model is fixed. Transactor i receives a constant endowment flow of good j, $w_j^i \geqslant 0$, normalized by $\sum_{j=1}^{n} w_j^i = 1$, $i, j = 1, \ldots, n$. Assume w_j^i is time-invariant and known.

Let $d_j^i(t)$ be the notional flow demand i.e., the purchase plan at time t of trader i for the jth good, $i, j = 1, \ldots, n$. We drop the time argument from now on to simplify notation.

Suppose trader i's utility function is given by

$$U^i = \prod_{j=1}^{n} (d_j^i)^{\alpha_j^i}, \qquad \alpha_j^i > 0$$

The budget constraint is

$$I_i = \sum_j p_j d_j^i$$

where

$$I_i = \delta_i \left(M_i + \gamma_i \sum_k p_k w_k^i \right), \qquad 0 < \delta_i < 1$$

Trader i's optimal notional demands are given by

$$\mathrm{Max}\left(U^i \,\middle|\, \sum_j p_j d_j^1 \leqslant I_i \right)$$

or equivalently

$$\mathrm{Max}\left(\ln U^i \,\middle|\, \sum_j p_j d_j^i \leqslant I_i \right)$$

---

†See footnote, page 203.

The constrained maximization may be carried out by Lagrange multiplier rule. As the result, we obtain d_j^i as

$$d_j^i = a_{ij}(\delta_i M_i + \eta_i \Sigma_k p_k w_k{}^i)/p_j \tag{7.2-1}$$†

where

$$a_{ij} = \alpha_{ij}/\Sigma_k \alpha_{ik} \quad \text{and} \quad \eta_i = \delta_i \gamma_i$$

Note that $\Sigma_j a_{ij} = 1$ by definition. The constants a_{ij}, δ_i and η_i are all positive and less than one. In case all goods are equally desirable so that α_j^i is the same for all j, then a_{ij} becomes $1/n$ for all i and j. This special case will be discussed later. Another special case we will discuss later is that in which $a_{ij}\eta_j$ is a constant independent of i and j, i.e., $a_{ij}\delta_i = 1/c\gamma_i$ where c is some number greater than one and such that $c\gamma_i > 1$ for all i. We shall refer to δ's and η's as the liquidity parameters. Note that d_j^i expresses the ith trader's notional real flow demand since doubling the money stocks held by transactors will result in doubling p_j's for all j, leaving d_j^i invariant, as will be seen later.

Equation (7.2-1) is economically reasonable: d_j^i satisfies the laws of demand such as being downward sloping with respect to p_j, being increasing with the money stock at trader i's command and with the flow endowments, and demand for own goods being decreasing when $w_k{}^i = 0$, $k \neq i$, as p_i increases.

---

†An alternate model formulation could be the following: Under static price expectation, the current value to trader i of the infinite sequence of endowments of goods j, $w_j{}^i$ each period, is $\gamma_j w_j{}^i p_j$, where p_j is the price of goods j, γ_j being determined by the relative "ease" with which the transactions in good j take place. Then, $M_i + \Sigma_j \gamma_j w_j{}^i p_j$ is the current wealth of trader i and he is willing to spend δ_i of it on current consumption. Then the budget constraint becomes $I_i = \delta_i(M_i + \Sigma_k \gamma_k w_k{}^i p_k)$. The notional demand is then given by

$$d_j^i = a_{ij}\left(\delta_i M_i + \delta_i \sum_k \gamma_k w_k{}^i p_k\right) \Big/ p_j \tag{7.2-1'}$$

When endowments are specialized to $w_k{}^i = \delta_{ik}$, no change in analysis is needed in using (7.2-1′) rather than (7.2-1) since the notional demand in both cases reduces to

$$d_j^i = a_{ij}(\delta_i M_i + \eta_i p_i)/p_j$$

where

$$\eta_i = \delta_i \gamma_i$$

Reverting back to the discrete-time description for a moment, if M_i^- is the money stock at the beginning of a market day, then he expects, if all his plans are carried out, his money stock at the end of the day to be

$$M_i^+ = M_i^- + \sum_k p_k(w_k^i - d_k^i)$$

where

$$\Sigma_j p_j d_j^i = (\delta_i M_i^- + \eta_i \Sigma_k p_k w_k^i).$$

Thus, in the equilibrium $M_i^+ = M_i^- = M_i^e$ or

$$M_i^e = \{1 - \eta_i\} \Sigma_k p_k^e w_k^i / \delta_i$$

This equation shows how the equilibrium money stock of trader i is determined by the equilibrium prices of the goods. Equilibrium prices will be determined after we derive the differential equations for p's and M's in disequilibrium. The superscript e refers to equilibrium.

We drop superscripts to indicate market variables: For example,

$$d_j = \sum_{i=1}^n d_j^i$$

and

$$w_j = \sum_{i=1}^n w_j^i$$

Denote trader i's net notional excess demand for good j by

$$x_j^i = d_j^i - w_j^i, \qquad i, j = 1, \ldots, n \tag{7.2-2}$$

We then denote by x_j the market net excess demand for good j

$$x_j = \sum_{i=1}^n x_j^i = d_j - w_j$$

Traders are generally rationed and their notional demands and supplies are not their realized demands and supplies. We assume the traders are rationed proportionally only if they are on the "long" side of the markets, i.e., only if $x_j^i \cdot x_j > 0$, $i, j = 1, \ldots, n$.

Denoting by k_j^i the proportion of the notional transaction that is realized, the differential equation for the money stock then can be written as

$$\begin{aligned}\dot{M}_i &= \sum_{j=1}^{n} p_j(w_j^i - d_j^i)k_j^i \\ &= -\sum_{j=1}^{n} p_j x_j^i k_j^i \end{aligned} \tag{7.2-3}$$

If trader i is on the short side of Market j, then $k_j^i = 1$. If trader i is on the long side of Market j, then k_j^i = (short side of Market j)/(long side of Market j) < 1. See Appendix 7-A for the analytical expression of k_j^i.

We assume that prices adjust according to

$$\begin{aligned}\dot{p}_j &= \beta_j p_j x_j \\ &= \beta_j z_j, \qquad j = 1, \ldots, n\end{aligned} \tag{7.2-4}$$

where

$$z_j = p_j x_j$$

is the nominal net excess demand for good j.

Equations (7.2-1), (7.2-3), and (7.2-4) determine completely the dynamics of the model. We explore their implications in the rest of this section. We assume from now on that $w_i^i = 1$, $w_j^i = 0$, $j \neq i$, i.e., assume that only trader i receives a unit of good i. This assumption permits us to examine the dynamic behavior of the model without unduly complicating the analysis and without losing the essential features of the model.

When we specialize the flow endowments to $w_i^i = 1$, $w_j^i = 0$, $i \neq j$, we have from (7.2-1) and (7.2-2) for all i,

$$\left.\begin{aligned} p_i x_i^i &= a_{ii}(\delta_i M_i + \eta_i p_i) - p_i \\ &= a_{ii}\delta_i M_i - (1 - a_{ii}\eta_i)p_i \\ p_i x_j^i &= a_{ij}\delta_i M_i + a_{ij}\eta_i p_i \quad, j \neq i \end{aligned}\right\} \tag{7.2-1''}$$

Hence (7.2-3) becomes

$$\dot{M}_i = k_i^i p_i - \left(\Sigma_j k_j^i a_{ij}\right)(\delta_i M_i + \eta_i p_i) \tag{7.2-3'}$$

It is convenient to express these equations using vector-matrix notations.

For that purpose, define the matrices of rationing proportions by

$$K(z) = \operatorname{diag}(k_1^{\ 1}, \ldots, k_n^{\ n})$$

$$\overline{\Lambda}(z) = \operatorname{diag}(\bar{\mu}_1, \ldots, \bar{\mu}_n)$$

where

$$\bar{\mu}_i = \sum_{j=1}^{n} k_j^{\ i}(z_j)a_{ij}$$

and the matrices of the liquidity parameters by

$$\Delta = \operatorname{diag}(\delta_1, \ldots, \delta_n) \qquad \text{and} \quad N = \operatorname{diag}(\eta_1, \ldots, \eta_n)$$

Then in the vector notation, we can write (7.2-3′) as

$$\dot{\boldsymbol{M}} = K(z)\boldsymbol{p} - \overline{\Lambda}(z)(\Delta \boldsymbol{M} + N\boldsymbol{p}) \tag{7.2-5}$$

where $\boldsymbol{M}$ is the n-dimensional column vector with components $M_1, \ldots, M_n$ and $\boldsymbol{p}$ is similarly defined.

The nominal excess demand z_j is given, from (7.2-1) and (7.2-2), by

$$\begin{aligned} z_j &= p_j x_j \\ &= \Sigma_i a_{ij}(\delta_i M_i + \eta_i p_i) - p_j \end{aligned}$$

or

$$z = A^{\mathrm{t}}(\Delta M + N\boldsymbol{p}) - \boldsymbol{p} \tag{7.2-6}$$

where

$$A = (a_{ij})$$

When α_{ij} in the utility function is independent of j, we know that $a_{ij} = 1/n$. Suppose that this is true for all i. In this special case, (7.2-5) reduces to

$$\dot{\boldsymbol{M}} = K(z)\boldsymbol{p} - \Lambda(z)(\,\Theta\, \boldsymbol{M} + N\boldsymbol{p}/n) \tag{7.2-5′}$$

where

$$\Lambda(z) = \operatorname{diag}(\mu_1, \ldots, \mu_n)$$

with

$$\mu_i = \Sigma_j k_j^{\,i}(z_j)$$

and where

$$\Theta = \mathrm{diag}(\theta_1, \ldots, \theta_n)$$

with

$$\theta_i = \delta_i / n \tag{7.2-5$'$a}$$

Also, in this special case $A = \boldsymbol{e}\boldsymbol{e}^{\mathrm{t}}/n$, where $\boldsymbol{e}^{\mathrm{t}} = (1, \ldots, 1)$. (7.2-6) reduces to

$$z = \boldsymbol{e}\boldsymbol{\theta}^{\mathrm{t}}\boldsymbol{M} + \left(\frac{1}{n}\boldsymbol{e}\boldsymbol{\eta}^{\mathrm{t}} - I\right)\boldsymbol{p} \tag{7.2-6$'$}$$

where

$$\boldsymbol{\theta} = (\theta_1, \ldots, \theta_n)^{\mathrm{t}} \quad \text{and} \quad \boldsymbol{\eta}^{\mathrm{t}} = \boldsymbol{e}^{\mathrm{t}}N$$

In the special case

$$a_{ij}\delta_i = 1/c\gamma_i$$

Redefine θ_i by

$$\theta_i = 1/c\gamma_i$$

Then (7.2-5) reduces to

$$\dot{\boldsymbol{M}} = K(z)\boldsymbol{p} - \Lambda(z)(\Theta\boldsymbol{M} + \boldsymbol{p}/c) \tag{7.2-5$''$}$$

and (7.2-6) becomes simplified to

$$z = \boldsymbol{e}\boldsymbol{\theta}^{\mathrm{t}}\boldsymbol{M} + (\boldsymbol{e}\boldsymbol{e}^{\mathrm{t}}/c - I)\boldsymbol{p} \tag{7.2-6$''$}$$

Since $K(z)$ and $\overline{\Lambda}(z)$ depend on z, (7.2-3′) or (7.2-5) clearly show that the differential equations for M_i are nonlinear. Equation (7.2-4) becomes

$$\dot{\boldsymbol{p}} = Bz \tag{7.2-7}$$

where

$$B = \mathrm{diag}(\beta_1, \ldots, \beta_n)$$

7.2.3 Equilibrium State

In equilibrium, since none of the prices and money stocks is changing, we have from (7.2-3) and (7.2-4)

$$\dot{p}_j = 0, \qquad \dot{M}_i = 0, \quad i, j = 1, \ldots, n$$

and consequently from (7.2-4),

$$z_j = 0, \qquad j = 1, \ldots, n$$

These conditions imply that $k_j^i = 1$ for all i and j (see Appendix 7-A). In other words, all the markets clear in equilibrium. In (7.2-5), $K(0) = I$ and $\overline{\Lambda}(0) = I$, hence $\dot{\boldsymbol{M}} = 0$ shows that

$$(I - N)\bar{\boldsymbol{p}} = \Delta\overline{\boldsymbol{M}}$$

or

$$\bar{p}_i = \delta_i \overline{M}_i / (1 - \eta_i) \tag{7.2-8}$$

where the overbar denotes equilibrium quantities. From the zero excess demand condition (7.2-6), we obtain another relation between $\bar{\boldsymbol{p}}$ and $\overline{\boldsymbol{M}}$,

$$A^{\mathrm{t}}\Delta\overline{\boldsymbol{M}} = (I - A'N)\bar{\boldsymbol{p}} \tag{7.2-8$'$}$$

Equations (7.2-8) and (7.2-8′) together determine the equilibrium prices and money stocks. We next derive their expressions explicitly for the two special cases we considered earlier and show that $\bar{\boldsymbol{p}} > 0$ and $\overline{\boldsymbol{M}} > 0$.

Case 1: $a_{ij} = 1/n$ for all i and j
By definition, $\Lambda(0) = nI$. From (7.2-5′) and (7.2-6′) we see that

$$\begin{aligned} \boldsymbol{e}\boldsymbol{\theta}^{\mathrm{t}}\overline{\boldsymbol{M}} &= (I - \boldsymbol{e}\boldsymbol{\eta}^{\mathrm{t}}/n)\bar{\boldsymbol{p}} \\ &= (nI - \boldsymbol{e}\boldsymbol{\eta}^{\mathrm{t}})(I - N)^{-1}\Theta\overline{\boldsymbol{M}} \end{aligned}$$

or

$$\hat{\theta}_i M_i = \text{constant} \qquad \text{for all } i,$$

where

$$\hat{\theta}_i = \theta_i / (1 - \eta_i)$$

From the assumption of the constant total money stock M we obtain

$$\overline{M}_i = \frac{M/\hat{\theta}_i}{\sum_k 1/\hat{\theta}_k} > 0, \qquad i = 1, \ldots, n \tag{7.2-9}$$

From (7.2-8) and (7.2-9) recalling (7.2-5′a)

$$\bar{p}_i = nM/\Sigma_k 1/\hat{\theta}_k > 0 \tag{7.2-9′}$$

Case 2: $a_{ij}\delta_i = 1/c\gamma_i$ for all i and j
From (7.2-5″) and (7.2-6″)

$$\overline{M}_i = \gamma_i M/\Sigma_k \gamma_k = M/\theta_i \Big/ \Sigma_k 1/\theta_k > 0$$

and

$$\bar{p}_i = \frac{ncM}{(c-n)\sum_k 1/\theta_k} > 0 \tag{7.2-9″}$$

We note that in both cases the equilibrium prices of all goods are constant. From now on, for the sake of simpler presentation of the main ideas, we consider only the special case, Case 2, in which c is taken to be $n + 1$.† For easier references later, we list the differential equations and their equilibrium values here:

$$\left.\begin{aligned} \dot{\boldsymbol{M}} &= K(z)\boldsymbol{p} - \Lambda(z)(\Theta \boldsymbol{M} + \boldsymbol{p}/(n+1)) \\ \overline{M}_i &= M/\theta_i \Big/ \Sigma_k 1/\theta_k \\ \bar{p}_j &= n(n+1)M \Big/ \Sigma_k 1/\theta_k \\ &\overset{\text{df}}{=} p_e \end{aligned}\right\} \tag{7.2-10}$$

7.2.4 Differential Equations Near the Equilibrium

We have obtained (7.2-3) and (7.2-4) or its vector matrix version (7.2-5) and (7.2-7) as the differential equations governing the changes in prices and the money stocks in §7.2.2.

†The model with this value is considered by Howitt (1973).

Since they are nonlinear, we linearize them about the equilibrium values to pave the way for stability analysis of the adjustment mechanisms near the equilibrium. We follow the procedures described in §2.6, in deriving the linearized equations. We measure prices and money stocks from their equilibrium levels,

$$p_j = p_e + \delta p_j$$

and

$$M_i = \overline{M}_i + \delta M_i$$

and rewrite differential equations (7.2-5) and (7.2-7) in terms of δp_j's and δM_i's. Note that the equilibrium levels of $\boldsymbol{x}$ and z are zero so we do not write $\delta \boldsymbol{x}$ nor δz, but use $\boldsymbol{x}$ and z.

Since the rationing makes the elements of $K(z)$ and $\Lambda(z)$ nonlinear we continue to investigate the special case where $w_j^i = \delta_{ij}$ where δ_{ij} is the Kronecker's delta, i.e., trader i receives one unit of flow of good i and nothing else. This assumption simplifies the analysis due to the fact that trader i then becomes a net supplier of good i and all other traders become net demanders of good i near equilibrium, since the individuals' excess demands in equilibrium are

$$\bar{x}_j^i = \begin{cases} (1-n)/n < 0, & i = j \\ 1/n > 0, & i \neq j \end{cases}$$

Thus, near the equilibrium the sign of $x_j^j \cdot x_j$ is the negative of the sign of x_j, and the sign of $x_j^i \cdot x_i$ is that of x_i $i \neq j$. This permits us to write the k_j^i explicitly as functions of z's.

Near the equilibrium, we obtain the expression of the rationing proportions as (see Appendix 7-A)

$$\left.\begin{aligned} k_j^i &= 1 - \gamma z_j u(z_j) + o(z_j) \\ k_i^i &= 1 + \gamma z_i u(-z_i) + o(z_i) \end{aligned}\right\}, \qquad \frac{1}{\gamma} = \frac{n^2-1}{(n+1)^2}\frac{M}{\Sigma_i \theta_i} \tag{7.2-11}$$

We now can derive approximate expressions for $K(z)$ and $\Lambda(z)$ in (7.2-5″) using (7.2-11).

Since $z_i = 0$ at the equilibrium and the dynamics of the p's are governed by the z's, it is convenient now to change variables from p's and M's to z's and M's. In other words, we use the differential equation for $\boldsymbol{M}$ and the nominal excess demand vector z rather than the price vector. Since the equilibrium prices are not zero these two sets of variables are equivalent in describing the dynamic behavior of the model near equilibrium.

To derive differential equations for z we can solve (7.2-6″) for $\boldsymbol{p}$ as†

$$\boldsymbol{p} = \left(I - \frac{1}{n+1}\boldsymbol{e}\boldsymbol{e}^{\mathrm{t}}\right)^{-1}(\boldsymbol{e}\boldsymbol{\theta}^{\mathrm{t}}\boldsymbol{M} - z)$$

Since

$$\left(I - \frac{1}{n+1}\boldsymbol{e}\boldsymbol{e}^{\mathrm{t}}\right)^{-1} = I + \boldsymbol{e}\boldsymbol{e}^{\mathrm{t}}$$

(this can be verified directly) we see the relation between $\boldsymbol{p}$, z and $\boldsymbol{M}$ is given by‡

$$\boldsymbol{p} = (n+1)\boldsymbol{e}\boldsymbol{\theta}^{\mathrm{t}}\boldsymbol{M} - (I + \boldsymbol{e}\boldsymbol{e}^{\mathrm{t}})z \qquad (7.2\text{-}12)$$

Denoting the vectors of deviations from equilibrium values by $\delta\boldsymbol{p}$ and $\delta\boldsymbol{M}$, (7.2-7) and (7.2-12) are valid with $\boldsymbol{p}$ and $\boldsymbol{M}$ replaced by $\delta\boldsymbol{p}$ and $\delta\boldsymbol{M}$. Equation (7.2-12) shows that nonzero excess demands show up generally in prices being different from the equilibrium prices but stocks of money that are not at their equilibrium values may not show up in prices if $\boldsymbol{\theta}^{\mathrm{t}}\delta\boldsymbol{M} = 0$. The differential equation for z can be obtained now by differentiating (7.2-12) and making use of (7.2-7)

$$\dot{z} = -Bz + \frac{1}{n+1}\boldsymbol{e}\boldsymbol{\beta}^{\mathrm{t}}z + \boldsymbol{e}\boldsymbol{\theta}^{\mathrm{t}}\delta\dot{\boldsymbol{M}} \qquad (7.2\text{-}13)$$

where $\boldsymbol{\beta}^{\mathrm{t}} = \boldsymbol{e}^{\mathrm{t}}B$. From (7.2-5″), (7.2-11), (7.2-12), (7.2-13), and noting that $z_i(u(z_i) + u(-z_i)) = z_i$ for all i, we derive the differential equation for $\delta\boldsymbol{M}$,

$$\delta\dot{\boldsymbol{M}} = \overline{\Theta}\delta\boldsymbol{M} + \frac{1}{n+1}(nI - \boldsymbol{e}\boldsymbol{e}^{\mathrm{t}})z - \frac{1}{n-1}(nU - \boldsymbol{e}\boldsymbol{v}^{\mathrm{t}})z \quad (7.2\text{-}14)$$

where

$$\boldsymbol{v} = (u(z_1), \ldots, u(z_n))^{\mathrm{t}}$$

$$U = \mathrm{diag}(u(z_1), \ldots, u(z_n))$$

$$\overline{\Theta} = \boldsymbol{e}\boldsymbol{\theta}^{\mathrm{t}} - n\Theta$$

Equation (7.2-14) is not yet the desired linearized differential equation since it contains $u(\cdot)$. We see from the dependence of U and $\boldsymbol{v}$ on z that

†Recall $c = n + 1$.

§In the state space (internal) description of this model, z and $\boldsymbol{M}$ constitute the state vector and $\boldsymbol{p}$ is the vector that is observed by the traders. The net excess demand and the money stocks held by other traders are *not* observed.

adjustment behavior depends on the signs of z_j, $j = 1, \ldots, n$, i.e., on which goods are in excess demands, which goods are in excess supply and which markets for the goods momentarily clear. Therefore, it is necessary to partition z by the signs of the components to examine the effects of noncleared markets.

Define the index sets J_+ and J_- by

$$J_+ = \{i | z_i > 0, i = 1, \ldots, n\}$$

and

$$J_- = \{i | z_i < 0, i = 1, \ldots, n\}$$

Partition z into

$$z = \begin{pmatrix} z_+ \\ z_- \end{pmatrix}$$

where z_+ is the n_+-dimensional subvector composed of all z_i, $i \in J_+$ and z_- is the n_--dimensional vector similarly defined. We show later that $z_j \neq 0$ unless $z_j = 0$ for all j so that we do not have to consider separately the subvector z_0 with all components zero. Partition $\delta \boldsymbol{M}$ and $\boldsymbol{e}$ conformably, where $\boldsymbol{e}$ is the n-dimensional vector with all components 1.

The perturbation equation, i.e., the differential equation near the equilibrium becomes from (7.2-13) and (7.2-14)

$$\begin{pmatrix} \dot{z} \\ \delta \dot{\boldsymbol{M}} \end{pmatrix} = F \begin{pmatrix} z \\ \delta \boldsymbol{M} \end{pmatrix} + \boldsymbol{q}(z, \delta \boldsymbol{M}) \tag{7.2-15a}$$

where

$$\boldsymbol{q}(0, 0) = 0, \ \|\boldsymbol{q}(z, \delta \boldsymbol{M})\| = o(z, \delta \boldsymbol{M}) \text{ and where}$$

$$F = \begin{bmatrix} -B + \dfrac{\boldsymbol{e}\boldsymbol{\delta}^{\mathrm{t}}}{n+1} + \boldsymbol{e}\boldsymbol{\theta}^{\mathrm{t}}[E, 0], & \boldsymbol{e}\boldsymbol{\eta}^{\mathrm{t}} \\ [E, 0] + \dfrac{1}{n+1}(nI - \boldsymbol{e}\boldsymbol{e}^{\mathrm{t}}), & \overline{\Theta} \end{bmatrix} \tag{7.2-15b}$$

where we define

$$E = \frac{1}{n-1}\boldsymbol{e}\boldsymbol{e}_+{}^{\mathrm{t}} - \frac{n}{n-1}\begin{pmatrix} I_+ \\ 0 \end{pmatrix} : n \times n_+$$

$$\boldsymbol{\delta} = \boldsymbol{\beta} + n\boldsymbol{\theta} - (\boldsymbol{\theta}^{\mathrm{t}}\boldsymbol{e})\boldsymbol{e}$$

that is,

$$\delta_i = \beta_i + n\theta_i - \sum_i \theta_i, \qquad i = 1, \ldots, n \tag{7.2-15c}$$

and where

$$\boldsymbol{\eta} = (\boldsymbol{\theta}^{\mathrm{t}}\boldsymbol{e})\boldsymbol{\theta} - n\boldsymbol{\theta}^2$$

that is,

$$\eta_i = \left(\sum_j \theta_j\right)\theta_i - n\theta_i^2, \qquad i = 1, \ldots, n$$

Note that

$$\boldsymbol{e}\boldsymbol{\theta}^{\mathrm{t}}\overline{\Theta} = \boldsymbol{e}\boldsymbol{\eta}^{\mathrm{t}}$$

See Appendix 7-B for details of the derivation of (7.2-15). Equation (7.2-15) is the set of linearized differential equations near the equilibrium for a given configuration of excess demands and supplies in the various markets specified by the index sets J_+ and J_-. Different F obtain for different J sets. We see the condition $\Sigma\delta M_i = 0$ is automatically satisfied since $\boldsymbol{e}^{\mathrm{t}}E = 0$, $\boldsymbol{e}^{\mathrm{t}}(nI - \boldsymbol{e}\boldsymbol{e}^{\mathrm{t}}) = 0$ and $\boldsymbol{e}^{\mathrm{t}}\overline{\Theta} = 0$. This implies at least one of the eigenvalues of F is zero.

7.2.5 Stability Analysis Near the Equilibrium State

SUFFICIENT CONDITION Since we have derived the differential equations (7.2-15) for the time paths of the economy in disequilibrium near the equilibrium state of the model, we now investigate the stability of the adjustment processes, the speeds of adjustments of prices and money stocks and how they depend on system parameters.

As time passes, some components of z will change sign and the forms of F and $\boldsymbol{q}$ in (7.2-15) change accordingly. There are finite number (2^n) of such possible F and $\boldsymbol{q}$'s, all individually satisfying the following two properties:

(1) The nonlinear term $\boldsymbol{q}(z, \boldsymbol{M})$ is continuous and satisfies the Lipschitz condition in a neighborhood of (0, 0) and $q(0, 0) = 0$.
(2) As will be shown later (see next subsection for eigenvalue computation), all the eigenvalues of the matrix F given by (7.2-15b) have negative real parts. Between switching, i.e., during the time interval over which the index sets J_+ and J_- remain the same, we can write the time path of $z(\cdot)$ as

$$z(t) \cong \phi_{11}(t - t_k)z(t_k) + \phi_{12}(t - t_k)\delta\boldsymbol{M}(t_k), \qquad t_k \leqslant t < t_{k+1}$$

where t_k is the last time J_+ and J_- have changed (i.e., some goods switched from being in excess demand to excess supply or vice versa),

and where

$$e^{Ft} = \begin{pmatrix} \phi_{11}(t), & \phi_{12}(t) \\ \phi_{21}(t), & \phi_{22}(t) \end{pmatrix}$$

in which $\phi_{ij}(t)$ is composed of linear combinations of terms such as $t^l e^{\lambda_f t}$, $l = 0, 1, \ldots, m_f$ where m_f is the multiplicity of the eigenvalue λ_f of F.

The next switching will occur when some components of $z(t)$ reach zero, i.e., some goods momentarily clear before becoming in excess demand (supply) from excess supply (demand).

Thus, if there are only a finite number of switrchings of signs of z, then a standard stability theorem can be invoked to show asymptotic stability of (7.2-15a) in a small neighborhood of the origin (0, 0). For this purpose, we have the following Lemma.

Lemma 1 (Pontryagin) *Let $\lambda_1, \ldots, \lambda_k$ be distinct real values and let $\phi_1(t), \ldots, \phi_k(t)$ be real polynomials in t of degrees $m_1, \ldots, m_k$, respectively. Then the function*

$$\sum_{i=1}^{k} \phi_i(t) e^{\lambda_i t}$$

has at most $\sum_{i=1}^{k} m_i + k - 1$ real roots.

For proof, see page 122 of Pontryagin (1962).

We say a matrix is stable if the real parts of its eigenvalues are all nonpositive. If they are all negative, we call the matrix asymptotically stable. The (asymptotic) stability of the system (7.2-15a) is determined by that of F. This is due to the following theorem.

Theorem 1 *The nonlinear system* (7.2-15) *is* (*asymptotically*) *stable if the linear part F has* (*negative*) *nonpositive real eigenvalues for all possible assignments of components of z into z_+ and z_-.*

Proof Since there are only a finite number of switchings of signs of the components of z by Lemma 1, there exists a finite time T after which no switching takes place. Thus, by choosing a sufficiently small neighborhood of the origin, a standard proof can be applied. For example, see page 88 of Lefschetz (1962).

Before we proceed to verify that eigenvalues of F are all negative real numbers (except one which is zero) in the rest of this section, we interpret the implication of Theorem 1. If the eigenvalues of F are all negative then excess demands for all the goods change sign at most a finite number of times. This implies that the prices (and the excess demands) may increase or decrease exponentially, changing the rates of increase or decrease at most a finite number of times before finally approaching the equilibrium values exponentially. Similarly, money stocks of individual traders may increase or decrease at different rates at most a finite number of times before they approach the equilibrium stock levels exponentially. (What could happen when some of the eigenvalues are complex is that the components of z could change signs infinitely often and the stability of adjustments may not follow.)

Equation (7.2-15b) shows also that if $\boldsymbol{\eta}^t\delta \boldsymbol{M} = 0$, for example if $\theta_1 = \cdots = \theta_n$, then small deviations of the trader's money balances from the equilibrium values do not affect $\dot{z}$. We discuss this case later.

EIGENVALUE COMPUTATIONS We want to know not only whether F is a stable matrix or not, but also how the eigenvalues of F depend on the parameters in the model, namely on θ_i's and β_j's. Although we can write down the characteristic function of F in (7.2-15) by using Lemma 2 in Appendix C, it is not possible generally to obtain its roots explicitly in analytically closed forms.

We therefore confine ourselves to obtaining the roots of the characteristic equation of F when $\boldsymbol{\theta}$ and/or $\boldsymbol{\beta}$ have special features. We know, from the continuity property of the eigenvalue, that the stability condition obtained for these special cases remain valid when $\boldsymbol{\theta}$ and $\boldsymbol{\beta}$ deviate from postulated forms slightly (Bellman 1953, Wilkinson 1965). We carry out the eigenvalue computation of F for the cases:

(1a) identical liquidity preference parameter, $\theta_i = \alpha$, all i.
(1b) similar liquidity preference parameter, $\theta_i = \alpha + \epsilon_i$, where

$$\alpha = \sum_{i=1}^{n} \theta_i/n, \qquad |\epsilon_i| \text{ small compared with } \alpha$$

(2) large number of traders.
(3) all net excess demands are negative and β_i = constant, all i.

In all these cases, in view of Theorem 1 we verify that the eigenvalues are real and negative.

CASE OF IDENTICAL LIQUIDITY PREFERENCE Assume θ_i to be the same for all traders. Denote it by $\alpha > 0$, so that $\boldsymbol{\theta} = \alpha \boldsymbol{e}$. From (7.2-15c), we see that $\boldsymbol{\delta} = \boldsymbol{\beta}$ and $\boldsymbol{\eta} = 0$ in this case. Then F of (7.2-15b) reduces to a lower triangular matrix

$$F_0 = \begin{bmatrix} -B + \dfrac{e\beta^{\mathrm{t}}}{n+1} & 0 \\ [E, 0] + \dfrac{1}{n+1}(nI - ee^{\mathrm{t}}) & (ee^{\mathrm{t}} - nI)\alpha \end{bmatrix}$$

Proposition 1 *When all transactors have the identical liquidity preference parameters, the differential equation governing the nominal net excess demands becomes separate from that for the money stock. The differential equation for z is stable near equilibrium and its eigenvalues are all negative real numbers,*

$$0 > \lambda_1 > -\beta_1 > \lambda_2 > -\beta_2 > \cdots > \lambda_n > -\beta_n$$

where we can take $\boldsymbol{\beta}$'s to be arranged as $0 < \beta_1 < \cdots < \beta_n$. The differential equation for $\delta \boldsymbol{M}$ has eigenvalues $\lambda = 0$ and $(n-1)$-multiple roots at $-n\alpha$.

Proof The statement of the first paragraph is obvious, since

$$\begin{aligned} 0 &= |\lambda I - F_0| \\ &= \left| \lambda I + B - \frac{e\beta^{\mathrm{t}}}{n+1} \right| \cdot |\lambda I - (ee^{\mathrm{t}} - nI)\alpha| \end{aligned}$$

The rest follows by computing the eigenvalues explicitly (see Appendix 7-C).

In this case, the adjustment process for z is independent of that for $\delta \boldsymbol{M}$ and z approaches zero with exponential speed $\sum_{i=1}^{n} \gamma_i e^{-\lambda_i (t - t_k)}$, the λ_i's are as stated in Proposition 1 and where t_k is the last switching time, and where γ_i are some linear combination of $z_j(t_k)$, $j = 1, \ldots, n$.

The cross-coupling between $\boldsymbol{M}$ and z is in one direction only from the real to the money sectors and is provided by

$$\left\{ [E, 0] + \frac{1}{n-1}(nI - ee^{\mathrm{t}}) \right\} z = \frac{1}{n-1} \begin{bmatrix} -e_+ e_-{}^{\mathrm{t}} z_- \\ e_- e_-{}^{\mathrm{t}} z_- + n z_- \end{bmatrix}$$

where

$$e^{\mathrm{t}} z = e_+{}^{\mathrm{t}} z_+ + e_-{}^{\mathrm{t}} z_-$$

is used.

In other words, the cross-coupling between the money stock vector δM and z is actually provided by z_- only, i.e., those goods which are in net excess supply,

$$\delta M(t) = \sum_{l=1}^{n} \left[\delta_l t^l e^{-n\alpha(t-t_k)} + \int_{t_k}^{t} \epsilon_l(\tau)(t-\tau)^l e^{-n\alpha(t-t_k-\tau)}\, d\tau \right]$$

where $\epsilon_l(t)$ is some linear combination of components of $z_-(\tau)$.

Thus if J_- is empty (all goods are in excess demand) then δM adjusts as $\Sigma_l \delta_l t^l e^{-n\alpha(t-t_k)}$.

Thus for larger β's and larger $n\alpha$ the markets reach the equilibrium state faster stably.

Case of Traders with Similar Liquidity Preference Parameters We assume that traders are not too different from each other in their liquidity preferences so that their θ_i are nearly the same for all i,

$$\theta_i = \alpha + \epsilon_i, \qquad i = 1, \dots, n$$

where

$$\alpha = \sum_{i=1}^{n} \theta_i / n$$

and

$$|\epsilon_i| \ll \alpha$$

We can write

$$\boldsymbol{\theta} = \alpha \boldsymbol{e} + \boldsymbol{\epsilon}$$

with

$$\boldsymbol{\epsilon} = (\epsilon_1, \dots, \epsilon_n)^{\mathrm{t}}$$

Note that $\boldsymbol{\epsilon}^{\mathrm{t}}\boldsymbol{e} = 0$.

We perform the eigenvalue analysis up to $O(\epsilon)$. Then we can see that eigenvalues can be determined up to $O(\epsilon)$ by ignoring $o(\epsilon)$—terms in the matrix F of (7.2-15b). (See, for example, Wilkinson, Chapter 2 (1965).)

With this restriction on $\boldsymbol{\theta}$, we see from (7.2-15a,b,c) that

$$\boldsymbol{\delta} = \boldsymbol{\beta}$$

The lower triangular form of F we saw in F_0 persists even when traders are not identical since $\boldsymbol{\eta} = 0(\epsilon^2) \simeq 0$. We also have

$$\overline{\Theta} = \alpha(\boldsymbol{e}\boldsymbol{e}^{\mathrm{t}} - nI) + \boldsymbol{e}\boldsymbol{\epsilon}^{\mathrm{t}} - n\boldsymbol{\epsilon}$$

Note that

$$\boldsymbol{e\theta}^{\mathrm{t}}(E \quad 0) = \boldsymbol{e\epsilon}^{\mathrm{t}}(E \quad 0)$$

where

$$\boldsymbol{\epsilon}^{\mathrm{t}}(E \quad 0) = -\frac{n}{n-1}(\boldsymbol{\epsilon}_{+}{}^{\mathrm{t}}, 0)$$

Thus, F becomes block lower triangular matrix up to $0(\epsilon)$ so that

$$|\lambda I - F| = \left|\lambda I + B - \frac{\boldsymbol{e\beta}^{\mathrm{t}}}{n+1} - \boldsymbol{e\epsilon}^{\mathrm{t}}(E, 0)\right| \cdot |\lambda I + \alpha(nI - \boldsymbol{ee}^{\mathrm{t}}) - \boldsymbol{e\epsilon}^{\mathrm{t}} + n\boldsymbol{\epsilon}|$$

$$= \prod_{i=1}^{n}(\lambda + \beta)\cdot\left\{1 - \sum_{i=1}^{n} g_i/\lambda + \beta_i\right\}$$

$$\cdot \prod_{i=1}^{n}(\lambda + n\alpha + n\epsilon_i)\cdot\left\{1 - \sum_{i=1}^{n}\frac{\alpha + \epsilon_i}{\lambda + n\alpha + n\epsilon_i}\right\}$$

where

$$g_i = \begin{cases} \beta_i/n + 1 - \dfrac{n}{n-1}(1 + \epsilon_i), & \text{if } i \in J_+ \\ \beta_i/n + 1, & \text{if } i \in J_- \end{cases}$$

Note that

$$\sum_i \frac{\alpha + \epsilon_i}{\lambda + n\alpha + n\epsilon_i} = \frac{n\alpha}{\lambda + n\alpha} + o(\epsilon)$$

Thus

$$|\lambda I - F| = \frac{\lambda}{\lambda + n\alpha}\Pi(\lambda + n\alpha + n\epsilon_i)\cdot \Pi_i(\lambda + \beta_1)$$

$$\cdot\left\{1 - \frac{1}{n+1}\sum_i \frac{\beta_i}{\lambda + \beta_i} + \frac{n}{n-1}\sum_{+}\frac{1 + \epsilon_i}{\lambda + \beta_i}\right\}$$

where

$$\sum_{+} \text{ stands for } \sum_{i \in J_+}.$$

Proposition 2 *When $\boldsymbol{\theta} = \alpha \boldsymbol{e} + \boldsymbol{\epsilon}$, F is a stable matrix with negative real eigenvalues. One eigenvalue is at zero, $n - 1$ of them are clustered about $-n\alpha$ and the remaining n of them are near $-\beta_i$, $i = 1, \ldots, n$.*

Again z adjusts independently of the money stock deviation $\delta \boldsymbol{M}$. The larger the $\boldsymbol{\beta}$'s, the faster the adjustment towards 0. Since $\lambda_i < 0$, the adjustments are exponentially fast and stable. Since the coupling term from z to $\delta \boldsymbol{M}$ is the same as in the previous section, only $\delta \boldsymbol{M}$ and z_- effect $\delta \boldsymbol{M}$ adjustments.

Case of Large Number of Traders It is possible to analyze the stability property of F without any special assumptions on $\boldsymbol{\theta}$ and $\boldsymbol{\beta}$ by performing the usual perturbation analysis, provided n is large, since

$$\theta_i = O(1/n)$$

$$\delta_i/n = O(1/n)$$

$$\eta_i = O(1/n)$$

if we assume $h_i = O(1)$ for all i.

The detailed analysis is omitted since it is routine. Standard perturbation analysis shows that the system is stable.

7.2.6 Short-Run and Long-Run Adjustments

Implication of Fix-Price and Archibald-Lipsey Analysis It is particularly simple to point out implications of the so-called fix-price method of analysis in this model as well as the Archibald-Lipsey type analysis. See Barrow-Grossman and Grossman for examples of the application of the fix-price method (Archibald and Lipsey 1969, Barro and Grossman 1971, Grossman 1969).

Suppose the eigenvalues associated with the differential equation for $\boldsymbol{p}$ are much larger in magnitude than those associated with $\boldsymbol{M}$. Then $\boldsymbol{p}$ adjusts more quickly than $\boldsymbol{M}$ as we discussed them in §7.2.5. At the extreme, for each $\boldsymbol{M}$ regarded as fixed, prices adjust instantaneously to clear all markets at each instant in time, i.e., there exists $\boldsymbol{p}$ such that $\dot{\boldsymbol{p}} = 0$, or $z = 0$. Denote this price vector by $\hat{\boldsymbol{p}}$. In other words, we see from (7.2-12) that this short run equilibrium price is given by $\hat{\boldsymbol{p}} = (n+1)\boldsymbol{e}\boldsymbol{\theta}^t\boldsymbol{M}$ or $\hat{p}_j = (n+1)\theta^t M$, for all j, $j = 1, \ldots, n$. Note that this short-run equilibrium price is the same as the long-run equilibrium price given (7.2-10), if $M_i = \overline{M}_i$ for all i or if $\boldsymbol{\theta}^t\delta\boldsymbol{M} = 0$.

Since $M_i = M_i + \delta M_i$ by definition, we see from (7.2-10) that

$$\hat{p}_j = p_e + \boldsymbol{\theta}^t\delta\boldsymbol{M}, \tag{7.2-16}$$

where p_e is the long-run equilibrium price for the good j. It is the same for all the goods as we have seen in §7.2.3. The short run equilibrium prices are also the same for all goods and they differ from the long-run equilibrium price p_e given by (7.2-10) by $\boldsymbol{\theta}^t\delta\boldsymbol{M}$. This difference in the prices is caused by the total effect of the deviations of all traders' current money stocks from their long run equilibrium levels. In other words, it is the total effect, $\boldsymbol{\theta}^t\delta\boldsymbol{M}$, not the individual deviation δM_i that determines $\hat{p}_j - p_e$. After $\boldsymbol{p}$ reaches the short-run equilibrium, the distribution of money holdings among transactors will be adjusted according to (7.2-5) (or equivalently (7.2-3′)) (there is no rationing since all markets clear),

$$\dot{M}_i = -n\theta_i M_i + \frac{1}{n+1}\hat{p}_i, \qquad i = 1, \ldots, n$$

or

$$\begin{aligned}\dot{\boldsymbol{M}} &= -n\Theta M + \frac{1}{n+1}\hat{\boldsymbol{p}} \\ &= (-n\Theta + \boldsymbol{e}\boldsymbol{\theta}^t)\boldsymbol{M} \qquad (7.2\text{-}17)\end{aligned}$$

where $\hat{\boldsymbol{p}}/(n+1) = \boldsymbol{e}\boldsymbol{\theta}^t\boldsymbol{M}$ has been substituted. Equivalently, we can express (7.2-17) as

$$\delta\dot{\boldsymbol{M}} = (-n\Theta + \boldsymbol{e}\boldsymbol{\theta}^t)\delta\boldsymbol{M}$$

When all markets clear, there is no cross coupling between traders or prices. Trader i's money stock changes as a function of his own money stock only and of $\hat{p}_i$, i.e., the model reduces to n isolated markets with a single trader in each market. Equation (7.2-17) is the long-run motion of $\boldsymbol{M}$. The eigenvalues governing the long-run motion of $\boldsymbol{M}$ are the roots of

$$\begin{aligned}0 &= \det(\lambda I + n\Theta - \boldsymbol{e}\boldsymbol{\theta}^t) \\ &= \Pi(\lambda + n\theta_i)\cdot\left(1 - \sum_i \frac{\theta_i}{\lambda + n\theta_i}\right)\end{aligned}$$

It is interesting to note that in this Archibald-Lipsey type analysis, it does not make any difference whether the demand for own goods is included in the model or not, because all markets are assumed to clear continuously and any difference in the rationing mechanism disappears. From $\dot{\hat{\boldsymbol{p}}} = (n+1)\boldsymbol{e}\boldsymbol{\theta}^t\dot{\boldsymbol{M}}$, the short-run equilibrium price vector $\hat{\boldsymbol{p}}$ changes according to

$$\dot{\hat{\boldsymbol{p}}} = (n+1)\boldsymbol{e}\boldsymbol{\theta}^t(\boldsymbol{e}\boldsymbol{\theta}^t - n\Theta)\boldsymbol{M}$$

where $\boldsymbol{M}$ is given by (7.2-17). When $\boldsymbol{p}$ adjusts faster than $\boldsymbol{M}$ but not instantaneously, the short-run motion of $\boldsymbol{p}$ is given from (7.2-6″) and (7.2-7) by

$$\dot{\boldsymbol{p}} = \left(-B + \frac{1}{n+1}\,\boldsymbol{\beta}\boldsymbol{e}^{\mathrm{t}}\right)\boldsymbol{p} + \boldsymbol{\beta}\boldsymbol{\theta}^{\mathrm{t}}\boldsymbol{M}$$

where $\boldsymbol{M}$ is taken to be a constant vector.

The above is the Archibald-Lipsey analysis. Note that the short-run and long-run models are both asymptotically stable.

The other extreme is the fix-price analysis in which $\boldsymbol{M}$ is assumed to adjust much more quickly than $\boldsymbol{p}$, i.e., β_j, $j = 1, \ldots, n$ are all much smaller than θ_i, $i = 1, \ldots, n$.

Thus, in the fix-price analysis, $\delta\boldsymbol{M}$ first adjusts to zero quickly i.e., the assumption of fix-price analysis implies that every trader has his equilibrium money stock.

Then $\delta\boldsymbol{p} = \boldsymbol{p} - \boldsymbol{p}_e$ adjust more slowly according to

$$\begin{aligned}\delta\dot{\boldsymbol{p}} &= B\left(\frac{1}{n+1}\,\boldsymbol{e}\boldsymbol{e}^{\mathrm{t}} - I\right)\delta\boldsymbol{p}\\ &= \left(\frac{1}{n+1}\,\boldsymbol{\beta}\boldsymbol{e}^{\mathrm{t}} - B\right)\delta\boldsymbol{p}\end{aligned}$$

since $\delta\dot{\boldsymbol{p}} = B\boldsymbol{z}$ and from (7.2-7)

$$\boldsymbol{z} = \left(\frac{1}{n+1}\,\boldsymbol{e}\boldsymbol{e}^{\mathrm{t}} - I\right)\delta\boldsymbol{p}, \text{ with } \delta\boldsymbol{M} = 0.$$

This, as we have seen, is stable.

From (7.2-12) and (7.2-13) the short-run movement of $\boldsymbol{M}$ near the equilibrium is governed by the differential equation

$$\delta\dot{\boldsymbol{M}} = \left[\frac{1}{n-1}\begin{pmatrix}(n_+ - 1)\boldsymbol{e}_+\\(n_- - 1)\boldsymbol{e}_-\end{pmatrix}\boldsymbol{\theta}^{\mathrm{t}} - n\Theta\right]\delta\boldsymbol{M} + (\text{terms involving } \boldsymbol{p} \text{ only})$$

From Lemma 2

$$\begin{aligned}0 &= \left|\lambda I + n\Theta - \frac{1}{n-1}\,\boldsymbol{f}\boldsymbol{\theta}^{\mathrm{t}}\right|\\ &= |\lambda + n\Theta|\left(1 - \frac{1}{n-1}\,\boldsymbol{\theta}^{\mathrm{t}}(\lambda I + n\Theta)^{-1}\boldsymbol{f}\right)\\ &= \Pi_i(\lambda + n\theta_i)\left(1 - \frac{1}{n-1}\sum\frac{\theta_i f_i}{\lambda + n\theta_i}\right)\end{aligned}$$

where

$$f = \begin{pmatrix} (n_+ - 1)e_+ \\ (n_- - 1)e_- \end{pmatrix}$$

Thus

$$0 = \Pi_i(\lambda + n\theta_i)\left[1 - \frac{n_+ - 1}{n - 1}\sum_{i\in J_+}\frac{\theta_i}{\lambda + n\theta_i} - \frac{n_- - 1}{n - 1}\sum_{i\in J_-}\frac{\theta_i}{\lambda + n\theta_i}\right]$$

showing that the short-run movement of δM is stable.

7.2.7 Adjustments with Price Control

The differential equation (7.2-15) can be used to answer some important questions related to the modified behavior of the model due to changes in some of the system parameters or the manner in which the markets operate. For example, suppose that the goods are classified into two groups and partition the price vector into $\boldsymbol{p}^1$ and $\boldsymbol{p}^2$ subvectors such that $\boldsymbol{p}^1$ is held fixed at some arbitrarily level $\hat{\boldsymbol{p}}^1$. In other words, $\boldsymbol{p}^1$ does not adjust even when $z^1 \neq 0$, where z^1 is the subvector of z corresponding to $\boldsymbol{p}^1$. What are the effects of this price freeze on other prices and money stocks? Would the markets eventually clear? Does this make the system unstable? We outline how to answer the first question here.

The price adjustment can be modeled conveniently by setting those β's corresponding to $\boldsymbol{p}^1$ to zero since the excess demands for goods have no effect on prices. Thus, instead of (7.2-7), we have

$$\dot{\boldsymbol{p}} = \hat{B}z \tag{7.2-7'}$$

where

$$\hat{B} = \begin{pmatrix} 0 & 0 \\ 0 & B_2 \end{pmatrix}$$

where $\hat{B}$ is partitioned conformally with $\boldsymbol{p}$.

The frozen price $\hat{\boldsymbol{p}}^1$ may or may not be near the equilibrium. Thus we cannot use (7.2-15) directly.

From (7.2-5″), recalling that $c = n + 1$ and partitioning $\boldsymbol{M}$ conformally with $\boldsymbol{p}$ we have

$$\begin{aligned} \dot{\boldsymbol{M}}^1 &= -\Lambda_1\Theta_1\boldsymbol{M}^1 + \left(K_1 - \frac{1}{n+1}\Lambda_1\right)\hat{\boldsymbol{p}}_1 \\ \dot{\boldsymbol{M}}^2 &= -\Lambda_2\Theta_2\boldsymbol{M}^2 + \left(K_2 - \frac{1}{n+1}\Lambda_2\right)\boldsymbol{p}^2 \end{aligned} \tag{7.2-18}$$

From (7.2-6″)

$$
\begin{aligned}
z^1 &= e^1\boldsymbol{\theta}^{\mathrm{t}}\boldsymbol{M} + \left(\frac{1}{n+1}e^1(e^1)^{\mathrm{t}} - I_1\right)\hat{\boldsymbol{p}}^1 + \frac{e^1(e^2)^{\mathrm{t}}}{n+1}\boldsymbol{p}^2 \\
z^2 &= e^2\boldsymbol{\theta}^{\mathrm{t}}\boldsymbol{M} + \frac{e^2(e^1)^{\mathrm{t}}}{n+1}\hat{\boldsymbol{p}}^1 + \left(\frac{e^2(e^2)^{\mathrm{t}}}{n+1} - I_2\right)\boldsymbol{p}^2
\end{aligned}
\tag{7.2-19}
$$

and differentiating the above

$$
\begin{aligned}
\dot{z}^1 &= e^1\boldsymbol{\theta}^{\mathrm{t}}\dot{\boldsymbol{M}} + \frac{e^1(e^2)^{\mathrm{t}}}{n+1}B_2 z^2 \\
\dot{z}^2 &= e^2\boldsymbol{\theta}^{\mathrm{t}}\dot{\boldsymbol{M}} + \left\{\frac{e^2(e^2)^{\mathrm{t}}}{n+1} - I_2\right\}B_2 z^2
\end{aligned}
\tag{7.2-20}
$$

At equilibrium, $\dot{\boldsymbol{M}} = 0$ and $\dot{\boldsymbol{p}}^2 = 0$. Hence from (7.2-7) or (7.2-20), $z^2 = 0$, i.e., the markets with no price control clear.

The equilibrium value of z^1, $\bar{z}^1$, is no longer zero generally.

We show in Appendix 7-D that the equilibrium state exists and that

$$\bar{z}^1 = H(\bar{z}^1)\hat{\boldsymbol{p}}^1$$

where

$$H(\bar{z}^1) = \frac{n+1}{n_1 + 1 + (e^2)^{\mathrm{t}}\left(I_2 - (n+1)\bar{\Lambda}_2^{-1}\right)e^2}\left(e^1(e^1)^{\mathrm{t}}\bar{\Lambda}_1^{-1}\bar{K}_1 - I_1\right)$$

Since $\bar{\Lambda}_1$, $\bar{\Lambda}_2$ and $\bar{K}_1$ are functions of $\bar{z}^1$, we cannot express $\bar{z}^1$ explicitly. However, we see from above that $\bar{z}^1 \neq 0$ since

$$e^1(e^1)^{\mathrm{t}}\bar{\Lambda}_1^{-1}\bar{K}_1 \neq I_1$$

Consequently, unless the frozen price happens to satisfy $H(\bar{z}^1)\hat{p}^1 = 0$, the equilibrium state with $\boldsymbol{p}^1 = \hat{\boldsymbol{p}}^1 = \text{const}$ is different from that of no price control given by (7.2-10).

Next we outline a procedure to answer the question: "Are the adjustments near the equilibrium stable?" From (7.2-12)

$$\boldsymbol{p}^2 = (n+1)e^2\,\boldsymbol{\theta}^{\mathrm{t}}\boldsymbol{M} - \left(I_2 + e^2 e^{\mathrm{t}}\right)z$$

Substituting this into the right-hand side of $\dot{M}^2$, we obtain

$$\frac{d}{dt}\begin{bmatrix} \boldsymbol{M}^1 \\ \boldsymbol{M}^2 \\ z \end{bmatrix} = F\begin{bmatrix} \boldsymbol{M}^1 \\ \boldsymbol{M}^2 \\ z \end{bmatrix} + \begin{bmatrix} K_1 - \Lambda_1/n + 1 \\ 0 \\ 0 \end{bmatrix} \boldsymbol{p}^1$$

where

$$F = \begin{bmatrix} F_{11} & 0 & 0 \\ F_{21} & F_{22} & F_{23} \\ F_{31} & F_{32} & F_{33} \end{bmatrix}$$

where

$$F_{11} = -\Lambda_1\Theta_1, \qquad F_{21} = (n+1)(K_2 - \Lambda_2/(n+1))\boldsymbol{e}^2(\boldsymbol{\theta}^1)^{\mathrm{t}}$$

$$F_{21} = (n+1)\left(K_2 - \frac{1}{n+1}\Lambda_2\right)e^2(\theta^1)^{\mathrm{t}}$$

$$F_{22} = (n+1)(K_2 - \Lambda_2/(n+1))\boldsymbol{e}^2(\boldsymbol{\theta}^2)^{\mathrm{t}} - \Lambda_2\Theta_2$$

$$F_{23} = (K_2 - \Lambda_2/(n+1))(I_2 + \boldsymbol{e}^2\boldsymbol{e}^{\mathrm{t}})$$

$$F_{31} = \boldsymbol{e}(\boldsymbol{\theta}^1)^{\mathrm{t}}, \qquad F_{32} = \boldsymbol{e}(\boldsymbol{\theta}^2)^{\mathrm{t}}, \qquad F_{33} = \begin{bmatrix} 0 & \frac{1}{n+1}\boldsymbol{e}^1(\boldsymbol{b}^2)^{\mathrm{t}} \\ 0 & \frac{1}{n+1}\boldsymbol{e}^2(\boldsymbol{b}^2)^{\mathrm{t}} - B_2 \end{bmatrix}$$

Thus, F is a stable matrix if

$$\begin{pmatrix} F_{22} & F_{23} \\ F_{32} & F_{33} \end{pmatrix} \text{ is stable, since } F_{11} \text{ is stable.}$$

We omit the details of the stability analysis, since it is quite analogous to that carried out earlier in §7.2.5. We merely mention that because the northwest corner submatrix of F_{33} is zero, it is conceivable that $\begin{pmatrix} F_{22} & F_{23} \\ F_{32} & F_{33} \end{pmatrix}$ has some eigenvalues with positive real roots. If so, the price control is destabilizing.

7.2.8 Partial Equilibrium Analysis

Suppose the initial conditions of the markets are such that $z^1 = 0$ and $\boldsymbol{p}^1 = p_e \boldsymbol{e}^1$ for some subvector of z where p_e is the equilibrium price given by (7.2-10).† In other words, assume that goods are momentarily cleared in a subset of markets at the correct equilibrium prices.

If $z^1 = 0$, $\boldsymbol{p}^1 = p_e \boldsymbol{e}^1$, then from (7.2-12) we have

$$\boldsymbol{p}^2 = p_e \boldsymbol{e}^2 - z^2$$

or the percentage deviation from the equilibrium price in the rest of the markets is directly proportional to the excess supply $-\boldsymbol{x}^2$,

$$\delta \boldsymbol{p}^2 = -z^2$$

The excess demand z^2 is related to $\delta \boldsymbol{M}$ as follows. Noting that (7.2-6″) is valid when $\boldsymbol{M}$ and $\boldsymbol{p}$ are replaced by $\delta \boldsymbol{M}$ and $\delta \boldsymbol{p}$, we obtain

$$\frac{(\boldsymbol{e}^2)^t z^2}{n_2 + 1} = (n+1)\boldsymbol{\theta}^t \delta \boldsymbol{M}, \qquad \text{where } \dim \boldsymbol{e}^2 = n_2$$

In other words $z^1 = 0$ and $\boldsymbol{p}^1 = p_e \boldsymbol{e}^1$ does not imply necessarily $z^2 = 0$ unless $\boldsymbol{\theta}^t \delta \boldsymbol{M} = 0$. One sufficient condition is of course that $\delta \boldsymbol{M}^1 = 0$ and $\delta \boldsymbol{M}^2 = 0$. The condition $\delta \boldsymbol{M}^1 = 0$ is not sufficient. As we show in detail presently, the condition "$z^1 = 0$, $\boldsymbol{p}^1 = p_e \boldsymbol{e}^1$, $\delta \boldsymbol{M}^1 = 0$" does not mean that the subset of markets and traders drop out of the dynamic adjustment processes of the rest of the model. They still interact with the rest of the traders.

We assumed on the other hand that

$$z^1 = 0, \qquad \boldsymbol{p} = p_e \boldsymbol{e}$$

These conditions imply from (7.2-6″) that $\boldsymbol{\theta}^t \delta \boldsymbol{M} = 0$ and $z^2 = 0$.

However, the condition $z = 0$, $\boldsymbol{p} = p_e \boldsymbol{e}$, i.e., that the real sector of the model is momentarily in equilibrium does not mean that money stocks are unimportant. They still interact with the real sector and upset the real

†The subvectors of this section are defined independently of other sections' definitions.

equilibrium. For example, this possibility suggests itself from (7.2-13) and (7.2-14),

$$\delta\dot{\boldsymbol{M}}(0) = (\boldsymbol{e}\boldsymbol{\theta}^{\mathrm{t}} - \Theta)\delta\boldsymbol{M}(0)$$

$$\begin{aligned} \dot{z}(0) &= \boldsymbol{e}\boldsymbol{\theta}^{\mathrm{t}}\delta\dot{\boldsymbol{M}}(0) \\ &= \boldsymbol{e}\boldsymbol{\theta}^{\mathrm{t}}(\boldsymbol{e}\boldsymbol{\theta}^{\mathrm{t}} - \Theta)\delta\boldsymbol{M}(0) \\ &= \boldsymbol{e}\boldsymbol{\eta}^{\mathrm{t}}\delta\boldsymbol{M}(0) \end{aligned}$$

Thus unless $\boldsymbol{\eta}^{\mathrm{t}}\delta\boldsymbol{M}(0) = 0$, z will not remain at zero. We have discussed some special cases in §7.2.5. For example, if θ_i is the same for all i, then $\boldsymbol{\eta}^{\mathrm{t}}\delta\boldsymbol{M}(0) = 0$.

More general cases may be examined using the observability arguments as illustrated in §3.6.4.

We have now illustrated in sufficient detail various control and system theoretic procedures that can be employed to analyze the adjustment processes of economic models going from disequilibrium states back to equilibrium states.

APPENDIX 7-A: RATIONING

Introduce a step function $u(\cdot)$ (a nonlinear function of the argument)

$$u(x) = \begin{cases} 0, & \text{if } x \leqslant 0 \\ 1, & \text{if } x > 0 \end{cases}$$

We can express $k_j^{\,i}$ that appears in (7.2-3) conveniently by

$$k_j^{\,i} = \frac{\sum_{k=1}^{n} p_k u\left(-x_j^{\,k} \cdot x_j\right) x_j^{\,k}}{\left[\sum_{k=1}^{n} p_j u\left(-x_j^{\,k} \cdot x_j\right) x_j^{\,k} - z_j u\left(x_j^{\,i} \cdot x_j\right)\right]}$$

where we define nominal net excess demand for good j as

$$z_j = p_j x_j, \qquad k, j = 1, \ldots, n$$

We interpret $0/0$ to be 1. The above expression can be seen as follows.

Define two index sets by

$$J_+(j) = \{i | x_j^i > 0, i = 1, \ldots, n\} : \left\{\begin{array}{l}\text{the set of traders with ex-}\\ \text{cess demand for good } j\end{array}\right\}$$

and

$$J_-(j) = \{i | x_j^i < 0, i = 1, \ldots, n\} : \left\{\begin{array}{l}\text{the set of traders with ex-}\\ \text{cess supply of good } j\end{array}\right\}$$

We have

$$x_j = \sum_{k \in J_+(j)} x_j^k + \sum_{k \in J_-(j)} x_j^k$$

Suppose $x_j > 0$. Then only demanders are rationed. Thus, for trader i, $i \in J_+(j)$, we have

$$k_j^i = \text{supply/demand}$$

$$= -\frac{\sum_{k \in J_-(j)} x_j^k}{\sum_{k \in J_+(j)} x_j^i} \tag{7-A-1}$$

Suppose $x_j < 0$. Only suppliers are rationed. We have

$$k_j^i = \text{demand/supply}$$

$$= -\frac{\sum_{k \in J_+(j)} x_j^k}{\sum_{k \in J_-(j)} x_j^k} \tag{7-A-2}$$

for all $i \in J_-(j)$.

With $w_j^i = \delta_{ij}$, at the equilibrium trader i is the net supplier of good i and demander of all other goods, i.e., $\bar{x}_j^i > 0, j \neq i$ and $\bar{x}_i^i < 0$. Thus, near the equilibrium we know the signs of z's and k_j^i is expressible as (7-A-3),

$$\begin{aligned}
k_j^i &= p_j x_j^j / \left[p_j x_j^j - z_j u(z_j) \right] \\
&= 1 + z_j u(z_j) / \left[\theta_j M_j - \frac{n}{n+1} p_j - z_j u(z_j) \right], \qquad i \neq j \\
k_i^i &= \left[p_i x_i^i - z_i u(-z_i) \right] / p_i x_i^i \\
&= 1 - z_i u(-z_i) / \left(\theta_i M_i - \frac{n}{n+1} p_i \right)
\end{aligned} \tag{7-A-3}$$

Equation (7-A-3) shows that $k_j^i = 1$ for all i and j if $z_j = 0$ for all j.

Near the equilibrium, Equation (7-A-3), shows that when good j is in excess demand then trader i can satisfy his notional demand. For his endowed good, i.e., good i, the same equation shows that if it is in excess supply he cannot sell it as much as he wishes, although he can fulfil his notional supply if it is in excess demand.

APPENDIX 7-B: DYNAMIC EQUATION FOR NOMINAL EXCESS DEMAND

From the definitions when z is partitioned into z_+ subvector and z_- subvector, we see that the partition induces the conformal partition of U and $\boldsymbol{v}$ which are

$$U = \begin{pmatrix} I_+ & 0 \\ 0 & 0 \end{pmatrix}, \qquad \boldsymbol{v}^{\mathrm{t}} = (\boldsymbol{e}_+{}^{\mathrm{t}}, 0)$$

Thus,

$$\frac{-1}{n-1}(nU - \boldsymbol{e}\boldsymbol{v}^{\mathrm{t}}) = \frac{1}{n-1}\begin{pmatrix} \boldsymbol{e}_+\boldsymbol{e}_+{}^{\mathrm{t}} - nI_+ & 0 \\ \boldsymbol{e}_-\boldsymbol{e}_+{}^{\mathrm{t}} & 0 \end{pmatrix}$$

Define this matrix to be $(E \quad 0)$, i.e.,

$$E = \frac{1}{n-1}\begin{pmatrix} \boldsymbol{e}_+\boldsymbol{e}_+{}^{\mathrm{t}} - nI_+ \\ \boldsymbol{e}_-\boldsymbol{e}_+{}^{\mathrm{t}} \end{pmatrix} : n \times n_+ \text{ matrix}$$

Then (7.2-14) can be expressed as

$$\delta\dot{\boldsymbol{M}} = \overline{\Theta}\delta\boldsymbol{M} + (E \quad 0)z + \frac{1}{n+1}(nI - \boldsymbol{e}\boldsymbol{e}^{\mathrm{t}})z + o(z, \delta\boldsymbol{M}) \qquad \text{(7-B-1)}$$

Rewrite (7.2-13) as

$$\dot{z} = -Bz + (\boldsymbol{e}\boldsymbol{\theta}^{\mathrm{t}} - \Theta)\delta\dot{\boldsymbol{M}} + \frac{\boldsymbol{e}\boldsymbol{\beta}^{\mathrm{t}}}{n+1}z + \Theta\delta\dot{\boldsymbol{M}} + o(z, \delta\boldsymbol{M}) \qquad \text{(7-B-2)}$$

One can show that when the "own" demands are not included in the model, the differential equation for $\delta\dot{\boldsymbol{M}}$ and $\dot{z}$ are represented by the first two terms in (7-B-1) and (7-B-2). The third term in (7-B-1) and the last two terms in (7-B-2) are therefore dynamic effects due to "own" demands.

From (7-B-2) then

$$\dot{z} = \left(-B + \frac{\boldsymbol{e}\boldsymbol{\delta}^{\mathrm{t}}}{n+1}\right)z + \boldsymbol{e}\boldsymbol{\theta}^{\mathrm{t}}Ez_+\boldsymbol{\eta} + \boldsymbol{e}\boldsymbol{\delta}^{\mathrm{t}}\delta\boldsymbol{M} + o(z, \delta\boldsymbol{M}) \qquad \text{(7-B-3)}$$

where $\boldsymbol{\delta}$ and $\boldsymbol{\eta}$ are defined as in (7.2-15c).

Equations (7-B-1) and (7-B-2) when written together give (7.2-15).

APPENDIX 7-C: COMPUTATION OF EIGENVALUES OF F

Applying Lemma 2, we see that

$$0 = \left|\lambda I + B - \frac{e\beta^{\mathrm{t}}}{n+1}\right| = |\lambda I + B|\left(1 - \beta^{\mathrm{t}}\frac{(\lambda I + B)^{-1}e}{n+1}\right)$$

$$= \prod_{i=1}^{n}(\lambda + \beta_i)\cdot\left[1 - \frac{1}{n+1}\sum_{i=1}^{n}\frac{\beta_i}{\lambda+\beta_i}\right] \tag{7-C-1}$$

We show in Appendix C that the roots of (7-C-1) are such that $0 > \lambda_1 > -\beta_1 > \cdots > \lambda_n > -\beta_n$. Also, by Lemma 2, we have

$$0 = |(\lambda + n\alpha)I - \alpha e e^{\mathrm{t}}|$$

$$= |(\lambda + n\alpha)I|\left(1 - \alpha e^{\mathrm{t}}(\lambda + n\alpha)^{-1}e\right)$$

$$= (\lambda + n\alpha)^n\left(1 - \alpha\frac{n}{\lambda + n\alpha}\right)$$

$$= \lambda(\lambda + n\alpha)^{n-1}$$

or $\lambda = 0$ and $-n\alpha$.

APPENDIX 7-D: NONZERO EQUILIBRIUM EXCESS DEMANDS UNDER PRICE CONTROL

From (7.2-19)

$$0 = e^2\theta^{\mathrm{t}}\overline{M} + \frac{e^2(e^1)^{\mathrm{t}}}{n+1}\hat{p}^1 + \left(\frac{e^2(e^2)^{\mathrm{t}}}{n+1} - I_2\right)\bar{p}^2$$

or

$$\bar{p}^2 = \frac{n+1}{n_1+1}e^2\theta^{\mathrm{t}}\overline{M} + \frac{1}{n_1+1}e^2(e^1)^{\mathrm{t}}\hat{p}^1 \tag{7-D-1}$$

where use is made of

$$\left(I_2 - \frac{e^2(e^2)^{\mathrm{t}}}{n+1}\right)^{-1} = I_2 + \frac{e^2(e^2)^{\mathrm{t}}}{n_1+1}, \qquad n = n_1 + n_2$$

From (7.2-19) and (7-D-1),

$$\bar{z}^1 = \frac{n+1}{n_1+1}e^1\theta^{\mathrm{t}}\overline{M} + \left(\frac{1}{n_1+1}e^1(e^1)^{\mathrm{t}} - I_1\right)\bar{p}^2 \tag{7-D-2}$$

To see that $\bar{z}^1 \neq 0$ generally, use (7.2-5), written for the subvectors $\boldsymbol{M}^1$ and $\boldsymbol{M}^2$ in the conformal partition of $\boldsymbol{M}$.

We have from (7.2-18)

$$\boldsymbol{\theta}^t\overline{\boldsymbol{M}} = (\boldsymbol{e}^1)^t\left(\overline{\Lambda}_1{}^{-1}\overline{K}_1 - \frac{1}{n+1} I_1\right)\hat{\boldsymbol{p}}^1 + (\boldsymbol{e}^2)^t\left(\overline{\Lambda}_2{}^{-1} - \frac{1}{n+1} I_2\right)\bar{\boldsymbol{p}}^2$$

where we use the fact $\overline{K}_2 = I$ from the definition just below (7.2-3′), since $\bar{z}_2 = 0$.

Substituting $\boldsymbol{\theta}^t\overline{\boldsymbol{M}}$ obtained above into (7-D-1) we solve for $\bar{\boldsymbol{p}}^2$ in terms of $\hat{\boldsymbol{p}}^1$ as

$$\bar{\boldsymbol{p}}^2 = \frac{n+1}{n_1+1}\left(I_2 + \boldsymbol{e}^2\boldsymbol{\zeta}^t\right)^{-1}\boldsymbol{e}^2(\boldsymbol{e}^1)^t\overline{\Lambda}^{-1}\overline{K}_1\hat{\boldsymbol{p}}^1 \tag{7-D-3}$$

where

$$\boldsymbol{\zeta} = \left(I_2 - (n+1)\overline{\Lambda}_2{}^{-1}\right)\boldsymbol{e}^2/(n_1+1)$$

Substituting $\boldsymbol{\theta}^t\overline{\boldsymbol{M}}$ and (7-D-3) into (7-D-2), we finally obtain

$$\bar{z}^1 = \frac{n+1}{n_1 + 1 + (\boldsymbol{e}^2)^t\left(I_2 - (n+1)\overline{\Lambda}_2{}^{-1}\right)\boldsymbol{e}^2}\left(\boldsymbol{e}^1(\boldsymbol{e}^1)^t\overline{\Lambda}_1{}^{-1}\overline{K}_1 - I_1\right)\hat{\boldsymbol{p}}^1$$

Clearly $\bar{z}^1 \neq 0$ unless $\hat{\boldsymbol{p}}^1$ happens to satisfy

$$\hat{\boldsymbol{p}}^1 = \boldsymbol{e}^1(\boldsymbol{e}^1)^t\overline{\Lambda}_1{}^{-1}\overline{K}_1\hat{\boldsymbol{p}}^1$$

8

Policy Mixes, Noninteracting Control and the Assignment Problem†

8.1 INTRODUCTORY EXAMPLES

It is a common experience in a study of macroeconomic policy implications that we find actions taken to bring about a wanted change in one endogenous macroeconomic variable introduce an unwanted change in another. Policymakers must, then, become involved in evaluations and rankings of various tradeoffs, either directly or indirectly through choice of objective (social welfare) functions that specify desired time paths for the target variables. Policymakers need to know what range of a target value is achievable under a set of specified circumstances while all the other target variables are held at some specified levels, or which of the target values can be modified with "least" adverse effects on other target values and so on. For example, policymakers are probably sure that it is impossible to slow the increase of wage rate while maintaining or decreasing the unemployment rate. They may not be sure, however, whether wage rate and interest rate can be controlled independently or not. When two (or more) target variables are such that one of them can be modified by suitable change of instruments while the other is not affected by the change, we say that these two target variables can be controlled or modified in a decoupled or noninteractive manner.‡ More precise descriptions are given later. Policymakers need to know, also, if instruments at their disposal are numerous and powerful enough to allow them to modify a given pair (set) of target variables in decoupled manner or not. If not, policymakers need information on the relevant tradeoff curves (surfaces), or must introduce additional instruments that will make possible independent control of the targets.

---

†Based in part on (Aoki 1974c).

‡Decoupling and noninteractions are used synonymously.

If it is possible to modify several target variables in noninteractive manner, at least in some neighborhood of a "state" in which the economy finds itself as approximately represented by some macroeconomic model, then policymakers need not face difficult trade-off deliberations regarding these target variables. Thus they will be less constrained in their consideration of policy implementations. Therefore, it is important to assess macroeconomic models for their capability of noninteractive control of target variables, i.e., to have some operational criteria by which policymakers can ascertain whether a given set of target variables can be controlled in decoupled manner with the instruments at their disposal and to discover all possible trade-offs implied by the use of the macroeconomic models, when the targets cannot be controlled noninteractively.

When instruments are versatile and numerous enough to give policymakers the capability of decoupled controls, they must find the appropriate policy-mixes to attain decoupling. As we shall see later, decoupling, if it is possible, is accomplished by tying parts of changes in the instruments to suitable linear combinations of variables describing the "macro-state" of the dynamic economic model (automatic adjustment rules) to cancel out the cross-coupling effects or interactions, and use suitable "discretionary" mixes of policy instruments.

Setting aside the question of how accurately such cancelling can be accomplished, this procedure can be applied to achieve other features that may be regarded desirable by some. For example, a set of policy mixes may be defined in such a way that one policy mix is assigned to one target variable in the Mundellian sense. There is a body of literature that discusses ways of achieving some sort of division of labor or assignment between policy instruments based on some "measure" of effectiveness or impacts of the instruments on one or several of target variables (McFadden 1969, Mundell 1962, 1968). For example, in the areas of macroeconomic models of open economy, Mundell advocates assigning each instrument to the target variable on which it has relatively the strongest effect. Noninteractive controls correspond to the extreme case of what Mundell calls comparative advantage in which each newly defined policy mix affects one and only one target variable.†

†The assignment problem is concerned with stabilization when a subset of instruments is assigned to a subset of targets, usually in a one-to-one fashion. It is a special case of noninteractive control problems in the sense that some additional constraint on the structure of the feedback matrix is imposed due to information decentralization that is inherent in the assignment problem (see Chapter 6). When decoupling is possible, it is done by assigning each newly introduced policy mix to one and only one target variable. Stabilization is one of the important considerations of noninteractive control systems. There are other problems such as distur-

Let us begin by discussing a couple of simple macroeconomic models. The first example is used by McFadden (1969). The second one is that of Example 2, §2.6.

Example 1 A highly aggregated macroeconomic model will be used to illustrate an assignment of instruments. The model is the one used by McFadden. Define the following macroeconomic variables to be as follows:

Y_t: Domestic U.S. production (= income of consumers)
X_t: U.S. aggregate expenditure
C_t: U.S. aggregate consumption
S_t: U.S. aggregate saving
I_t: U.S. aggregate domestic investment
M_t: U.S. imports of foreign goods and services
K_t: Net capital outflows from the U.S.
T_t: Taxes
G_t: U.S. government expenditures for goods and services
B_t: U.S. net surplus in international balance of payments
E: U.S. exports of goods and services, assumed constant
r_t; U.S. domestic interest rate
r_f: Foreign interest rate, assumed constant

The accounting identities connecting these variables are

$$Y_t = C_t + S_t + T_t \tag{8.1-1}$$

$$X_t = C_t + I_t + K_t + G_t \tag{8.1-2}$$

$$B_t = E - M_t - K_t \tag{8.1-3}$$

$$B_t = Y_t - X_t \tag{8.1-4}$$

Assume the behavioral relations

$$S_t = -\alpha_0 + \alpha_1 Y_t, \qquad \alpha_1 > 0 \tag{8.1-5}$$

$$M_t = -\beta_0 + \beta_1 Y_t, \qquad \beta_1 > 0 \tag{8.1-6}$$

$$I_t = \gamma_0 - \gamma_1 r_t, \qquad \gamma_1 > 0 \tag{8.1-7}$$

$$K_t = \delta_0 - \delta_1 r_t, \qquad \delta_1 > 0 \tag{8.1-8}†$$

---

bance localization or pole-assignment which are not considered here. We must caution the reader that these considerations generally presuppose accurate knowledge of the economies, more accurate than available to policymakers in the present state of economic science.

†r_f, being a constant, is included in δ_0.

Define the state vector

$$z_t = \begin{pmatrix} B_t \\ Y_t \end{pmatrix}$$

and the control instrument

$$x_t = \begin{pmatrix} \Delta r_t \\ \Delta D_t \end{pmatrix}$$

where

$$\Delta r_t = r_{t+1} - r_t$$

$$\Delta D_t = D_{t+1} - D_t$$

and where $D_t = G_t - T_t$ is the domestic deficit. The instrument Δr_t is taken to represent monetary measure, while ΔD_t represents the fiscal measure.

Substituting (8.1-5) through (8.1-8) into (8.1-1) through (8.1-4), we have from (8.1-4),

$$B_t = -\{\alpha_0 + \gamma_0 + \delta_0\} + \{\gamma_1 + \delta_1\}r_t + \alpha_1 Y_t - D_t \qquad (8.1\text{-}9)$$

and from (8.1-3)

$$B_t = E + \beta_0 - \delta_0 - \beta_1 Y_t + \delta_1 r_t$$

Solving the above two expressions for Y_t, we obtain

$$Y_t = \mu E + \mu\{\alpha_0 + \beta_0 + \gamma_0\} - \mu\gamma_1 r_t + \mu D_t \qquad (8.1\text{-}10)$$

where $\mu = \{\alpha_1 + \beta_1\}^{-1}$.
Substituting this into (8.1-9),

$$\begin{aligned} B_t = & -\mu\beta_1\{\alpha_0 + \gamma_0 + \delta_0\} + \mu\alpha_1 E \\ & + \{\delta_1 + \mu\beta_1\gamma_1\}r_t - \mu\beta_1 D_t \end{aligned}$$

From this equation and (8.1-10), the dynamic equation for z_t is obtained

$$z_{t+1} = z_t + Hx_t \qquad (8.1\text{-}11)$$

where

$$H = \begin{pmatrix} \delta_1 + \mu\beta_1\gamma_1 & -\mu\beta_1 \\ -\mu\gamma_1 & \mu \end{pmatrix}$$

H^{-1} exists if and only if $\delta_1 \neq 0$, in other words if and only if capital outflow is sensitive to changes in the difference of the interest rates. Equation (8.1-11) shows that the increase in the international balance of payment surplus and the domestic output are both affected by the two instruments even though their effects on them are in opposite directions. Increase in domestic deficit decreases the international balance of payments surplus (since $\mu\beta_1 > 0$), while it increases the domestic output. Raising the interest rate has a positive effect on the balance of payment even though it depresses the domestic output. The impact of Δr_t and ΔD_t can be decoupled by

$$\boldsymbol{x}_t = H^{-1}(\boldsymbol{d}_{t+1} - z_t)$$

where $\boldsymbol{d}_{t+1}$ is the desired state at time $t + 1$ since

$$\begin{aligned} z_{t+1} &= z_t + H \cdot H^{-1}(\boldsymbol{d}_{t+1} - z_t) \\ &= \boldsymbol{d}_{t+1} \end{aligned}$$

This is essentially a static example in that we could solve the $\boldsymbol{d}_{t+1} = z_t + H\boldsymbol{x}_t$ explicitly for $\boldsymbol{x}_t$.†

Our next example is a little more dynamic, even though it is equally as simple as Example 1.

Example 2 The model is the same as in Example 2 of §2.6 and has two endogenous variables

$$z_1 = \ln \frac{r}{\overline{w}_1 e^{\mu t}}$$

and

$$z_2 = \ln \frac{Y}{\overline{w}_2 e^{\mu t}}$$

†The situation is somewhat analogous to the algebraic equation for the instruments that appears in Tinbergen's theory of policy as compared with the more general output controllability condition.

where $\overline{w}_1$, $\overline{w}_2$ and μ are some constants as in Example 2, §2.6. Consider a monetary instrument x that is defined by

$$x = \ln \frac{M^s}{M^* e^{mt}}$$

We have discussed in Example 2, §2.6, that $M^* e^{mt}$ is the equilibrium growth money supply that makes r grow as $\overline{w}_1 e^{\mu t}$ and Y as $\overline{w}_2 e^{\mu t}$, where $\mu = (\alpha + \beta - 1)/m$.

Suppose we change the money supply by

$$M^s = M^* e^{mt} e^x$$

so that r stays at some specified level r^*, while leaving Y as is, i.e. keeping $z_2 = 0$. Can this be done? The answer is "no," as we see below.

The dynamic equation has been derived in §2.6 as

$$\begin{aligned} \dot{z}_1 &= h\beta z_1 + h(\alpha - 1)z_2 - hx \\ \dot{z}_2 &= -\delta z_1 + \delta z_2 \end{aligned} \tag{8.1-11'}$$

From the second equation, $z_2 = 0$ and $\dot{z}_2 = 0$ implies

$$z_1 = z_2 = 0$$

or $\dot{z}_1 = 0$, implying that x must be zero. But then r is growing as $\overline{w}_1 e^{\mu t}$ and not a constant, a contradiction.

To state the matter in another way, maintaining $r = r^*$ implies from the first equation of (8.1-11′) that x must be given by

$$x = \beta\left(\ln \frac{r^*}{\overline{w}_1} - \mu t\right) + (\alpha - 1)z_2 + \mu/h$$

From the second equation, however, z_2 is changing according to

$$\dot{z}_2 = -\delta\left(\ln \frac{r^*}{\overline{w}_1} - \mu t\right) + \delta z_2$$

or

$$
\begin{aligned}
z_2 &= e^{\delta t} z_2(0) - \delta \ln\left(\frac{r^*}{\bar{w}_1}\right) \int_0^t e^{\delta(t-\tau)}\, d\tau \\
&\quad + \delta\mu \int_0^t \tau e^{\delta(t-\tau)}\, d\tau \\
&= e^{\delta t} z_2(0) + \ln\left(\frac{r^*}{\bar{w}_1}\right)(1 - e^{\delta t}) - \mu t \\
&\quad - \frac{\mu}{\delta}(1 - e^{\delta t}) \\
&= e^{\delta t} z_2(0) + \left[\ln \frac{r^*}{\bar{w}_1} - \frac{\mu}{\delta}\right](1 - e^{\delta t}) - \mu t \\
&\neq 0
\end{aligned}
$$

unless $z_2(0) = 0$ and $\ln (r^*/\bar{w}_1) = \mu/\delta$.

In such a situation, we need an additional instrument to control these two variables in an independent manner.

In the case of the above model, we could define a set of auxiliary instruments ω_1 and ω_2 by

$$
\begin{pmatrix} x_1 \\ x_2 \end{pmatrix} = -B^{-1}Az + \begin{pmatrix} \omega_1 \\ \omega_2 \end{pmatrix}
$$

Then the dynamic equation becomes

$$
\dot{z} = \begin{pmatrix} \omega_1 \\ \omega_2 \end{pmatrix}
$$

or

$$
\dot{z}_i = \omega_i, \qquad i = 1, 2
$$

showing that z_1 and z_2 are controlled independently.

This means that the monetary variable x_1 is now tied to z by

$$
x_1 = -\beta z_1 - (\alpha - 1) z_2 + \omega_1
$$

or

$$
\begin{aligned}
M^s &= M^* e^{mt} e^{-\beta z_1 - (\alpha-1) z_2} e^{\omega_1} \\
&= M^* e^{mt} \left(\bar{w}_1 e^{\mu t}\right)^{\beta} \left(\bar{w}_2 e^{\mu t}\right)^{\alpha-1} e^{\omega_1} / r^{\beta} Y^{\alpha-1}
\end{aligned}
$$

where ω_1 is the new monetary instrument.

From (8.1-12) of §2.6, we see that the money supply is determined by

$$M^s = \frac{M^* e^{mt} M_e^d}{M^d} e^{\omega_1}$$

where $pY = \gamma$ is assumed, and where the subscript e denotes variables with equilibrium growth rate.

Similarly, the fiscal instrument x_2 is now generated as the sum of the automatic expenditure tied to r and Y and the additional expenditure variable ω_2 by

$$x_2 = \frac{\delta}{\lambda}(z_1 - z_2) + \omega_2$$

or

$$G/Y = \nu^* + \frac{\delta}{\lambda}(z_1 - z_2) + \omega_2$$

This expenditure rate means that

$$G/Y = \nu^* + \frac{\delta}{\lambda} \ln\left(\frac{r\overline{w}_2}{Y\overline{w}_1}\right) + \omega_2$$

Then

$$z_1(t) = \int_0^t \omega_1(\tau)\, d\tau$$

or r changes as

$$r(t) = \overline{w}_1 e^{\mu t} \exp \int_0^t \omega_1(\tau)\, d\tau$$

and

$$z_2(t) = \int_0^t \omega_2(\tau)\, d\tau$$

or

$$Y(t) = \overline{w}_1 e^{\mu t} \exp \int_0^t \omega_2(\tau)\, d\tau$$

For example, the constant rate

$$\omega_1 = -\mu, \qquad \omega_2 = 0$$

will keep

$$r(t) = \overline{w}_1$$

while Y grows as

$$Y(t) = \overline{w}_1 e^{\mu t}$$

8.2 CRITERIA OF NONINTERACTION

We now turn our attention to deriving a necessary and sufficient condition that a dynamic macroeconomic (econometric) model must satisfy in order that individual target variables can be modified in noninteractive manner, i.e., in such a way that one target variable can be changed without affecting the values of the rest of the target variables. Needless to say, we must remember that there is a wide gap between the policy recommendations and the results of the analysis using a model that is (1) linear, (2) deterministic and (3) that assumes that all parameters are known. Since the criterion of noninteraction is essentially that certain rank conditions be met, parameters of macroeconomic models need not be fixed exactly to determine when a given set of targets can be controlled in a noninteracting manner. A procedure to assess the effects of imprecisely known parameter values is discussed later.

Let a dynamic linear macroeconomic model be given as

$$\dot{z} = Az + B\boldsymbol{q}$$
$$y = Cz \tag{8.2-1}$$

where A, B and C are constant matrices. The vector z is the n-dimensional state vector, $\boldsymbol{q}$ is the control vector and $\boldsymbol{y}$ is the target vector. We assume that $\boldsymbol{y}$ and $\boldsymbol{q}$ are both m-vectors.† Individual components of q; q_i, $i = 1, \ldots, m$ are the individual control instruments $m < n$. The components of y; y_i, $i = 1, \ldots, m$ are the target variables.

†A more general case with dim $\boldsymbol{q} >$ dim $\boldsymbol{y}$ can also be discussed analogously. This case is not discussed in order to explain the main idea unencumbered with technical complications.

We can intuitively motivate the definition of decoupling to be introduced shortly by using the representation of (8.2-1) in the Laplace transforms. We return to the time domain description of the model when we discuss target time paths. As in §2.3, s is the Laplace transform variable, and we use "^" to denote Laplace transform so that $\hat{x}(s) = \int_0^\infty e^{-st} x(t)\, dt$, for example.

The target vector is related to the instruments by

$$\hat{y}(s) = H(s)\hat{q}(s) + C(sI - A)^{-1} z(0) \tag{8.2-2}$$

where

$$H(s) = C(sI - A)^{-1}B : m \times m$$

is the transfer matrix introduced in §2.4.4, and where I is the $n \times n$ identity matrix. The second term in (8.2-2) may be ignored if A is a stable matrix, i.e., if eigenvalues of A all have negative real parts. Each element of $H(s)$ is some rational function in the complex variable s.

The model (8.2-1) is said to be decoupled if $H(s)$ is nonsingular (except at a finite number of s values) and if $H(s)$ is diagonal.† The meaning of this definition is intuitively clear since $\hat{q}_i(s)$ appears only in $\hat{y}_i(s)$, $i = 1, \ldots, m$, if $H(s)$ is diagonal. Thus, $y_1(t), \ldots, y_m(t)$ are clearly controlled in a noninteractive manner, if effects of the initial condition are neglected.

We know from the final value theorem of Laplace transform of §2.3 that

$$\lim_{t \to \infty} f(t) = \lim_{s \to 0} s\hat{f}(s)$$

if $\hat{f}(s)$ and the limit exist.

This motivates a useful specialization of the decoupling concept introduced here, namely that of asymptotic decoupling‡ that may be of some economic interest: Say that a system is asymptotically decoupled if $H(s)$ is stable and $H(0) = -CA^{-1}B$ is nonsingular and diagonal. In other words, a system is asymptotically decoupled if the effects of a sudden finite change in the instrument q_i disappear in all target variables except in y_i as $t \to \infty$, $i = 1, \ldots, m$.

---

†Alternatively the system (8.2-2) may be defined to be decoupled if $H(s)$ is block-diagonal rather than diagonal. Modifications required to accommodate this weaker definition of decoupling are straightforward.

‡A concept of impact decoupling may be introduced also. Unless the targets are related to z and q by $y = Cz + Eq$, where E is nonsingular and diagonal, there is no possibility for impact decoupling. This topic is not discussed here.

According to the definition we introduced here, we need to compute the transfer matrix to see if a given system is controlled in this manner. It is desirable if we can avoid computing $H(s)$ that involves inverting the $n \times n$ matrix $(sI - A)$. Such an operationally more convenient criterion of decoupling is available. This approach is due to E. Gilbert (1969).

By the Cayley-Hamilton Theorem, we can express A^n as

$$A^n = \sum_{i=1}^{n} \alpha_i A^{n-i}$$

where α's are those that appear in the characteristic polynomial of A, $\mu(s) = |sI - A| = s^n - \alpha_1 s^{n-1} - \cdots - \alpha_n$.

In §2.3 we showed that

$$(sI - A)^{-1} = \operatorname{adj}(sI - A)/|sI - A|$$

where

$$\operatorname{adj}(sI - A) = Is^{n-1} + R_1 s^{n-2} + R_2 s^{n-3} + \cdots + R_{n-1}$$

with

$$R_k = A^k - \alpha_1 A^{k-1} - \cdots - \alpha_k I, \qquad k = 1, \ldots, n-1 \tag{8.2-3}$$

Therefore, we can write the transfer matrix in (8.2-2) as

$$H(s) = C(sI - A)^{-1}B = G(s)/\mu(s) \tag{8.2-4}$$

where

$$G(s) = s^{n-1}CB + s^{n-2}CR_1B + \cdots + CR_{n-1}B$$

Substituting (8.2-3) into (8.2-4),

$$\begin{aligned} G(s) = \mu(s)H(s) &= \left(s^{n-1} - \alpha_1 s^{n-2} - \cdots - \alpha_{n-1}\right)CB \\ &\quad + \left(s^{n-2} - \alpha_1 s^{n-3} - \cdots - \alpha_{n-2}\right)CAB \\ &\quad + \cdots \\ &\quad + \left(s^{n-j-1} - \alpha_1 s^{n-j-2} - \cdots - \alpha_{n-j-1}\right)CA^jB \\ &\quad + \cdots \\ &\quad + CA^{n-1}B \end{aligned} \tag{8.2-5}$$

We are now ready to characterize the diagonality of $H(s)$. Denote the row vectors of C by $c_j, j = 1, \ldots, m$.

For each i, define an integer d_i by

$$d_i = \begin{cases} 0, & c_i B \neq 0 \\ j, & c_i B = 0 \end{cases} \tag{8.2-6}$$

where j is the least positive integer such that $c_i A^j B \neq 0, j = 1, \ldots, n-1$.

In other words for each i, d_i is such that $c_i B = c_i AB = \cdots = c_i A^{d_i - 1} B = 0$ and $c_i A^{d_i} B \neq 0$. Take d_i to be $n-1$ if $c_i A^j B = 0$ for all $j = 0, 1, \ldots, n-1$.

Define

$$D_i = c_i A^{d_i} B, \qquad i = 1, \ldots, n \tag{8.2-7}$$

Form an auxiliary $m \times m$ matrix by

$$B^* = \begin{bmatrix} c_1 A^{d_1} \\ c_2 A^{d_2} \\ \vdots \\ c_m A^{d_m} \end{bmatrix}, \qquad B = \begin{bmatrix} D_1 \\ D_2 \\ \vdots \\ D_m \end{bmatrix} \tag{8.2-8}$$

We establish next that if B^* is diagonal and if $c_i A^{d_i+1} = 0$, $i = 1, \ldots, m$, then $H(s)$ is a diagonal matrix also.

Let $B^* = \operatorname{diag}(\gamma_1, \ldots, \gamma_m)$ by assumption. Then from (8.2-8) $D_i = c_i A^{d_i} B = \gamma_i e_i^t$ where $e_i^t = (0 \quad 1 \quad 0 \quad \ldots \quad 0)$ where only the ith element is nonzero. Denoting the ith row vector of $H(s)$ by $H_i(s)$, we have from (8.2-4) and (8.2-5) and the assumptions,

$$\begin{aligned} H_i(s) &= \mu(s)^{-1}\left(s^{n-1} c_i B + s^{n-2} c_i R_1 B + \cdots + c_i R_{n-1} B\right) \\ &= \mu(s)^{-1}\left(s^{n-1-d_i} - \alpha_1 s^{n-2-d_i} - \cdots - \alpha_{n-1-d_i}\right)\gamma_i e_i^t \end{aligned}$$

Since $\mu(s)$ is a polynomial of degree n in s, we have another characterization of D_i (Gilbert, 1969),

$$D_i = \lim_{s \to \infty} s^{j+1} H_i(s) \neq 0 \qquad \text{with } j = d_i$$

where $H_i(s)$ is the ith row vector of $H(s)$. From the Cayley-Hamilton Theorem $c_i A^{n+j} B = \sum_{k=1}^{n} \alpha_k c_i A^{n+j-k} B$ holds for any nonnegative integer

j. Taking $j = 0$ and using the assumption $c_i A^k B = 0$ unless $k = d_i$, we see that $\alpha_{n-d_i} = 0$. Taking $j = 1, 2, \ldots, d_i$ respectively, $\alpha_{n-d_i+1} = \cdots = \alpha_n = 0$ follows. In other words, the characteristic polynomial of this matrix A is

$$\mu(s) = s^n - \alpha_1 s^{n-1} - \cdots - \alpha_{n-1-d_i} s^{n-1-d_i}$$

Thus

$$H_i(s) = \gamma_i e_i^t / s^{1+d_i}, \qquad i = 1, \ldots, m \tag{8.2-9}$$

establishing the diagonality of $H(s)$.

If, in addition, B^* is nonsingular, then $H(s)$ is also nonsingular, since $\gamma_i \neq 0$ for $i = 1, \ldots, m$.

Thus we have established a criterion for decoupling.

Criterion 1 *The system* (8.2-1) *is decoupled if* (1) B^* *is nonsingular and diagonal and* (2) $A^* = 0$, *where*

$$A^* = \begin{bmatrix} c_1 A^{d_1+1} \\ \vdots \\ c_m A^{d_m+1} \end{bmatrix}$$

In general, $H(s)$ will not be diagonal. Then, we consider whether it is possible to decouple the system (8.2-1) by introducing some mixes of instruments ω, plus adjustment rules tied to z (feedback) so that the transfer function matrix between the targets and ω becomes nonsingular and diagonal. We confine our search for the feedback and policy mixes to the form

$$\boldsymbol{q} = Kz + Q\omega \tag{8.2-10†}$$

where K is a constant $m \times n$ matrix and Q is a constant nonsingular matrix,‡ and where ω is an m-vector which is introduced into the model as a set of policy mixes so that y_i responds to ω_i but not to ω_j, $j \neq i$. In other

†Since Q is nonsingular, (8.2-10) may be written as $\omega = Q^{-1}\boldsymbol{q} - Q^{-1}Kz$. This shows that the variable ω_i is being introduced by (8.2-10) as a mixture of the instruments $q_1, \ldots, q_m$ and a term representing a state variable feedback.

‡When only a subset of the components of z is available to policymakers (due to delay in compiling necessary data, for example), (8.2-10) is replaced with $\boldsymbol{q} = \tilde{F}z + Q\omega$, where $\tilde{F}$ is an $m \times n$ matrix of a particular form.

words, we let some of the adjustments be generated automatically according to Kz and let the rest be given by discretionary or conscious efforts of the policymakers who now focus their attention on ω. The variables ω_1 and ω_2 of Example 2, §8.1 correspond to ω. For example, the money supply there equals the equilibrium growth money supply times e^{ω_1}. Thus, if the policymaker lets the markets adjust by themselves, $\omega_1 = 0$ and the economy grows at the equilibrium growth rate. The policy action $\omega_1 \neq 0$ represents some conscious effort of the policymakers to let the money supply grow at a rate different from the equilibrium rate. Similar remarks may be made about $\omega_2 \neq 0$ of Example 2.

Use of the policy (8.2-10) modifies the dynamics from those of (8.2-1) to

$$\dot{z} = (A + BK)z + BQ\omega \tag{8.2-1$'$}$$

The target vector is still related to z by $y = Cz$. In other words, the new transfer matrix between the newly defined instruments ω and the target vector is given by

$$H(s; K, Q) = C(sI - A - BK)^{-1}BQ \tag{8.2-11}$$

Note that $H(s; K, Q)$ expresses the effects of ω on y, rather than of q on y as $H(s)$ does. Note also that $H(s; K, Q) = H(s; K, I)Q$. From Criterion 1 we can state the following Proposition.

Proposition 1 *The target vector* y *is controlled in a decoupled manner by the instruments* ω *if and only if* $H(s; K, Q)$ *is diagonal and nonsingular (except at a finite number of* s *in the complex plane). In other words, the condition that the transfer matrix* (8.2-11) *be nonsingular and diagonal is necessary and sufficient for the dynamics of* (8.2-2) *to be modified by the rule* (8.2-10) *in such a way that* ω *affects* y *noninteractively.*

We next derive an operationally more convenient criterion for the existence of K and Q to achieve the decoupling conditions along the line of Criterion 1.

Define $\alpha_i(K)$, $i = 1, \ldots, n$ by

$$\begin{aligned}\mu(s; K) &= |sI - A - BK| \\ &= s^n - \alpha_1(K)s^{n-1} - \cdots - \alpha_n(K)\end{aligned}$$

Then, from (8.2-11),

$$H(s; K, Q) = C(sI - A - BK)^{-1}BQ = G(s; K)Q/\mu(s; K)$$

where

$$G(s;K) = \left(s^{n-1}CB + s^{n-2}CR_1(K)B + \cdots + CR_{n-1}(K)B\right)$$

where

$$R_0(K) = I$$

$$R_i(K) = (A + BK)R_{i-1}(K) - \alpha_i(K)I$$

$$= (A + BK)^i - \alpha_1(K)(A + BK)^{i-1}$$

$$-\alpha_2(K)(A + BK)^{i-2} - \cdots - \alpha_i(K)I$$

Note that

$$\alpha_i(0) = \alpha_i$$

and

$$R_i(0) = R_i, \qquad i = 1, \ldots, n$$

Now, define $d_i(K, Q)$ and $D_i(K, Q)$ analogously with d_i and D_i of (8.2-6) and (8.2-7).

Perhaps a remarkable fact is that

$$d_i(K, Q) = d_i$$

i.e., d_i is invariant with respect to K and Q, or a change of instruments (8.2-10) does not change d_i defined by (8.2-6). This is easily verified directly. Then we also see easily that

$$D_i(K, Q) = D_iQ$$

Criterion 2 *The matrices K and Q to decouple* (8.2-1) *exist if and only if B^* of* (8.2-8) *is nonsingular. One such set of K and Q is given by*

$$K^* = -B^{*-1}A^*, \qquad Q^* = B^{*-1}$$

This form of the criterion is due to Falb and Wolovich (1967).

The necessity is easily established. Suppose that $H(s; K, Q)$ is decoupled. Then it is nonsingular and its ith row vector is

$$H_i(s; K, Q) = h_i(s; K, Q)e_i^t$$

where

$$h_i(s;\ K, Q) \neq 0$$

Then

$$D_i Q = \lambda_i e_i{}^{t}$$

where

$$\lambda_i = \lim_{s \to \infty} s^{d_i+1} h_i(s;\ K, Q)$$

where λ_i are all nonzero, for if $\lambda_k = 0$, for some k, then $D_k Q = 0$. Since Q is nonsingular, this implies $D_k = 0$ and $d_k = n - 1$, i.e., $c_k(A + BK)^j B = 0, j = 0, \ldots, n - 1$ which implies that $H_i(s;\ K, Q) = 0$ contradicting the nonsingularity of $H(s;\ K, Q)$.

Sufficiency is established with the aid of the following two facts which show that Criterion 1 is now applicable to $H(s;\ K^*, Q^*)$.

Fact 1

$$\begin{aligned} c_i(A + BK^*)^{d_i} BQ^* &= c_i(A + BK^*)^{d_i} BB^{*-1} \\ &= B_i^* B^{*-1} \\ &= e_i{}^{t} \end{aligned}$$

where B_i^* is the tth *row of* B^* *and where e_i is the m-dimensional vector with all components zero except the* ith *which is* 1.

Note that the choice of $Q^* = B^{*-1}$ is not essential. Any Q such that $B^* Q$ is diagonal is satisfactory for noninteraction.

Fact 2

$$c_i(A + BK^*)^{d_i + j} = 0 \quad \textit{for all positive integer } j.$$

To see this, it suffices to see $c_i(A + BK^*)^{d_i+1} = 0$. This follows from

$$\begin{aligned} c_i(A + BK^*)^{d_i+1} &= c_i A^{d_i+1} + c_i A^{d_i} BK^* \\ &= c_i A^{d_i+1} - c_i A^{d_i} BB^{*-1} A^* \\ &= 0 \end{aligned}$$

since

$$c_i A^{d_i} BB^{*-1} A^* = e_i{}^{t} A^* = c_i A^{d_i+1}$$

From (8.2-9) in terms of the time domain, the target variables are related to ω's by

$$y^{(d_i+1)} = \gamma_i \omega_i, \qquad i = 1, \ldots, m \tag{8.2-12}$$

when effects of the initial condition terms are ignored (see Appendix 1 for detail).

As for asymptotic decoupling, recall that the definition requires that (8.2-1) is asymptotically stable when the instruments are generated by (8.2-10) and that $C(-A - BK)^{-1}B$ is nonsingular and diagonal.

Proposition 2 *A pair of matrices K and Q can be found to decouple asymptotically the system* (8.2-1) *if and only if* (1) *a matrix K can be found to make $A + BK$ an asymptotically stable matrix (i.e. has eigenvalues with negative real parts) and* (2) *rank* $\begin{pmatrix} A & B \\ C & 0 \end{pmatrix} = n + m$.

The matrix identity

$$\begin{pmatrix} I & 0 \\ C(-A - BK)^{-1} & I \end{pmatrix} \cdot \begin{pmatrix} A & B \\ C & 0 \end{pmatrix} \cdot \begin{pmatrix} I & 0 \\ K & I \end{pmatrix} = \begin{pmatrix} A + BK & B \\ 0 & C(-A - BK)^{-1}B \end{pmatrix}$$

is used to show (2) is necessary and sufficient for $C(-A - BK)^{-1}B$ to be nonsingular and diagonal (see (Wolovich, 1973) for further details of the proof).

8.3 IMPERFECTLY KNOWN DYNAMIC SYSTEM

Since decoupling is achieved by subtracting coupling effects between the targets it is not practical to rely heavily on this type of policy, since we do not know the relevant system parameters to carry out this cancellation operation perfectly. In practice no parameters will be known exactly; hence the effects of the other control instruments will only be approximately nullified in the target variables.

It is therefore essential to have some idea of the effects of imprecisely known parameters. We show how to assess approximately the effects of the imperfect knowledge of the parameter values in decoupling.

Let us now return to the time-domain analysis of (8.2-1) to assess the effects of nonexact cancellation of the other control instruments on a given output y_i. We do this by deriving the differential equation that y_i satisfies following Falb and Wolovich (1967) and see ω_j, $j \neq i$ appear as perturbation terms.

Denote as before the ith row vector of C by c_i. We have

$$y_i = c_i z$$

Differentiating this once, and noting that $\dot{z} = (A + BK)z + BQ\omega$,

$$\dot{y}_i = c_i \dot{z} = c_i(A + BK)z + c_i BQ\omega$$

If d_i defined by (8.2-6) is not zero, then we can write it as

$$\dot{y}_i = c_i A z, \qquad \text{since } c_i B = 0$$

Differentiate $\dot{y}_i$ once more to obtain

$$\begin{aligned} \ddot{y}_i &= c_i A \dot{z} \\ &= c_i A(A + BK)z + c_i ABQ\omega \end{aligned}$$

If $d_i > 1$, then this expression simplifies to

$$\ddot{y}_i = c_i A^2 z$$

since $c_i AB = 0$. Thus, by repeated differentiation we obtain

$$\begin{aligned} y_i &= c_i z = c_i(A + BK)^0 z \\ \dot{y}_i &= c_i A z = c_i(A + BK)z \\ &\vdots \\ y_i^{(d_i)} &= c_i A^{d_i} z = c_i(A + BK)^{d_i} z \\ y_i^{(d_i+1)} &= c_i A^{d_i}(A + BK)z + c_i A^{d_i} BQ\omega \\ &= c_i(A + BK)^{d_i+1} z + c_i(A + BK)^{d_i} BQ\omega \\ y_i^{(d_i+2)} &= c_i(A + BK)^{d_i+2} z + c_i(A + BK)^{d_i+1} BQ\omega + c_i(A + BK)^{d_i} BQ\dot{\omega} \\ &\vdots \\ y_i^{(n)} &= c_i(A + BK)^n z + c_i(A + BK)^{n-1} BQ\omega \\ &\quad + \cdots + c_i(A + BK)^{d_i} BQ\omega^{(n-d_i-1)} \end{aligned}$$

Note in the above ω is absent from the expressions for $y_i, \ldots, y_i^{(d_i)}$. Eliminate z from the above using the Cayley–Hamilton Theorem.

We write $(A + BK)^n$ as

$$(A + BK)^n = \sum_{k=0}^{n-1} \beta_k (A + BK)^k$$

where β_k is a real number, $k = 0, 1, \ldots, n - 1$. Note that β_k depends on K.

Multiply $y_i^{(k)}$ by β_k, $k = 0, 1, \ldots, n - 1$, and substract it from $y_i^{(n)}$ to obtain the differential equation for y_i,

$$y_i^{(n)} - \sum_{k=0}^{n-1} \beta_k y_i^{(k)} = \boldsymbol{l}_{n-1}\omega + \cdots + \boldsymbol{l}_{n-d_i}\omega^{(n-d_i-1)} \qquad (8.3\text{-}1)\dagger$$

where $\boldsymbol{l}_{n-1}$, etc., are m-dimensional row vectors defined by

$$\boldsymbol{l}_{n-1} = c_i\Big((A + BK)^{n-1} - \beta_{n-1}(A + BK)^{n-2} - \cdots - \beta_{d_i+1}(A + BK)^{d_i}\Big)BQ$$

$$\boldsymbol{l}_{n-2} = c_i\Big((A + BK)^{n-2} - \beta_{n-1}(A + BK)^{n-3} - \cdots - \beta_{d_i+2}(A + BK)^{d_i}\Big)BQ$$

$$\vdots$$

$$\boldsymbol{l}_{n-d_i} = c_i(A + BK)^{d_i}BQ$$

A necessary and sufficient condition that the right-hand side contains only ω_i and its derivatives and not $\omega_j, j \neq i$ nor its derivative for $i \neq j$ for some choice of K and Q (this is the precise meaning of noninteraction) is that B^* is nonsingular. Note that assuming that B^* is nonsingular, the choice of $K = -B^{*-1}A^*$ and $Q = B^{*-1}$ reduces $c_i(A + BK)^{d_i}BQ$ to $e_i^t = (0, \ldots, 0, 1, 0, \ldots, 0)$. Then the matrix K is computed approximately as $\hat{K} = -\hat{B}^{*-1}\hat{A}^*$. Consequently, the right-hand side of (8.3-1) contains more than ω_i or its derivatives. The proof of this assertion is given in Falb and Wolovich (1967) in the time domain and is equivalent to the transform domain proof given here. We now proceed to derive the perturbation term due to $\hat{A} \neq A$ and $\hat{B} \neq B$.

---

†The right-hand side is assumed to be not zero identically.

Suppose that $\hat{B}^*$ is the matrix computed using the model, while B^* is the corresponding quantity for the exact (unknown) parameter case.

Associate d_i with the true system with A and B.

Let $\hat{B}^* = B^* + \Delta$ where Δ is an $m \times m$ matrix, the magnitudes of the elements of which reflect the imprecision of the model.

Then instead of $B_i^* B^{*-1}$ giving e_i^{t}, we have

$$B_i^* \hat{B}^{*-1} = e_i^{\mathrm{t}} + \delta_i^{\mathrm{t}}$$

where $\delta_i^{\mathrm{t}} = B_i^* \hat{B}^{*-1} - \tilde{e}_i^{\mathrm{t}}$ has the only nonzero component at the ith position, and where the elements of this m-dimensional vector δ_i^{t} will be smaller in magnitude for a model with better known parameter values.

Similarly, $c_i(A + B\hat{K})^{d_i+1}$ is no longer zero but will be

$$c_i(A + B\hat{K})^{d_i+1} - c_i(A + BK)^{d_i+1} = c_i A^{d_i+1} - c_i A^{d_i} B \hat{B}^{*-1} \hat{A}^*$$

$$= -\delta^{\mathrm{t}} \hat{A}^* - c_i(A^{d_i+1} - \hat{A}^{d_i+1})$$

The expression $c_i(A + BK^*)^{d_i+j} - c_i(A + B\hat{K})^{d_i+j}$ can similarly be computed. Denote it as $\mu_j + \nu_j$ where μ_j is the row vector with the only nonzero entry at the ith position.

From (8.3-1) it is clear that

$$y_i^{(n)} - \sum_{k=0}^{n-1} \alpha_i y_i^{(k)} = L_1 + L_2$$

where L_1 contains only terms involving ω_1 or its derivatives and where

$$L_2 = (\nu_{n-d_j-1} - \alpha_{n-1}\nu_{n-d_j-2} - \cdots - \alpha_{d_j+1}\nu_0)\omega$$

$$+ (\nu_{n-d_j-2} - \alpha_{n-1}\nu_{n-d_j-3} - \cdots - \alpha_{d_j+2}\nu_0)\dot{\omega}$$

$$+ \cdots + \nu_0 \omega^{(n-d_i-1)}$$

The term L_2 is the interaction term, containing all $\omega_j, j \neq i$ and their derivatives.

This equation can be used to estimate the effect of ω_k on y_i, $k \neq i$, which is intended for influencing y_k only.

By assuming certain bounds on the unknown system parameters, an upper bound on the magnitude of L_2 is obtainable for any proposed policy. Thus the effect of the interaction from components of ω other than ω_i can be estimated, using standard techniques of the perturbation analysis.

Finally, we note that $B^{*-1}D$, where D is any nonsingular diagonal matrix, can be chosen to be Q^*, instead of B^{*-1}.

The choice of

$$K = B^{*-1}\left(\sum_{k=0}^{\delta} M_k C A^k - A^*\right)$$

$$Q = B^{*-1}$$

with $\delta = \operatorname{Max} d_i$, $M_k = \operatorname{diag}(m_k^{\ 1}, m_k^{\ 2}, \ldots, m_k^{\ m})$ results in

$$y_i^{d_i+1} = \sum_{j=0}^{d_i} m_j^i y_k^{(j)} + \omega_i, \qquad i = 1, \ldots, m$$

while the choice

$$K = -B^{*-1}A^*$$

$$Q = B^{*-1}$$

results in

$$y_i^{(d_i+1)} = \omega_i$$

A suitable choice of M_k can modify some of the eigenvalues of $(A + BK^*)$ (see Falb and Wolovich, 1967, for details).

8.4 EXAMPLE

We discuss an example of a closed macroeconomic model. This model is a slight modification of models discussed by Bergstrom (1967).

In this example the equation for Y_t (national income) is nonlinear. We apply the criterion of decoupling to the linearized part of the dynamic equation, as will be shown shortly.

Thus, instead of (8.2-1), we have for a vector y the following equation indicating deviation from a reference growth path,

$$\dot{y} = Ay + Bx + f(y)$$

where $f(y) = o(\|y\|)$, i.e., $f(y)$ goes to zero faster than $y \to 0$.

Then, we show local noninteracting control for this example (which is valid in a short-run).

The model is composed of

$$C_t = (1 - s)Y_t \tag{8.4-1}$$

$$\frac{\dot{K}_t}{K_t} = \gamma \ln \frac{p_t Y_t - w_t L_t}{(1 + c)r_t p_t K_t} \tag{8.4-2}$$

$$\dot{Y}_t = \lambda(C_t + \dot{K}_t + G_t - Y_t) \tag{8.4-3}$$

$$L_t = \bar{B}e^{-\rho t}Y_t^{\,b}K_t^{\,1-b}, \qquad b > 1 \tag{8.4-4}$$

$$\frac{\dot{w}_t}{w_t} = \beta \ln \frac{L_t}{L_o e^{lt}} + a \tag{8.4-5}$$

$$M_t^{\,d}/p_t = \bar{A}Y_t^{\,u}r_t^{\,-v} \tag{8.4-6}$$

$$\dot{r}_t/r_t = h \ln M_t^{\,d}/M_t^{\,s} \tag{8.4-7}$$

$$p_t = \frac{b(1 + \pi)w_t L_t}{Y_t} \tag{8.4-8}$$

where

C_t: real consumption
Y_t: real output
s: constant propensity to save
K_t: amount of physical capital stock
G_t: government expenditure on goods and services
L_t: employed labor force; note we use Cobb-Douglas production function with constant returns to scale for simplicity
w_t: wage rate
r_t: interest rate
p_t: price level
$M_t^{\,d}$: demand for money
$M_t^{\,s}$: supply of money

γ, λ, c, $\bar{B}$, ρ, b, β, l, a, $\bar{A}$, u, v, h and π are constant parameters.

For this model the supply of money and G_t are taken to be monetary and fiscal instruments, respectively.

Dropping the subscript t for ease of notation, define

$$y_1 = \ln K, \quad y_2 = \ln r, \quad y_3 = \ln w \quad \text{and} \quad y_4 = \ln Y \tag{8.4-9}$$

Substituting (8.4-8) into (8.4-2)

$$\dot{y}_1 = \gamma(-y_1 - y_2 + y_4) + \delta_1 \tag{8.4-10}$$

where

$$\delta_1 = \gamma \ln \frac{b(1+\pi) - 1}{b(1+\pi)(1+c)}$$

Substituting (8.4-4), (8.4-6), (8.4-8) into (8.4-7)

$$\dot{y}_2 = h\{(u + b - 1)y_4 - vy_2 - (b-1)y_1 + y_3 - \rho t\} \\ - h \ln M^s + \delta_2 \tag{8.4-11}$$

with

$$\delta_2 = h \ln \overline{A}\overline{B}\,(1+\pi)b$$

Substituting (8.4-4) into (8.4-5)

$$\dot{y}_3 = -\beta(\rho + l)t + \beta b y_4 - \beta(-1)y_1 + \delta_3 \tag{8.4-12}$$

where

$$\delta_3 = \beta \ln \overline{B}/L_0 + a$$

From (8.4-1) and (8.4-3)

$$\dot{Y} = \lambda(-sY + \dot{K} + G)$$

which can be written as

$$\dot{y}_4 = \lambda(-s + e^{y_1 - y_4}\dot{y}_1) + \lambda\nu_t \tag{8.4-13}$$

where

$$\nu_t = G/Y \tag{8.4-14}$$

With $M_t^s = M^* e^{\overline{m}t}$ and $\nu_t = \nu^*$, it is a straightforward calculation to see that the set of differential equations for $y_1 \sim y_4$ has the particular solution

$$\bar{y}_1 = \ln K^* + \mu_1 t$$

$$\bar{y}_2 = \ln r^*$$

$$\bar{y}_3 = \ln w^* + \mu_2 t$$

$$\bar{y}_4 = \ln Y^* + \mu_1 t$$

and where K^*, r^*, w^* and Y^* are given as solutions of the set of algebraic equations

$$\ln K^* + \ln r^* - \ln Y^* = (\delta_1 - \mu_1)/\gamma$$

$$(b-1)\ln K^* + v \ln r^* - (u+b-1)\ln Y^* - \ln w^* = -\ln M^* + \delta_2/h$$

$$(b-1)\ln K^* - b\ln Y^* = (\delta_3 - \mu_2)/\beta$$

$$\frac{K^*}{Y^*}(\rho + l) - s = -\nu^* + \frac{\mu_1}{\lambda}$$

Now define the state vector z with components

$$\begin{aligned}
z_1 &= \ln \frac{K}{K^* e^{\mu_1 t}} = y_1 - \ln K^* - \mu_1 t \\
z_2 &= \ln \frac{r}{r^*} = y_2 - \ln r^* \\
z_3 &= \ln \frac{w}{w^* e^{\mu_2 t}} = y_3 - \ln w^* - \mu_2 t \\
z_4 &= \ln \frac{Y}{Y^* e^{\mu_1 t}} = y_4 - \ln Y^* - \mu_1 t
\end{aligned} \tag{8.4-15}$$

with the control instruments

$$q_1 = \ln \frac{M^s}{M^* e^{\overline{m}t}} = \ln M^s - \ln M^* - \overline{m}t$$

$$q_2 = \nu - \nu^*$$

The set of differential equations for z's becomes

$$\begin{aligned}
\dot{z}_1 &= \gamma(-z_1 - z_2 + z_4) \\
\dot{z}_2 &= h[(u+b-1)z_4 - vz_2 - (b-1)z_1 - z_3] - hq_1 \\
\dot{z}_3 &= \beta[bz_4 - (b-1)z_1]
\end{aligned} \tag{8.4-16}$$

$$\begin{aligned}
\dot{z}_4 &= \lambda\left\{\mu_1 \frac{K^*}{Y^*}(e^{z_1 - z_4} - 1) + \frac{K^*}{Y^*} e^{z_1 - z_4}\dot{z}_1 + q_2\right\} \\
&= \lambda\left\{\mu_1 \frac{K^*}{Y^*}(e^{z_1 - z_4} - 1) + \gamma \frac{K^*}{Y^*} e^{z_1 - z_4}(-z_1 - z_2 + z_4) + q_2\right\} \\
&= \lambda\gamma \frac{K^*}{Y^*}(-z_1 - z_2 + z_4) + \lambda\mu_1 \frac{K^*}{Y^*}(z_1 - z_4) + \lambda q_2 + o(\|z\|)
\end{aligned}$$

where $o(\|z\|)$ indicates second and higher order terms in the components of z.

The linearized part of the differential equation for z is given then by

$$\dot{z} = Az + Bq \tag{8.4-17}$$

where

$$A = \begin{bmatrix} -\gamma & -\gamma & 0 & \gamma \\ -h(b-1) & -hv & -h & h(u+b-1) \\ -\beta(b-1) & 0 & 0 & \beta b \\ a_{41} & a_{42} & 0 & a_{44} \end{bmatrix}$$

with

$$a_{41} = \lambda \frac{K^*}{Y^*}(\mu_1 - \gamma)$$

$$a_{42} = -\lambda\gamma \frac{K^*}{Y^*}$$

$$a_{44} = \lambda \frac{K^*}{Y^*}(\gamma - \mu_1) = -a_{41}$$

$$B = \begin{bmatrix} 0 & 0 \\ -h & 0 \\ 0 & 0 \\ 0 & \lambda \end{bmatrix}$$

We now apply the criterion for noninteracting control to various possibilities.

8.4.1 Case 1

Can r_t and w_t be controlled in a noninteracting manner? From (8.4-15), take the target variables to be z_2 and z_3. The matrix C is therefore 2×4, with $c_1 = (0, 1, 0, 0)$ and $c_2 = (0, 0, 1, 0)$. We next calculate D_1 and D_2. First $c_1 B = (-h, 0)$. Since $c_2 B$ is zero, we next compute

$$c_2 AB = (-\beta(b-1), 0, 0, \beta b)\begin{bmatrix} 0 & 0 \\ -h & 0 \\ 0 & 0 \\ 0 & \lambda \end{bmatrix}$$

$$= (0, \beta b \lambda)$$

Thus

$$B^* = \begin{pmatrix} c_1 B \\ c_2 AB \end{pmatrix} = \begin{pmatrix} -h & 0 \\ 0 & \beta b \lambda \end{pmatrix}$$

From the criterion, we see that the answer is yes, r_t and w_t can be controlled noninteractively. We elaborate on this next.

POLICIES THAT ACHIEVE NON-INTERACTION To give an economic interpretation to the noninteracting control given by $\boldsymbol{q} = -B^{*-1}A^{*}\boldsymbol{x} + B^{*-1}\omega$, consider Case 1 as an illustration. The economic interpretation of ω in this case is similar to that given in connection with Example 2 of §8.1.

In this case

$$A^{*} = \begin{pmatrix} c_1 A \\ c_2 A^2 \end{pmatrix} = \begin{pmatrix} -h(b-1) & -hv & -h & h(u+b-1) \\ \beta l_1 & \beta l_2 & 0 & \beta l_4 \end{pmatrix}$$

with

$$l_1 = ba_{41} + \gamma(b-1), \quad l_2 = ba_{42} + \gamma(b-1)$$
$$l_4 = ba_{44} - \gamma(b-1)$$

The indicated control law modifies the system dynamics into

$$\dot{z} = (A - BB^{*-1}A^{*})z + BB^{*-1}\omega$$

where

$$A - BB^{*-1}A^{*} = \begin{bmatrix} -\gamma & -\gamma & 0 & \gamma \\ 0 & 0 & 0 & 0 \\ -\beta(b-1) & 0 & 0 & \beta b \\ -\mu & -\mu & 0 & \mu \end{bmatrix}$$

where

$$\mu = \gamma(1-b)/b$$

and

$$BB^{*-1} = \begin{bmatrix} 0 & 0 \\ 1 & 0 \\ 0 & 0 \\ 0 & \dfrac{1}{\beta b} \end{bmatrix}$$

It is easily seen then that

$$\dot{z}_2 = \omega_1$$

$$\ddot{z}_3 = \omega_2$$

This is a special case of the general result (8.2-9).

Of course, by choosing K^* slightly differently, different dynamics are obtainable as indicated at the end of §8.3.

In effect, the dependence of z_3 on the other components z_1, z_2 and z_4, hence on ω_1, is exactly cancelled. The variable ω_1, which is related to the money supply (measured from the level of automatic adjustment as indicated in Example 2 of §8.1), influences not the interest rate itself but the percentage rate of change of interest rate and ω_2 related to government expenditure (measured also from the automatically determined proportion of Y) influences not the wage rate itself but $d/dt\ (\dot{w}/w)$, i.e., the time derivative of the percentage rate of change of wage.

8.4.2 Case 2

Consider the price and the employment. Can p_t and L_t be independently controlled by monetary and fiscal policy?

From (8.4-4) and (8.4-8) (dropping the subscript t again),

$$\ln p = \ln w - \rho t + (b-1) \ln Y - (b-1) \ln K + \ln b(1+\pi)B$$

Define

$$y_1 = \ln \frac{p}{p^* e^{\lambda t}}$$

with

$$np^* = \ln w^* + (b-1) \ln \frac{Y^*}{K^*} + \ln b(1+\pi)\bar{B}$$

$$\lambda = m - u(\rho + l)$$

Then

$$y_1 = (-(b-1) \quad 0 \quad 1 \quad (b-1))z$$

Define

$$y_2 = \ln \frac{L}{L^* e^{lt}}$$

with

$$L^* = \bar{B}\,(Y^*)^b (K^*)^{1-b}$$

Then

$$y_2 = (-(b-1) \quad 0 \quad 0 \quad b)z$$

Thus, the outputs to be controlled independently are given by Cz with

$$C = \begin{pmatrix} c_1 \\ c_2 \end{pmatrix}$$

$$c_1 = (-(b-1) \quad 0 \quad 1 \quad (b-1)), \quad c_2 = (-(b-1) \quad 0 \quad 0 \quad b)$$

According to (2.8)

$$B^* = \begin{pmatrix} c_1 B \\ c_2 B \end{pmatrix} = \begin{pmatrix} 0 & \lambda(b-1) \\ 0 & \lambda b \end{pmatrix}$$

This is a singular matrix. Thus the levels of price and employment cannot be controlled independently.

8.4.3 Case 3

Let w_t and L_t be the target variables. This is an interesting case in view of the Philips curve, which seems to show that a definite relation exists between w_t and L_t.

We have the target vector Cz,

$$C = \begin{pmatrix} c_1 \\ c_2 \end{pmatrix}$$

with

$$c_1 = (0 \quad 0 \quad 1 \quad 0)$$

$$c_2 = (-(b-1) \quad 0 \quad 0 \quad b)$$

Thus, from (8.2-8),

$$c_1 B = (0 \quad 0)$$

$$c_1 AB = (0 \quad \lambda\beta b)$$

From Case 2, $c_2 B = (0 \quad \lambda b)$

Thus

$$B^* = \begin{pmatrix} c_1 AB \\ c_2 B \end{pmatrix} = \begin{pmatrix} 0 & \lambda\beta b \\ 0 & \lambda b \end{pmatrix}$$

Therefore, w_t and L_t cannot be controlled in a noninteracting way.

Before closing this chapter, we want to emphasize again a potential use of the decoupling criterion as a means of discovering hidden functional relations in a macroeconomic model. For example, when a macroeconomic model is specified by a set of equations (8.4-1) $\sim$ (8.4-8), the model builder may not have suspected that functional relations among some macroeconomic variables, such as between w_t and L_t, are implicitly specified. Decoupling criterion uncovered such implicit relations as in Cases 2 and 3. As a matter of practice, this use of the concept of decoupling to discover inherent limitations of macroeconomic models may be more important than designing policy mixes to actually achieve decoupling.

PART III
STOCHASTIC DYNAMIC ECONOMIC SYSTEMS

9

Autoregressive Moving Average Processes

9.1 INTRODUCTION

Often, economic variables are modeled as being generated by a difference equation of the form

$$y_t + B_1 y_{t-1} + \cdots + B_n y_{t-n} = n_t + A_1 n_{t-1} + \cdots + A_m n_{t-m} \quad (9.1\text{-}1)$$

where $\dim y_t = \dim n_t = l$, and where $\{n_t\}$ is a sequence of independent exogenous noises with mean zero and finite second-order moments.

Equations of this type are known as autoregressive moving average (ARMA) equations. From our work in §2.2 we know that ARMA models can be put into state-space representation. In this chapter, we show that they are actually equivalent. It is then not surprising to find that controllability and observability also provide key technical conditions in many discussions involving ARMA models.

9.2 ORTHOGONAL PROJECTIONS

Minimization of quadratic expressions naturally arises in such problems as a least squares problem or a maximum likelihood estimation problem of Gaussian random variables. It is easy to see that a quadratic form

$$E_\theta[(x-\theta)^t(x-\theta)|\mathfrak{B}]$$

is minimized with respect to x by setting it equal to the conditional expected value of θ given $\mathfrak{B}$, since

$$E_\theta[(x-\theta)^t(x-\theta)|\mathfrak{B}] = E_\theta[(x-\hat{\theta})^t(x-\hat{\theta}) + (\hat{\theta}-\theta)^t(\hat{\theta}-\theta)|\mathfrak{B}]$$

$$\geqslant E_\theta[(\hat{\theta}-\theta)^t(\hat{\theta}-\theta)|\mathfrak{B}]$$

for all $\boldsymbol{x}$ where

$$\hat{\boldsymbol{\theta}} = E_\theta(\boldsymbol{\theta} | \mathcal{B})$$

A slightly more general quadratic form $E_\theta[\boldsymbol{x} - \boldsymbol{\theta})^t Q(\boldsymbol{x} - \boldsymbol{\theta}) | \mathcal{B}]$ where Q is a symmetric positive definite matrix is still minimized by $\boldsymbol{x} = \hat{\boldsymbol{\theta}}$. A seemingly more general quadratic minimization problem involving $\boldsymbol{x}^t Q \boldsymbol{x} + 2\boldsymbol{x}^t T \boldsymbol{\theta} + \boldsymbol{\theta}^t R \boldsymbol{\theta}$ where $Q^t = Q$ is a positive definite matrix may be reduced to the above by completion of squares

$$\boldsymbol{x}^t Q \boldsymbol{x} + 2\boldsymbol{x}^t T \boldsymbol{\theta} + \boldsymbol{\theta}^t R \boldsymbol{\theta} = (\boldsymbol{x} - \boldsymbol{w})^t Q (\boldsymbol{x} - \boldsymbol{w}) + \boldsymbol{\theta}^t (R - T Q^{-1} T) \boldsymbol{\theta}$$

where

$$\boldsymbol{w} = Q^{-1} T \boldsymbol{\theta}$$

so that $\boldsymbol{x} = -Q^{-1} T \hat{\boldsymbol{\theta}}$ minimizes its conditional expectation.

In all three cases, the error of estimation, defined by $\boldsymbol{e} = \boldsymbol{\theta} - \hat{\boldsymbol{\theta}}$ satisfies

$$E_\theta(\boldsymbol{e}^t \hat{\boldsymbol{\theta}} | \mathcal{B}) = 0 \quad \text{or} \quad E_\theta(\boldsymbol{e}^t Q \hat{\boldsymbol{\theta}} | \mathcal{B}) = 0 \tag{9.2-1}$$

It is convenient and helps intuition to employ geometric terms and characterize the optimal conditional estimate of $\boldsymbol{\theta}$ being such that the estimation error becomes orthogonal to $\mathcal{B}$, when we define the orthogonality of any two vectors $\boldsymbol{x}$ and $\boldsymbol{y}$ with finite variances by $E(\boldsymbol{x}^t \boldsymbol{y} | \mathcal{B}) = 0$ in the first two cases and by $E(\boldsymbol{x}^t Q \boldsymbol{y} | \mathcal{B}) = 0$ in the last case. In other words, the best estimate of $\boldsymbol{\theta}$ is its orthogonal projection on $\mathcal{B}$. By making the collection of all random variables with finite second moments into a Hilbert space, the machinery of Hilbert space theory becomes available to us. The projection can thus be understood in the sense of conditional expectation or of mean square.

Orthogonal projection also has an important role to play in our discussion of ARMA processes. ARMA models involves two stochastic processes $\{\boldsymbol{n}_t\}$ and $\{\boldsymbol{y}_t\}$. Without too much loss of generality we take them to be zero-mean real valued stochastic processes with finite second-order moments.

Assume also that $R_t = E(\boldsymbol{y}_{s+t} \boldsymbol{n}_s^t)$. We denote by $\overline{S(t-)}$ the closure, in the sense of mean-square norm, of the linear space of all possible finite linear combinations of $\boldsymbol{n}_t, \boldsymbol{n}_{t-1}, \ldots$; all current and past exogenous noise realization. Similarly $\overline{S(t+)}$ denotes the closure of the subspace spanned by $\boldsymbol{n}_t, \boldsymbol{n}_{t+1}, \ldots$. Two random variables X and Y are equivalent or equal in the mean square if $E|X - Y|^2 = 0$. We incorporate into the space of all random variables this equivalence relation. The best s step ahead predictor of y_{t+s} at time t is then its projection on $\overline{S(t-)}$ and is denoted by $y_{t+s|t-}$. The error of prediction is then orthogonal to $\overline{S(t-)}$. In particular

$$E\left[(y_{t+s} - y_{t+s|t-})n_{t-m}^{\mathrm{t}}\right] = 0, \qquad m = 0, 1, \ldots \tag{9.2-2}$$

These relations correspond to (9.2-1). This is not surprising since (9.2-2) is the result of minimizing $E(y_{t+s} - x)^2$ with respect to $x \in \overline{S(t-)}$ hence the error must be orthogonal to $\overline{S(t-)}$.

From (9.2-2), we derive a set of relations

$$E\left[(y_{t+s|t-})n_{t-m}^{\mathrm{t}}\right] = R_{s+m}, \qquad s, m = 0, 1, 2, \ldots$$

9.3 STATE-SPACE REPRESENTATION

We call a Markovian representation of $\{y_t\}$ a state-space representation since it is a natural extension of the concept of state of deterministic dynamic systems introduced in Chapter 2 to stochastic system. We follow Akaike (1974, 1975) in this section to derive state-space representation of $\{y_t\}$ by choosing a basis in $\overline{S(t+|t-)}$, the space spanned by all the predictions $y_{t+s|t-}$, $s = 0, 1, \ldots$. There is a close relation with the concept of canonical correlation in multivariate statistics. See for example Anderson (1958) and Akaike (1975). The space $\overline{S(t+|t-)}$ is finite dimensional when $\{y_t\}$ is generated by (9.1-1).

Denote its basis by z_t. Since $y_{t|t-}$ is an element of $\overline{S(t+|t-)}$, we know that

$$y_t = y_{t|t-} + u_t$$

where

$$y_{t|t-} = Cz_t \tag{9.3-1}$$

and where u_t is orthogonal to, i.e., independent of, all random vectors in $\overline{S(t-)}$.

Let z_{t+1} be obtained analogously with z_t when t is replaced by $t+1$. In other words z_{t+1} is a basis of $\overline{S((t+1)-)}$, the closure of the subspace spanned by $n_{t+1}, n_t, n_{t-1}, \ldots$.

Define

$$w_{t+1} = n_{t+1} - n_{t+1|t-}$$

Since $\overline{S(t-)}$ is a subspace of $\overline{S((t+1)-)}$ and any element of $\overline{S((t+1)+|(t+1)-)}$ is expressible as a linear combination of an element of $\overline{S(t-)}$ and of w_{t+1}, we have

$$z_{t+1} = Az_t + Bw_{t+1} \tag{9.3-2}$$

The matrices A, B and C are constant since $\{y_t\}$ and $\{n_t\}$ are stationary and their correlations are also stationary. Equations (9.3-1) and (9.3-2) are in state-space representation. For systems governed by (9.1-1) $y_{t+s|t-}$ may be successively generated by

$$\begin{aligned} y_{t+s|t-} &+ B_1 y_{t+s-1|t-} + \cdots + B_n y_{t+s-n|t-} \\ &= n_{t+s|t-} + \cdots + A_m n_{t+s-m|t-} \end{aligned} \tag{9.3-3}$$

where

$$y_{t+s|t-} = y_{t+s}, \qquad \text{for } s = 0, -1, -2, \ldots$$

and

$$n_{t+s|t-} = 0, \qquad \text{for } s = 0, 1, 2, \ldots$$

For $s \geqslant m+1$, the right-hand side of (9.3-3) vanishes. For z_t of (9.1-1) we may take for some positive integer K

$$z_t^{\mathrm{t}} = (y_{t|t-}^{\mathrm{t}}, y_{t+1|t-}^{\mathrm{t}}, \ldots, y_{t+K-1|t-}^{\mathrm{t}})$$

Then

$$A = \begin{pmatrix} 0 & I & \cdots & 0 \\ 0 & 0 & \cdots & 0 \\ & & \vdots & \\ 0 & 0 & \cdots & I \\ -B_K & -B_{K-1} & \cdots & -B_1 \end{pmatrix}, \qquad B = \begin{pmatrix} W_0 \\ W_1 \\ \vdots \\ W_{K-1} \end{pmatrix}$$

$$C = (I \quad 0 \quad \ldots \quad 0)$$

where W's are as they appear in

$$y_t = \sum_{m=0}^{\infty} W_m n_{t-m}$$

W's are conveniently generated by the lag-transforms of (9.1-1), introduced in Chapter 2, as

$$\sum_{i=0}^{\infty} W_i L^i = (I + B_1 L + \cdots + B_n L^n)^{-1}(I + A_1 L + \cdots + A_m L^m)$$

or

$$W_i = A_i - B_1 W_{i-1} - \cdots - B_n W_{i-n} \qquad \text{for } i > 0$$

$$W_0 = I$$

$$W_i = 0 \qquad \text{for } i < 0$$

To convert a state-space representation into an autoregressive moving average form, we appeal to Cayley-Hamilton Theorem.

Let a model be given in state-space representation

$$z_{t+1} = A z_t + B \boldsymbol{n}_{t+1}$$

$$y_t = C z_t$$

where A is $p \times p$. The matrix A satisfies by Cayley-Hamilton Theorem

$$A^p + \sum_{i=1}^{p} a_i A^{n-i} = 0$$

where a's are those in $|\lambda I - A| = \lambda^p + \Sigma a_i \lambda^{p-i}$. Eliminate powers of A from

$$z_{t+i} = A^i z_t + A^{i-1} B \boldsymbol{n}_{t+1} + \cdots + B \boldsymbol{n}_{t+i}, \qquad i = 0, 1, \ldots$$

to obtain the autoregressive moving average form

$$y_{t+p} + a_1 y_{t+p-1} + \cdots + a_n y_t = CB\boldsymbol{n}_{t+p} + C_1 \boldsymbol{n}_{t+p-1} + \cdots + C_{n-1} \boldsymbol{n}_{t+1}$$

where

$$C_i = C\left(A^i + a_1 A^{i-1} + \cdots + a_i I\right)B, \qquad i = 1, \ldots, n-1$$

9.4 CONTROLLABILITY IN ARMA PROCESS

Controllability has implications on the probability distribution of endogenous variable $\boldsymbol{y}_t$ of (9.1-1). Consider the ARMA model in its state-space representation

$$z_{t+1} = Az_t + Bn_{t+1}$$

where $\{n_t\}$ is a sequence of independently and identically distributed noise processes with mean zero and positive definite variances. Suppose (A, B) is not a controllable pair. Then, there exists a nonzero vector a such that

$$a^t(B \quad AB \quad \dots \quad A^{n-1}B) = 0$$

This implies that

$$E(a^t z_t n_{t-i}) = 0, \qquad i = 0, 1, \dots, n-1$$

By Cayley–Hamilton Theorem, this means that $E(a^t z_t n_{t-i}) = 0$ for all $i \geqslant 0$. Since z_t is a linear combination of $n_t, n_{t-1}, \dots$, we see that $a^t z_t = 0$ in the mean square, i.e., the distribution of z_t is degenerate. On the other hand, if rank $\langle A|B \rangle = n$, then there is no nonzero vector a such that $E(a^t z_t n_{t-i}) = 0$, indicating that the distribution of z_t is not degenerate. Similar implication holds for systems governed by Itoh stochastic differential equations driven with Brownian process (Elliott 1969).

There are many computational algorithms to construct estimates of A, B and C matrices from the record of $\{y_t\}$ and $\{n_t\}$. In most algorithms, certain linearly independent vectors are suitably chosen from the controllability and observability matrices or their product matrix. We refer interested readers to Akaike (1975).

An obvious question of ARMA processes such as (9.1-1) is how to compute $y_{t+s|t-}$ or to construct equations to generate it. By putting ARMA models into state-space representation we can discuss it using Kalman filter theory (Aoki 1967). Convergence behavior of estimates depends crucially on the controllability and observability properties of the systems. Actually dynamics of Kalman filters is dual of regulator dynamics to be discussed in the next chapter (Aoki 1967). We do not pursue this topic in detail here.

9.5 EXAMPLES

Example 1 Even though our discussion was confined to stationary stochastic processes, various extension to nonstationary stochastic processes exist. The next example is indicative of the nature of such extensions. The technique is that proposed by Rissanen (1967).

Total output in a certain industry composed of several firms is supposed to be governed by

$$y_t + w_1 y_{t-1} + w_2 y_{t-2} = \epsilon_t \tag{9.5-1}$$

where all the variables are scalar, and where

$$E(\epsilon_t) = 0$$

$$R_{t,\tau} = E(\epsilon_t \epsilon_\tau)$$

where the subscript t refers to a suitable period such as a market day or a quarter. Suppose that $R_{t,\tau} = 0$ for $|t - \tau| > 2$.

Define the covariance matrix

$$R = \begin{bmatrix} & \gamma_{\tau t} & \cdots & \cdots \\ & & \cdots & \cdots \\ \vdots & & & \\ \cdots & \cdots & \gamma_{22} & \gamma_{20} \\ \cdots & \cdots & \gamma_{02} & \gamma_{00} \end{bmatrix} \tag{9.5-2}$$

It is a Toeplitz matrix and can be factored as

$$R = TT' \tag{9.5-3}$$

where T is an upper triangular matrix.

From (9.1-1) we see that $y_0, y_1, \epsilon_2, \ldots, \epsilon_t$ (or y_0, y_t if $t < 2$) generate the subspace $\overline{S(t-)}$. The dimension of the subspace is equal to the rank of $R^{(t)} = \{\gamma_{ij}\}$, $0 \leqslant i, j \leqslant t$, the truncated matrix of R. We assume it to be $(t+1)$.

Let $\{v_0, v_1, \ldots\}$ be an orthonormal basis in $\overline{S(t-)}$. With respect to this basis, we have

$$\begin{aligned} \epsilon_t &= a_0(t)v_t + a_1(t)v_{t-1} + a_2(t)v_{t-2}, \qquad && \text{for } t \geqslant 2, \\ y_t &= a_0(t)v_t + \cdots + a_t(t)v_0, && \text{for } 0 \leqslant t \leqslant 1 \end{aligned} \tag{9.5-4}$$

where from (9.5-1) and (9.5-3) we see that $a_0(t)$, $a_1(t)$, ... are the elements of T in (9.5-3).

The one-step ahead prediction of y_t made on $(t-1)$th day, denoted by $\hat{y}_t$ is then related to y_t by

$$y_t = \hat{y}_t + a_0(t)v_t$$

or more generally

$$v_\tau = (y_\tau - \hat{y}_\tau)/a_0(\tau), \qquad \tau = t, t-1, \ldots \tag{9.5-5}$$

From (9.5-1), (9.5-4) and (9.5-5) we obtain the equation for best one-day ahead prediction of the industry output

$$\hat{y}_t + \frac{a_1(t)}{a_0(t-1)}\,\hat{y}_{t-1} + \frac{a_{2(t)}}{a_2(t-2)}\,\hat{y}_{t-2}$$

$$= \left(\frac{a_1(t)}{a_0(t-1)} - w_1\right) y_{t-1} + \left(\frac{a_2(t)}{a_2(t-2)} - w_2\right) y_{t-2}$$

with initial conditions

$$\hat{y}_0 = \hat{y}_{-1} = 0$$

For recursive computational algorithm for computing a's, see (Rissanen (1967).

Example 2 (Model With Rational Expectation)† As Example 5 of Chapter 2 illustrates, some problems are better treated in autoregressive moving average representation even though state-space representation is generally preferable. As an example, consider a dynamic economy in which conditional expectations (rational expectation) of future values of some endogenous variables (such as future prices used to estimate inflation rate) appear in the dynamic equation. This gives rise to an equation of the form (in the deviational form)

$$z_t = Az_{t-1} + B_1 z_{t|t-1} + \cdots + B_k z_{t|t-k} + e_t$$

where $\{e_t\}$ are stationary serially independent mean zero with $Ee_t e_t^{\mathrm{t}} = I$,

$$z_{t|t-1} = E(z_t | \mathcal{G}_{t-1})$$

and where $\mathcal{G}_t$ is the σ-field generated by $z_0, e_0, \ldots, e_t$. Here we treat $k = 2$, and we assume that A, B_1 and B_2 are known. We also assume z_0 to be independent of e's. Hence, we can express z_t as

$$z_t = S_t z_0 + \sum_{s=0}^{t} W_{t-s} e_s \tag{E.1}$$

Substituting into the original equation, noting that $e_{t|t-1} = e_{t|t-2} = 0$, we

†The model is suggested by M. Canzoneri.

determine S's, and W's recursively where

$$S_s = \hat{A}S_{s-1}, \qquad s = 1, 2, \ldots, t$$

$$\hat{A} = (I - B)^{-1}A, \qquad \text{where } B = B_1 + B_2$$

$$S_0 = I$$

$$W_1 = \bar{A}, \qquad \text{where } \bar{A} = (I - B_1)^{-1}A$$

$$W_{s+1} = \hat{A}W_s, \qquad s = 1, \ldots, t - 1$$

An alternate expression for z_t is

$$z_t = \hat{A}z_{t-1} + e_t + (W_1 - \hat{A})e_{t-1}$$

where

$$W_1 - \hat{A} = \left[(I - B)^{-1} - (I - B_1)^{-1}\right]A$$
$$= -(I - B_1)^{-1}B_2(I - B)^{-1}A$$

Suppose now that the government utilizes instruments to modify the dynamic equation for z_t

$$z_t = Az_{t-1} + B_1 z_{t|t-1} + B_2 z_{t|t-2} + Cx_t + e_t$$

If the government employs $x_t = Gy_{t-1}$, then this is taken into account in computing the expectations. We see its effect is to modify A into $A + CG$ where G is a constant matrix, values of which the government can specify freely. Equation (E.1) tells us that

$$z_t = e_t + \hat{z}_{t-1}$$

where the government can affect the distribution of the random variable $\hat{z}_{t-1}$ with

$$\hat{z}_{t-1} = D^t z_0 + D^{t-1}He_0 + \cdots + DHe_{t-2} + He_{t-1}$$

where

$$D = (I - B)^{-1}(A + CG)$$

$$H = -(I - B_1)^{-1}B_2(I - B)^{-1}(A + CG)$$

Note that G enters into both D and H. How should G be chosen? Apparently a novel optimization problem results† if we try

$$\min_G E(\hat{z}_{t-1}{}^{t}\hat{z}_{t-1})$$

$$= \min_G \mathrm{tr}\left\{H^{t}H + H^{t}D^{t}DH + \cdots + H^{t}(D^{t})^{t-1}D^{t-1}H + (D^{t})^{t}D^{t}\right\}$$

where we assume

$$Ez_0z_0{}^{t} = I$$

A sufficient condition for this minimization is to choose G to make the eigenvalues of $A + CG$ all zero.‡ Such a G exists if (A, C) is a controllable pair.

Let

$$F = (I - B_1)^{-1}B_2(I - B)^{-1}$$

The effect of the most immediate past disturbance e_{t-1} is minimized by $\min_G \mathrm{tr}(E^{t}E)$ that is achieved by

$$G = -(C^{t}F^{t}FC)^{-1}C^{t}F^{t}FA, \qquad i = 1, \ldots, n$$

†More generally we may treat G to be time varying.
‡In this case $D = 0$ and $H = 0$, hence $z_t = e_t$.

10

Regulation and Modelling of Stochastic Dynamic Economies

A Prototype optimization problem for a linear deterministic dynamic economy has been posed and solved in §5.3 for discrete-time systems and in §5.4 for continuous time systems.

In both cases we observe optimal control policies are linear in the state vector as $\boldsymbol{x}_t = -K_t z_t$, where K_t is called feedback gain matrix. This form of policy tacitly implies that the whole state vector z_t is available at time t for the purpose of implementing control actions. This assumption may not be very realistic because data gathering may mean that the state vector becomes available only with some delay. Perhaps more serious defect is the fact that data, even though current, contain errors and omissions.

Errors in data are usually modeled as exogenous additive disturbances so that instead of $y_t = C_t z_t + D_t \boldsymbol{x}_t$ we use $y_t = C_t z_t + D_t \boldsymbol{x}_t + \boldsymbol{n}_t$ as the observation equation, where $\{\boldsymbol{n}_t\}$ is the random noise process.

This is one of several ways we come to model economic systems as stochastic systems. The others come from having additive disturbances in the state equation itself and/or parameters of the dynamic equations are random variables or stochastic processes.

It is perhaps the need for controlling dynamic macroeconomic models with randomly varying system parameters that first caused economists to turn their attention to stochastic control theory. Consequently, some aspects of the subject matter of this chapter may be rather well-known to economists compared with other aspects of control and system theory.

With stochastic models it is customary to formulate intertemporal optimization problem as minimizing expected value of given performance indices.†

†Their variances are usually ignored. In some problems, however, variances convey important information.

When dynamic economies are assumed to be linear with known system parameters (or system parameters with estimated values that are treated as the correct ones), and with additive rather than multiplicative random disturbances in the dynamic equations and/or in the data, and in some cases under less stringent conditions, it turns out that the solution of stochastic optimal control problems with quadratic performance index separates into two stages: one of computing $E(z_t|\mathcal{I}_t)$, where $\mathcal{I}_t$ is the information set available at time t, and the other of computing feedback gains (Holt 1962, Simon 1956, Theil 1957, Theil and Kloek, 1960). Thus, typically optimal control rules are linear in $E(z_t|\mathcal{I}_t)$ rather than in z_t.

This area more than any other economic problems has provided an interface between economists and control engineers. See also Peston (1973) for views of economists or econometricians on interaction with control engineers. Instead of following a standard route that would be a stochastic version of Chapter 5, therefore, we merely give for ease of reference an outline of the developments of a set of formulas for this standard case.† We devote most of our attention to less well-known cases, such as dynamic economies with randomly varying system parameters with uncertain statistics and the use of several aggregation models or observers to construct policies and make comments on some numerical problems.

10.1 KALMAN FILTER: KNOWN SYSTEM PARAMETERS CASE

We summarize a procedure for computing the conditional expected value of z_t, $E(z_t|\mathcal{I}_t)$ for the system governed by

$$\begin{aligned} z_{t+1} &= A_t z_t + B_t x_t + \xi_t \\ y_t &= C_t z_t + \eta_t \end{aligned} \tag{10.1-1}$$

We show in this section that under suitable assumptions $E(z_t|\mathcal{I}_t)$ can be computed recursively by

$$z_{t+1|t+1} = z_{t+1|t} + K_{t+1}(y_{t+1} - C_{t+1} z_{t+1|t})$$

where

$$z_{t+1|t} = A_t z_{t|t} + B_t x_t$$

and where

$$z_{t|t} = E(z_t|\mathcal{I}_t)$$

†Because of the large variations in some of the system parameters, use of certainty equivalence is less desirable in economic systems than in engineering systems in which the variations are usually smaller.

and where K_{t+1} is called the filter gain matrix (given by (10.1-3)).

We assume that A, B and C are known, that ξ's and η's have zero mean, that they are mutually and serially uncorrelated with known variances $Q_t = E\xi_t\xi_t^{\mathrm{t}}$ and $R_t = E\eta_t\eta_t^{\mathrm{t}}$. Recall our geometric picture developed in connection with autoregressive moving average models in Chapter 9.

We have

$$z_{t+1} = z_{t+1|t} + w_{t+1}$$

where w_{t+1} is orthogonal to $\overline{S(t-)}$, the closure (in the mean square) of the subspace generated by $z_0, \xi_0, \ldots, \xi_{t-1}, \xi_0, \ldots, \xi_t$ or equivalently by $\mathcal{G}_t = \{y_0, \ldots, y_t\}$. Since $\mathcal{G}_{t+1} = \{\mathcal{G}_t, y_{t+1}\}$, we have $\overline{S(t-)} \subset \overline{S((t+1)-)}$ and $y_{t+1} - y_{t+1|t}$ is the component orthogonal to $\overline{S(t-)}$, we see that the orthogonal projection of z_{t+1} on $\overline{S((t+1)-)}$ is given as

$$\begin{aligned} z_{t+1|t+1} &= z_{t+1|t} + w_{t+1|t+1} \\ &= z_{t+1|t} + K_{t+1}(y_{t+1} - y_{t+1|t}) \end{aligned} \tag{10.1-2}$$

where

$$y_{t+1|t} = Cz_{t+1|t}$$

since $\eta_{t+1|t} = 0$ by the independence assumption. The matrix K_{t+1} is called the filter gain matrix, and is chosen to make the error of estimation $z_{t+1} - z_{t+1|t+1}$ orthogonal to $\overline{S((t+1)-)}$, in particular K_{t+1} must be such that

$$E\{(z_{t+1} - z_{t+1|t})y_{t+1}^{\mathrm{t}}\} = K_{t+1}E\{(y_{t+1} - y_{t+1|t})y_{t+1}^{\mathrm{t}}\}$$

We can easily compute, remembering that η_{t+1} and $z_{t+1} - z_{t+1|t}$ are orthogonal to $\overline{S(t-)}$, that

$$E\{(y_{t+1} - y_{t+1|t})y_{t+1}^{\mathrm{t}}\} = CM_{t+1}C^{\mathrm{t}} + R_{t+1}$$

where we denote

$$M_{t+1} = E\{(z_{t+1} - z_{t+1|t})(z_{t+1} - z_{t+1|t})^{\mathrm{t}}\}$$

and

$$E\{(z_{t+1} - z_{t+1|t})y_{t+1}^{\mathrm{t}}\} = M_{t+1}C^{\mathrm{t}}$$

We thus computed the optimal gain

$$K_{t+1} = M_{t+1}C_{t+1}^{\mathrm{t}}(C_{t+1}M_{t+1}C_{t+1}^{\mathrm{t}} + R_{t+1})^{-1} \tag{10.1-3}$$

where we assume for simplicity that $R_t > 0$ for all t. From (10.1-1)

$$z_{t+1} - z_{t+1|t} = A_t(z_t - z_{t|t}) + \xi_t$$

hence

$$M_{t+1} = A_t \Gamma_t A_t^{\mathrm{t}} + Q_t$$

where

$$\Gamma_t = \operatorname{cov}(z_t | \mathcal{G}_t)$$

After a bit of matrix algebra the recursion formula for Γ_t is obtained as

$$\Gamma_{t+1} = M_{t+1} - M_{t+1} C_{t+1}{}^{\mathrm{t}} (C_{t+1} M_{t+1} C_{t+1}{}^{\mathrm{t}} + R_{t+1})^{-1} C_{t+1} M_{t+1} \quad (10.1\text{-}5)$$

or in its inverse form

$$\Gamma_{t+1}{}^{-1} = M_{t+1}{}^{-1} + C_{t+1}{}^{\mathrm{t}} R_{t+1}{}^{-1} C_{t+1} \quad (10.1\text{-}6)$$

See (5.3.10) and Section VII.3 of Aoki (1967) for alternate expressions for the filter gain and a discussion on a numerical sensitivity in recursive gain computation. Note that the calculation of covariance matrices, hence of the filter gain matrices are independent of actual measurement data and can be carried out separately. Note also that these matrices are independent of the actual controls.

DISCUSSION There is a close relation between the filter gain computation and that of the feedback control gain of §5.3. It is known as the duality principle in the engineering literature (Aoki, 1967, page 64). It is not surprising since both calculations, are tied to solutions of certain matrix Riccati equations. Their asymptotic behavior is determined by controllability and observability conditions. For example, $\Gamma_t \to 0$ as $t \to \infty$ if (A, C) is an observable pair.

10.2 LINEAR REGULATOR PROBLEMS WITH QUADRATIC COST FUNCTION

10.2.1 Known System Parameters

STANDARD PROBLEM We consider a stochastic version of the problem we discussed in §5.3. It is to minimize the expected value of

$$J_{0,T} = z_T{}^{\mathrm{t}} P z_T + \sum_{s=0}^{T-1} W_s \quad (10.2\text{-}1)$$

where $P = P^{\mathrm{t}}$ is positive semidefinite and where $W_s \geqslant 0$, subject to the dynamic equation

$$\begin{aligned} z_{s+1} &= A_s z_s + B_s x_s + \xi_s \\ y_s &= C_s z_s + D_s x_s + \eta_s \end{aligned} \tag{10.2-2}$$

The matrices A's, B's, C's and D's are assumed to be known deterministic matrices. The ξ 's and η's are as in §10.1. As we discussed in §5.3, we take W_s to be

$$W_s = y_s^{\prime} y_s$$

without any loss of generality.

Stochastic version of dynamic programming can be used to solve this problem. Because of random disturbances in (10.2-2) we do not know the values of the state vectors exactly. Given the information set $\mathcal{G}_t$ at time t we know its conditional mean $E(z_t|\mathcal{G}_t)$ and conditional covariance matrix $\operatorname{cov}(z_t|\mathcal{G}_t)$ which we denote by Γ_t. These moments suffice to evaluate the conditional expected value of a quadratic expression. If random noises are Gaussian as in §10.1, then we have shown there that Γ_t is a constant matrix, independent of controls.†

In stochastic control problems, it is very important to be specific and explicit about the information set (or pattern) available for a decision maker to perform these calculations. Different information sets would result in different optimal controls. As the concept of "physical" state vector is very important in deterministic control problems, so is the concept of "informational" state in stochastic control problems. To emphasize the fact, we call $\mathcal{G}_t$ information state from now on.

Normally, $\mathcal{G}_t$ consists of the current and past measurements and past controls, $y_0, \ldots, y_t$ and $x_0, \ldots, x_{t-1}$ plus whatever information is available *a priori* such as the statistical characteristic of noises. If we wish to examine effects of delays in acquiring data, then the information state consists of $y_0, \ldots, y_{t-d}, x_0, \ldots, x_{t-d-1}$, where d is the number of periods of delay. The conditional probability distribution function of z_t, if available, is the complete description of the informational state. So is a set of sufficient statistics, such as mean and variance in the case of Gaussian

---

†When we control stochastic systems with imperfectly known system parameters, $\operatorname{cov}(z_t|\mathcal{G}_t)$ becomes a function of past control values, even when ξ 's and η's are Gaussian random variables. Dependence of estimation error covariance matrices on the values of controls employed in the past clearly indicates the so-called dual control effects in the engineering literature, the existence of conflict between learning the incompletely known parameters quickly and currently more advantageous controls. Dual control theory has been developed to examine best compromises, see Aoki (1967) and Tse and Bar-Shalom (1973) for detail.

random variables, to summarize the information available to decision makers.

The functional equation is given by

$$J_{t,T}(\mathcal{I}_t) = \min_{x_t} E\{W_t + J_{t+1,T}[\mathcal{I}_{t+1}(\mathcal{I}_t, x_t)] | \mathcal{I}_t\},$$

$$\text{with} \quad t = t_0, t_0 + 1, \ldots, T - 2$$

$$J_{T-1,T}(\mathcal{I}_{T-1}) = \min_{x_{T-1}} (z_T{}'Pz_T + W_{T-1} | \mathcal{I}_{T-1}) \tag{10.2-3}$$

Note that the information state vector $\mathcal{I}_t$ as well as z_t undergoes transformation as functions of x_t. The equation (10.2-3) can now be solved by backward induction in exactly the same way as we solved (5.3-7). The only difference we must note is in evaluation of quadratic expressions in z's. For example,

$$E(z_t{}'\Pi z_t | \mathcal{I}_t) = \text{tr}(\Pi \Gamma_t) + z_{t|t}{}'\Pi z_{t|t}$$

where

$$\begin{aligned} \Gamma_t &= \text{cov}(z_t | \mathcal{I}_t) \\ &= E\{(z_t - z_{t|t})(z_t - z_{t|t})' | \mathcal{I}_t\} \end{aligned}$$

and where

$$z_{t|t} = E(z_t | \mathcal{I}_t)$$

With this change, the procedure for solving (10.2-3) is identical to that of §5.3.

We list the results for reference convenience.

OPTIMAL CONTROLS The following formulae give the optimal controls:

$$x^0_{T-1} = -\Lambda_{T-1} z_{T-1|T-1}$$

where

$$\Lambda_{T-1} = (B_{T-1}{}'PB_{T-1} + D_{T-1}{}'D_{T-1})^+ (B_{T-1}{}'PA_{T-1} + D_{T-1}{}'C_{T-1})$$

$$J_{T-1,T}{}^0(\mathcal{I}_{T-1}) = \|z_{T-1,T-1}\|^2_{\Pi_{T-1,T}} + \text{tr}\, \Pi_{T-1,T}$$

where

$$\Pi_{T-1,T} = (A_{T-1}{}'PA_{T-1} + C_{T-1}{}'C_{T-1})\Gamma_{T-1} + PQ_{T-1} + R_{T-1}$$

The matrix $\Pi_{T-1,T}$ is the one given in §5.3. Note that the optimal feedback gain matrix Λ_{T-1} is identical to the one for the deterministic problem (5.3-7′). In general

$$x_t^0 = -\Lambda_t z_{t|t}$$

where Λ_t is as given in (5.3.10)

$$J_{t,T}^0(\mathcal{I}_t) = \|z_{t,t}\|^2_{\Pi_{t,T}} + \operatorname{tr} \Pi_{t,T}$$

where $\Pi_{t,T}$ is as given by (5.3.9) and where

$$\Pi_{t,T} = (C_t{}^t C_t \Gamma_t + R_t) + \Pi_{t+1,T} \qquad t = t_0, \ldots, T-2$$

Thus, except for constant Π's, the only change in the control law is to replace z_t by $z_{t|t}$.

Some Variations of the Standard Problem It is possible to reduce to this standard formulation some seemingly different problems. We have already mentioned different information pattern due to delay. We can discuss the optimal control of (10.2-2) when some of the matrices A's, B's, C's and D's are random matrices with known mean and second moments rather than known constant matrices. The problem of following a reference time path (tracking problem) is obviously reducible to the regulator problem involving derivational variables by suitable change of variables.

The solution has been obtained assuming that the ξ's and η's are not serially nor mutually correlated. The solution can be extended without difficulty to problems in which ξ's and η's are serially and/or mutually correlated and to problems in which only some data are corrupted with noise and so on. For details of some of these extensions reader is referred to Aoki (1967, Chapters 2 and 5).

Examples of how the probability density $p(z_t|\mathcal{I}_t)$ is transformed into $p(z_{t+1}|\mathcal{I}_{t+1})$ may also be found in Section II.1 of Aoki (1967).

10.2.2 Unknown System Parameters

Optimal Closed-Loop Feedback Control When we drop the assumption that either the system parameters are known, or are random variables with a known joint probability distribution function or that noise statistics are known, we ran into analytical difficulties in deriving optimal control policies. The difficulties show up even in scalar problems.

For example, suppose a scalar model is described by

$$z_{t+1} = az_t + bx_t + \xi_t$$

where a is known but b is a random variable independent of ξ's with its density function given by $N(\theta, \sigma_1^2)$ with θ the unknown mean. In other words, we assume that b is normally distributed with unknown mean θ and known variance σ_1^2. Suppose further that we have an *a priori* information that θ is normally distributed with mean θ_0 and variance σ_2^2. Such assumptions could be appropriate for a model in which the effectiveness of the instrument, e.g., the multiplier associated with the instrument, is only vaguely known. Suppose z_t is observed without error. Take the objective function to be z_T^2 where T is the end of the planning horizon.

This problem has been treated in Aoki (1967, pages 111–113) to show that although the optimal control at $T-1$ is given by

$$x_{T-1}^* = -b_{T-1}az_{T-1}/\left(b_{T-1}^2 + \sigma_{T-1}^2\right)$$

where * indicates optimality, where

$$b_{T-1} = E(b|z_0, \ldots, z_{T-1})$$

and where

$$\sigma_{T-1}^2 = \text{var}(b|z_0, \ldots, z_{T-1})$$

and where b_t's and σ_t's for $t = 0, 1, \ldots, T-1$ can be computed recursively, $x_{T-2}^*, x_{T-3}^*, \ldots, x_0^*$ cannot be given in analytically closed form even though approximations to them are available ([Aoki 1967], pages 113–116).

This example is indicative of the nature of difficulties we encounter in treating the optimal controls of uncertain dynamic reduced form economic models by the Bayesian approach.

Open-Loop Feedback Control We cannot express optimal control laws in analytically closed form because we cannot evaluate explicitly expressions of 'cost-to-go' such as $E(\gamma_T^*|z_0, \ldots, z_{T-2}, x_0, \ldots, x_{T-2}, b)$ where $\gamma_T^* = \min E(z_T^2|z_0, \ldots, z_{T-1}, x_0, \ldots, x_{T-2})$.

One approximation to avoid this kind of difficulty is to compute the so-called open-loop feedback control law instead of optimal closed-loop control laws.

We approximate γ_t^*, the optimal expected cost over the period covering t to T (called cost to go) by its predicted cost based on the conditional probability distribution of everything, given the information set $\mathcal{I}_t$ at time t. After x_t is derived and applied, the joint probability distribution function for all random variables is updated and the conditional distribution function given $\mathcal{G}_{t+1}$ is used next to approximate γ_{t+1}^* (see [Aoki 1967, pages 241–245] for further explanation). The open-loop feedback rule has gained a certain amount of popularity due to its ease of computation and implementation.

ADAPTIVE CONTROL OR DUAL CONTROL LAWS Although the open-loop feedback control laws are easy to obtain, they tend to be too conservative in some applications where the available resources for control purposes are fixed in advance, since they ignore future learning that is expected to take place between t and T in approximating the optimal cost to go γ_t^*. By pursuing an active learning program, at least initially, unknown parameter values may be learned faster with better control results. To learn and to control are the familiar dual functions any controls must perform and they are not often compatible. Any policies that mix these two roles are variously called active learning, parameter adaptive or dual control laws. (See Fel'dbaum, 1960, 1961, Tse and Athans, 1972 for some engineering applications).

10.2.3 Structural Form

When economic theories or reasoning are applied to construct models, they are usually in structural form and not in reduced form such as (10.1-1).

An example of the structural form is given by

$$F_{t+1}z_{t+1} = H_{t+1}z_{t+1} + G_t z_t + B_t x_t + \rho_t \qquad (10.2\text{-}4)$$

$$y_t = C_t z_t + \eta_t \qquad (10.2\text{-}5)$$

If $(F - H)_{t+1}$ is invertible, then (10.2-4) can be reduced formally to (10.1-1).

This procedure is not very desirable when the matrices such as F, H, etc., contain some unknown parameters. These parameters usually have well-defined economic meanings which tend to be scrambled and lost in forming $(F_{t+1} - H_{t+1})^{-1}G_t$ and $(F_{t+1} - H_{t+1})^{-1}B_t$. Also, the computation of the estimation error covariance matrices for $(F - H)^{-1}G$, etc., become complicated and time-consuming.

Control algorithms suitable for models in structural form are yet to be developed.

10.3 AGGREGATION†

As we endeavor to reflect more economic realism in models that we build, they quickly become unmanageable analytically and computationally very expensive to simulate. Construction of simple models that nonetheless are useful for their theoretical insights still is an art. Suppose we are somehow given a large econometric model. We then need a systematic procedure for reducing some measure of complexity of the model without too much loss of information contained in the original model.

One approach is to borrow ideas from multivariate statistics such as clustering or factor analysis (Blin, 1973, Fisher, 1969). Another approach is to "aggregate" some variables in the models, thus reducing the number of economic variables, hence complexity of the models. There are two related approaches. One is to aggregate by similarities of speed of time responses. The other is to aggregate by multicollinearity such as similarity of tastes or proportionality of incomes and so forth. Of course, these two approaches should be combined in constructing aggregated dynamic economic models, whenever possible.

Dynamic economic models may be conveniently considered to be composed of several markets such as real markets, money markets, and bond markets and so forth. These markets are not independent. There are many transmission paths and/or feedback loops linking them.

Some markets have shorter time constants than the others, and markets with shorter time constants may reach equilibrium more quickly. Since large disaggregated models could be very complicated and it is very hard, if not impossible, to comprehend full range of implications of the models, it is sometimes useful to aggregate models by the speeds of dynamic responses of the various markets. This idea has been advanced by Simon and Ando (1961) and has been extended to quadratic regulation problems by Aoki (1971). We have also given an example of this approach in §7.2.6 when we discussed fix-price method of short-run analysis. Marshall's

†The material of this section is based in part on Aoki (1971), Aoki and Huddle (1967), Leondes and Novak (1972). When the dimension of z_t, n, is large, sequential estimation of z_t or computation of linear optimal feedback control requires a large amount of computation (of the order of n^3 multiplications) and could be quite time consuming and expensive. For large dimensional econometric models, therefore, it is of interest to construct one or more models of smaller dimension, the state vectors of which are linearly related to z_t, and use them to construct a reasonably good suboptimal control policy for the original system since they require less computation in total. For example, data processing for two models of dimension $n/2$ would approximately require $1/4$ of computation of the original model.

period analysis is also recognizable as an approximation method using aggregation by dynamic responses.

In constructing a macroeconomic model we may, for example, assume that markets for money and bonds are in equilibrium at each instant of time since responses of money markets are generally much faster than those in real sectors. This is equivalent to assuming that the time responses of markets for various paper assets as infinite and ignoring very short run fluctuations in economic variables. Then economic relations in the paper asset markets will hold as algebraic relations rather than as dynamic relations. This may significantly simplify analysis of models. The models become dynamic, for example, through investment equations that are not instantaneous, money stocks that grow over time, prices and real outputs that are slowly changing and/or growing labor population. Equilibrium growth models of economy usually are constructed in this manner.

Given a dynamic model of the economy and given a method for aggregating economic variables, we can ask how "best" to construct dynamic models of reduced dimension. The "best" criteria for aggregation depends on the intended use of the aggregated models. In some cases it may be to minimize the forecasting error of some endogenous variables using the aggregated models. In other cases it may be to minimize the differences in some performance index of the economy using control rules devised by using large economic models and control rules derived by using the aggregated models. This is the problem of choosing T in the diagram below. Some aspects of these problems have been discussed in Aoki (1971). Another question for which we need answers is the "best" way to aggregate i.e., how to choose T in the diagram. This question is still largely unresolved in dynamic models, although there are answers in the static context of aggregation (Chipman 1975, Fisher 1969).

These ideas are conveniently illustrated by the following diagram that commutes:

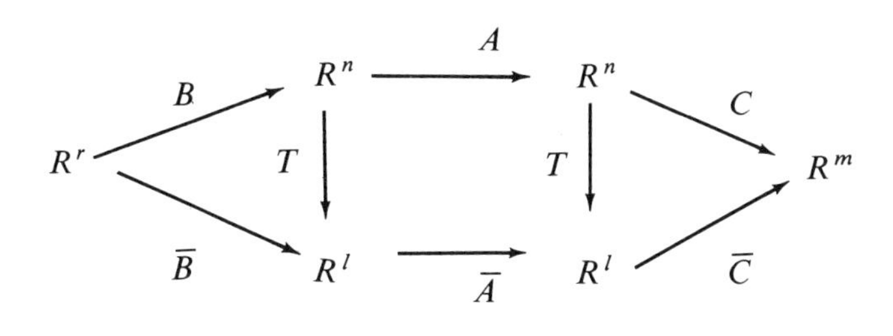

where

$$\dot{z} = Az + Bx, \qquad \dim z = n$$

$$y = Cz$$

is the original dynamic model and

$$\dot{s} = \bar{A}s + \bar{B}x \qquad \dim x = l < n$$
$$y = \bar{C}s$$

is the aggregated dynamic model, and where T in

$$x = Tz$$

is the aggregation matrix.

Let $\mathcal{X}$ be the largest A-invariant subspace in the null space of T. Then R^l may be taken to be $R^n/\mathcal{X}$, where $l = n - \dim \mathcal{X}$. We do not consider here the question of how to aggregate to minimize information loss in any generality. As usual what we can say about aggregating nonlinear relations are limited. These are isolated examples that are not easily generalizable. What we propose to discuss in this section is linear aggregation to give a systematic procedure of constructing an aggregated linear dynamic model with a given aggregation mapping and explore its potential use in estimation and control of original larger dynamic economic models.

10.3.1 State Vector Estimation via Aggregation

Given the dynamics of the economy and its measurement equation (10.1-1), we consider using an aggregated model of (10.1-1) to estimate the state vector z_t. Later we consider using two or more aggregated models of (10.1-1) to estimate z_t. These estimation schemes are of interest for linear economic models with a large number of variables since they are computationally less demanding than the estimates by the Kalman filter and are reasonably good when observation noises are "small."

Let T_t be an aggregation matrix so that

$$s_t = T_t z_t + \epsilon_t \tag{10.3-1}$$

where s_t is an aggregated state vector of z_t and ϵ_t is the error of aggregation. The variables z_t are inaccessible and s_t are available.

Let $\dim s_t = l$. We take $l = \dim z_t - \dim y_t$ to be specific. Suppose the aggregated state vector s_t is governed by the dynamic equation

$$s_{t+1} = F_t s_t + G_t x_t + D_t y_t \tag{10.3-2}$$

where F_t is an $l \times l$ matrix.

For (10.3-2) to be consistent with (10.3-1), it is necessary that the matrices F_t, G_t and D_t in (10.3-2) satisfy the relations in (10.3-3), that are obtained from

$$\begin{aligned} &T_{t+1}(A_t z_t + B_t x_t + \xi_t) + \epsilon_{t+1} \\ &\qquad = F_t(T_t z_t + \epsilon_t) + G_t x_t + D_t(C_t z_t + \eta_t) \end{aligned}$$

for all z_t and x_t, or

$$\begin{aligned} T_{t+1}A_t &= F_t T_t + D_t C_t \\ T_{t+1}B_t &= G_t \end{aligned} \tag{10.3-3}$$

and the aggregation error propagates with time according to the dynamic equation

$$\epsilon_{t+1} = F_t \epsilon_t + D_t \eta_t - T_{t+1}\xi_t \tag{10.3-4}$$

The equations (10.3-3) and (10.3-4) are also sufficient for (10.3-1) and (10.3-2) to be an aggregation. To see this, we have from (10.3-1) and (10.3-3)

$$\begin{aligned} s_{t+1} - T_{t+1}z_{t+1} &= F_t s_t + G_t x_t + D_t(C_t z_t + \eta_t) \\ &\quad - T_{t+1}(A_t z_t + B_t x_t + \xi_t) \\ &= F_t(s_t - T_t z_t) + D_t \eta_t - T_{t+1}\xi_t \end{aligned}$$

or

$$\epsilon_{t+1} = F_t \epsilon_t + D_t \eta_t - T_{t+1}\xi_t \tag{10.3-5}$$

We now consider choosing $\{T_t\}$ to minimize the error of estimating z_t. Write the first equation of (10.3-3) as

$$T_{t+1}A_t = (D_t \quad F_t)\begin{pmatrix} C_t \\ T_t \end{pmatrix} \tag{10.3-6}$$

We impose a condition that T_t is chosen to make $\begin{pmatrix} C_t \\ T_t \end{pmatrix}$ nonsingular. Let

$$\begin{pmatrix} C_t \\ T_t \end{pmatrix}^{-1} = (V_t \quad P_t) \tag{10.3-7}$$

where P_t is an $n \times l$ matrix and V_t is an $n \times m$ matrix, $l = n - m$. From (10.3-6) and (10.3-7) the matrices D and F in (10.3-2) are expressible as

$$\begin{aligned} F_t &= T_{t+1}A_tP_t \\ D_t &= T_{t+1}A_tV_t \end{aligned} \tag{10.3-8}$$

since

$$(D_t \quad F_t) = T_{t+1}A_t(V_t \quad P_t)$$

In other words, a choice of T_{t+1} determines the matrices in the dynamic equation for the aggregated state vector (10.3-2) uniquely, since G_t is determined also uniquely by T_{t+1}.

Let

$$E_t = \operatorname{cov} \epsilon_t$$

From (10.3-5), it propagates with time according to

$$E_{t+1} = T_{t+1}\Omega_t T_{t+1}{}^{\mathrm{t}} \tag{10.3-9}$$

where

$$\Omega_t = A_tP_tE_tP_t{}^{\mathrm{t}}A_t{}^{\mathrm{t}} + A_tV_tR_tV_t{}^{\mathrm{t}}A_t{}^{\mathrm{t}} + Q_t \tag{10.3-10}$$

and where we assume

$$\begin{aligned} &E\eta_t\eta_\tau{}^{\mathrm{t}} = R_t\delta_{t\tau} \\ &E\xi_t\xi_\tau{}^{\mathrm{t}} = Q_t\delta_{t\tau} \\ &E\xi_t\eta_\tau{}^{\mathrm{t}} = 0, \qquad \text{for all } t, \tau \end{aligned}$$

Augmenting the measurement vector y_t by the aggregated state vector s_t, we obtain an equation that can be used to estimate z_t,

$$\begin{pmatrix} y_t \\ s_t \end{pmatrix} = \begin{pmatrix} C_t \\ T_t \end{pmatrix} z_t + \begin{pmatrix} \eta_t \\ \epsilon_t \end{pmatrix} \tag{10.3-11}$$

or we estimate z_t by

$$\hat{z}_t = \begin{pmatrix} C_t \\ T_t \end{pmatrix}^{-1} \begin{pmatrix} y_t \\ s_t \end{pmatrix}$$

From (10.3-7), the estimate can be written as

$$\hat{z}_t = (V_t \quad P_t)\begin{pmatrix} y_t \\ s_t \end{pmatrix} \tag{10.3-12}$$

The estimation error is expressible from (10.3-11) as

$$\begin{aligned}\tilde{z}_t &= \hat{z}_t - z_t \\ &= -(V_t \quad P_t)\begin{pmatrix} \boldsymbol{\eta}_t \\ \boldsymbol{\epsilon}_t \end{pmatrix}\end{aligned}$$

Let $Z_t = \text{cov}\,(\tilde{z}_t)$. Then the estimation error covariance is related to the covariance matrices of the observation noise and the aggregation error and to the aggregation matrix T_{t+1} by

$$Z_t = (V_t \quad P_t)\begin{pmatrix} R_t & 0 \\ 0 & E_t \end{pmatrix}(V_t \quad P_t)^{\mathrm{t}} \tag{10.3-13}$$

since we see from (10.3-5) that $\boldsymbol{\epsilon}_t$ is uncorrelated with $\boldsymbol{\eta}_t$.

We now introduce some simplifying assumptions without loss of generality in order to exhibit an optimal aggregation matrix explicitly. First, we may assume that C_t in (10.1-1) is of the form

$$C_t = (I_m \quad 0)$$

To see this, we note that we can partition C_t as $C_t = (C_t^1 \quad C_t^2)$ where C_t^1 is a nonsingular $m \times m$ matrix by renumbering the components of the state vector if necessary. Then by transforming z_t by a nonsingular matrix to a new state vector $\boldsymbol{q}_t$ by

$$z_t = \begin{bmatrix} (C_t^1)^{-1} & -(C_t^1)^{-1} C_t^2 \\ 0 & I_l \end{bmatrix} \boldsymbol{q}_t$$

we can write the observation equation as

$$\begin{aligned} y_t &= C_t z_t + \boldsymbol{\eta}_t \\ &= (I_m \quad 0)\boldsymbol{q}_t + \boldsymbol{\eta}_t \end{aligned}$$

Secondly, we may assume T_t to be of the form $T_t = (K_t \quad I_l)$. This is because C_t with T_t must form a nonsingular matrix by choice, which necessitates that T_t^2 in $T_t = (T_t^1 \quad T_t^2)$ be nonsingular. Again by a nonsingular transformation of the state vector we can take $K_t = T_t^1(T_t^2)^{-1}$. In this form, the choice of the aggregation matrix amounts to that of the matrix K_t. Thus

$$\begin{pmatrix} C_t \\ T_t \end{pmatrix}^{-1} = \begin{pmatrix} I_m & 0 \\ -K_t & I_l \end{pmatrix}$$

or from (10.3-7)

$$P_t = \begin{pmatrix} 0 \\ I_l \end{pmatrix} \quad \text{and} \quad V_t = \begin{pmatrix} I_m \\ -K_t \end{pmatrix}$$

In the coordinate system in which C_t and T_t take the assumed simplified form, (10.3-12) shows that $\hat{w}_{1t} = y_t$, $\hat{w}_{2t} = -K_t y_t + s_t$ where w_t is the state vector in this new coordinate system.

Using these special structures of P_t and V_t, and from (10.3-13)

$$\begin{aligned} Z_{11,t+1} &= R_{t+1}, \qquad Z_{12,t+1} = R_{t+1}K_{t+1}{}^{\mathrm{t}} \\ Z_{22,t+1} &= -K_{t+1}R_t E_{t+1} + K_{t+1}R_{t+1}K_{t+1}{}^{\mathrm{t}} \end{aligned} \qquad (10.3\text{-}13')$$

Partitioning Ω_t of (10.3-10) comformably, we see from (10.3-9) that

$$E_{t+1} = K_{t+1}\Omega_{11t}K_{t+1}{}^{\mathrm{t}} + K_{t+1}\Omega_{12t} + \Omega_{21t}K_{t+1}{}^{\mathrm{t}} + \Omega_{22t}$$

Take the trace of Z_{t+1} in (10.3-13). It is a quadratic function of K_{t+1},

$$\begin{aligned} \operatorname{tr} Z_{t+1} &= \operatorname{tr} R_{t+1} \\ &\quad + \operatorname{tr}\{K_{t+1}(\Omega_{11t} + R_{t+1})K_{t+1}{}^{\mathrm{t}} + K_{t+1}\Omega_{12t} + \Omega_{21t}K_{t+1}{}^{\mathrm{t}} + \Omega_{22t}\} \end{aligned}$$

The best choice of the aggregation matrix T_{t+1} is therefore determined by

$$\min_{K_{t+1}} \operatorname{tr} Z_{t+1}$$

or

$$K_{t+1}{}^{*} = -\Omega_{21t}(\Omega_{11t} + R_{t+1})^{-1}$$

This determines the optimal T_{t+1}. The dynamics of this optimal aggregation model of the economy are therefore given by (10.3-2), where from (10.3-8)

$$\begin{aligned} F_t &= A_{22t} + K_{t+1}A_{12t} \\ D_t &= A_{21t} - A_{22}K_t + K_{t+1}(A_{11t} - A_{12t}K_t) \end{aligned}$$

The aggregation model (10.3-2) is known as the observer in engineering literature Luenberger (1971). The estimate of z_t given by (10.3-12), $\hat{z}_t = V_t y_t + P_t s_t$, is not a truly optimal estimate in the sense of the conditional expectation of z_t given $y_0, \ldots, y_t$ generated via the Kalman filter, since for example, the estimate uses only the current values of y_t and s_t not the past values. However, it is much simpler computationally and it is known that $\hat{z}_t$ obtained via the aggregation model is exact when observation noise is absent (Leondes and Novak, 1972).

For an alternate characterization of the aggregation of a dynamic economic model, see the Appendix at the end of this chapter.

10.3.2 Use of Two or More Aggregated Macroeconomic Models

It is often useful to employ two or more aggregated macroeconomic models in studying a large economic model. These aggregated models represent, so to speak, different aspects of the "true" economy represented by a disaggregated dynamic model.

Suppose a disaggregated model is given as

$$
\begin{aligned}
z_{t+1} &= A_t z_t + B_t x_t + \xi_t \\
y_t &= C_t z_t + \eta_t
\end{aligned}
$$

Typically, n the dimension of the state vector is in the order of several hundreds or more. Computations for optimal estimates or control laws require that we solve nonlinear difference equations of Riccati type discussed in §5.3 and in §10.1.

We consider using two aggregated models with the aggregation matrix T_t and S_t,

$$
\begin{aligned}
s_{t+1} &= F_t s_t + G_t x_t + D_t y_t \\
s_t &= T_t z_t + \zeta_t
\end{aligned}
\tag{10.3-14}
$$

and

$$
\begin{aligned}
w_{t+1} &= K_t w_t + L_t x_t + N_t y_t \\
w_t &= S_t z_t + n_t
\end{aligned}
\tag{10.3-15}
$$

From (10.3-3), the matrices in (10.3-14) and (10.3-15) are related by

$$
\begin{aligned}
T_{t+1} A_t &= F_t T_t + D_t C_t \\
T_{t+1} B_t &= G_t \\
S_{t+1} A_t &= K_t S_t + N_t C_t \\
S_{t+1} B_t &= L_t
\end{aligned}
$$

From the system equation and (10.3-14) and (10.3-15), z_t which is unknown, is related to y_t, s_t and w_t which are known by

$$\begin{pmatrix} y_t \\ s_t \\ w_t \end{pmatrix} = \begin{pmatrix} C_t \\ T_t \\ S_t \end{pmatrix} z_t + \begin{pmatrix} \eta_t \\ \zeta_t \\ n_t \end{pmatrix}$$

From this we can obtain an estimate of z_t if C_t, T_t and S_t are chosen to make the stacked matrix nonsingular. This estimate will generally be suboptimal. However, it is much simpler to compute and when there is no observation noise or aggregation error, it will be optimal.† We could use the estimate of z_t thus obtained to control the economy by

$$x_t = \Pi_t \hat{z}_t$$

where Π_t is a feedback gain matrix and where

$$\hat{z}_t = \begin{pmatrix} C_t \\ T_t \\ S_t \end{pmatrix}^{-1} \begin{pmatrix} y_t \\ s_t \\ w_t \end{pmatrix}$$

The importance of using such a suboptimal scheme lies in the saving of computation that results when n is very large, since the computation of optimal estimates requires computation which is roughly proportional to n^3. Thus, if dim s_t = dim w_t = $n/2$, approximately only a quarter of the computation will be required to construct $\hat{z}_t$ since $2(n/2)^3 = n^3/4$.

See Aoki (1971), for further details of such construction of suboptimal control laws via aggregated models. Much more work remains to be done to take advantages offered by aggregation models. For example, using the framework set up here we can discuss how to combine control laws generated using two or more aggregated models to improve accuracy of control and other related topics.

APPENDIX 10-A: AGGREGATION

A geometric treatment‡ of the concept of aggregation is given here. Heuristically, we regard a disaggregated model of the economy as generat-

---

†If there are no aggregation errors and if $\begin{pmatrix} T_t \\ S_t \end{pmatrix}$ is nonsingular, we can use just s_t and w_t in recovering z_t exactly.

‡Based in part on Bhattacharyya *et. al.* (1972).

ing a time path in an n-dimensional Euclidean space. We think of an aggregated model as describing a projection of the time path onto a suitably chosen subspace. We formalize this geometric concept by the following requirements.

System models obtained by aggregation can be characterized by the subspace $\mathcal{S}$ with the property that

(1) if $s_0 \in \mathcal{S}$, then $s(t) \in \mathcal{S}$ for all $t \geqslant 0$
(2) if $z_0 \notin \mathcal{S}$, then $|z(t) - s(t)| \to 0$ as $t \to \infty$

Property (1) means that the trajectory of the aggregated system model remains in $\mathcal{S}$, hence the model is of lower dimension than the system. In the presence of state feedback $x = Fz$ this requires that $(A + BF)\mathcal{S} \subset \mathcal{S}$ for some F, i.e., $\mathcal{S}$ is (A, B)-invariant (Wonham and Morse 1970).

The aggregation matrix is then a matrix representation of the projection map: $E^n \to \mathcal{S}$.

The induced dynamics $\bar{A} : E^n/\mathcal{S} \to E^n/\mathcal{S}$ are the error dynamics associated with this aggregation model, where $E^n/\mathcal{S}$ is the quotient space.

Property (2) expresses the requirement that the aggregation error goes to zero asymptotically. Therefore, $(\bar{A}, \bar{B})$ of the error dynamics has to be stabilizable.

Given A, B and a subspace $\mathcal{V} \subset E^n$, consider a class of feedback matrices F such that

$$\underline{F}(\mathcal{V}) = \{F|(A + BF)\mathcal{V} \subset \mathcal{V}\}$$

If we assume $\underline{F}(\mathcal{V}) \neq \varnothing$, and choose $F_0 \in \underline{F}(\mathcal{V})$, then $\mathcal{V}$ is $(A + BF_0)$-invariant. Let P be the canonical map $P : E^n \to E^n/\mathcal{V}$, so that Px is in the coset of $x(\mathrm{mod}\ \mathcal{V})$. We use P also to denote its matrix representation.

For any linear map $\bar{F} : E^n/\mathcal{V} \to E^r$, let

$$x = \left(F_0 + \bar{F}P\right)z$$

Then

$$\left(A + BF_0 + B\bar{F}P\right)\mathcal{V} = (A + BF_0)\mathcal{V} + B\bar{F}P\mathcal{V} \subset \mathcal{V}$$

We have the following proposition from Bhattacharyya *et. al.*

Proposition *If* $z_0 \in \mathcal{V}$, *then* $z(t) \in \mathcal{V}$ *for all* $t \geqslant 0$, *with* $x = (F_0 + \bar{F}P)z$ *for all* $F_0 \in \underline{F}(\mathcal{V})$ *and all* $\bar{F} : E^n/\mathcal{V} \to E^r$.

The trajectory of the system is therefore confined in $\mathcal{V}$ if the system initial condition is in $\mathcal{V}$ if the state feedback signal $x = (F_0 + \bar{F}P)z$ is applied to the system, and if the appropriate dynamics are $(A + BF_0)|\mathcal{S}$, i.e., the restriction of $(A + BF_0)$ to $\mathcal{S}$.

If $z_0 \not\subset \mathcal{V}$, then the trajectory does not in general remain in $\mathcal{V}$. In this case we require that the trajectory approaches $\mathcal{V}$ asymptotically.

Since $\bar{z} = Pz$ obeys the induced dynamic equation

$$\dot{\bar{z}} = \left(\overline{A + BF_0} + \bar{B}\bar{F}\right)\bar{z}$$

where

$$\bar{B} = PB$$

the requirement of asymptotic behavior of the trajectory becomes that of finding $\bar{F}$ such that $\overline{A + BF_0} + \bar{B}\bar{F}$ is stable. This can be done if $(\overline{A + BF_0}, \bar{B})$ is a stabilizable pair.

11

Random Multiplier—An Example of Control of A Stochastic Macroeconomic Model†

11.1 INTRODUCTION

In this chapter we continue our discussion of control of stochastic economic systems which began in Chapter 10. There seems to be considerable interest in applying, or in assessing the applicability of stochastic control techniques to econometric models, for the purpose of forecasting or planning economic activities. The models may range from single equation models to large scale macroeconomic models of national economies (Tinsley *et al.*, 1974). Parameters in these models are usually imperfectly known and must be estimated. In some cases parameters may vary randomly over time. In other words, random variables multiply some endogenous variable rather than appear additively in these economic dynamic models.

Much of the existing stochastic control literature is concerned with models in which random disturbances are modeled as additive disturbances. Control rules resulting from these two different specifications of stochastic disturbances are usually and sometimes substantially different. Caution is necessary in choosing stochastic specifications, since control rules based on models which incorrectly specify stochastic disturbances may destabilize systems rather than stabilize them. Although it is sometimes not clear, a priori, whether additive random disturbances are more plausible than multiplicative ones, we must understand the implications of multiplicative random disturbances in stochastic control problems.

We examine control of an economic model in which multiplicative random disturbances appear naturally. This model arises when we consider short-run control of an economy by a monetary instrument. Most existing treatment of monetary macroeconomic models have additive rather than

†Based in part on Aoki (1975b).

multiplicative stochastic disturbances under the assumption that the monetary authority can control the money supply and/or the interest rate exactly. In other words, the money supply and/or some interest rates are taken as the instruments of the monetary authority. In reality, they are not the instruments† (Andersen 1965, Brunner 1969, Federal Reserve Bank of Boston 1970, Gramlich 1971, Waud 1973). We treat the net source base as the instrument. As we show in the next section, it is related to the money supply via a multiplier that is randomly varying. There are many other examples in which multipliers should be treated as random. The model we discuss here illustrates how to treat such multiplicative random disturbances in a linear dynamic macroeconomic model.

11.2 RANDOM MONEY MULTIPLIER

Since we merely wish to illustrate how a multiplicative random variable may appear in monetary control problems rather than provide an accurate description of the institutional setup of monetary control mechanisms by central banks (The Federal Reserve System in the U.S.), we provide only enough details of the money supply process to suit our primary objective. Loosely speaking, the Federal Reserve changes the amount of money in the economy through its open market operations and by changes in the reserve requirements, the discount rate and ceilings on the interest rates (Regulation Q). The open market operations modify the amount of money by buying or selling government securities. The part of money thus directly affected by the open market operation is called the net source base. (See Burger, 1971, for a more detailed and accurate description of the net source base.)

We take the net source base to be the variable which is directly controlled by the monetary authority. Then the stock of money M_1 composed of the cash and the demand deposits held by the public is expressible by‡

$$M = mB$$

†A great deal of controversy exists regarding choices of appropriate monetary policy variables, and proximate target variables. Some argue for control of some aggregate reserve measures such as total reserves, monetary base, etc., while others argue for control of variables that reflect money and capital market conditions such as free reserves.

‡We drop the subscript from M_1 since it is the only kind of money stock discussed here.

where B stands for the net source base. The multiplicative factor m is called the (money) multiplier.

The multiplier m is expressible as

$$m = \frac{1 + k}{(r - b)(1 + t + d) + k} \tag{11.2-1}$$

where

k: ratio of the amount of cash over the demand deposit, (the cash ratio)

t: ratio of the time deposit over the demand deposit (the time deposit ratio)

d: ratio of the deposit by the Treasury over the demand deposit (the treasury deposit ratio)

r: amount of required reserve of the banks over the demand deposit (the reserve ratio)

b: amount borrowed by the member banks over the demand deposit (borrowing ratio)

These ratios all vary with time. We drop the time subscript to simplify notations. The time subscript refers to such a basic time unit as a month or a quarter. We take it as a quarter to be definite. Various elasticities of ratios appearing in m have been derived, and regression analyses have been run on m in an effort to predict future m values (Andersen 1965, Burger 1971). We do not follow this approach. Instead, we examine the nature of $r - b$ (the unborrowed reserve ratio) and obtain a formula for predicting m_{t+1} from known values of the various ratios in m_t and a random noise term. We know that $r - b$ is expressible as (see Burger 1971)

$$\begin{aligned} r - b &= r^d \delta u + r^t \tau (1 - u) + e + v - b \\ &= \alpha_0 + \alpha_1 u + f + v \end{aligned} \tag{11.2-2}$$

where

$u = \dfrac{1 + d}{1 + t + d}$

v: vault cash

r^t: required reserve ratio on time deposits

r^d: required reserve ratio on demand deposits

e: excess reserve ratio

δ = (member bank demand deposit)/(total bank demand deposit)

τ = (member bank time deposit)/(total bank time deposit)

$f = e - b$: free reserve ratio

$\alpha_0 = r^t \tau$ and $\alpha_1 = r^d \delta - r^t \tau$

The factors δ and τ are necessary because member and nonmember banks of the Federal Reserve System have different required reserve ratios.

It seems reasonable to assume that δ and τ do not change as much or as rapidly as the other ratios. Due to the recent change in regulations governing member banks, the excess reserve ratio may be considered very small. We assume that changes in u are known more accurately and timely than those in f or v.

We assume that the vault cash ratio and the free reserve ratio respond in an unpredictable manner and represent them by a sequence of random variables $\{\xi_t\}$. Therefore, we model $r - b$ as

$$r - b = \alpha_0 + \alpha_1 u + \xi \tag{11.2-2'}$$

where ξ is a random variable.

Thus, according to our assumption,

$$\Delta(r - b)_t = \alpha_1 \Delta u_t + \Delta\xi_t$$

i.e., the change in $(r - b)_t$ from one quarter to the next is composed of the two parts: Δu_t which is assumed to be known and $\Delta\xi_t$ which is unknown at the beginning of the tth quarter.

To lighten notation, we introduce n defined by

$$n = \frac{mk}{1 + k} \tag{11.2-3}$$

Then from (11.2-1) and (11.2-3), we have

$$\frac{\Delta n}{(1 - n)n} = \frac{\Delta k}{k} - \frac{\Delta(r - b)}{(r - b)} - \frac{\Delta(t + d)}{1 + t + d} \tag{11.2-4}$$

where

$$\frac{\Delta(r - b)}{r - b} + \frac{\Delta(t + d)}{1 + t + d} = \frac{\Delta d}{1 + t + d}\left(\alpha_1 \frac{t}{1 + t + d}\,\frac{1}{r - b} + 1\right) + \frac{\Delta t}{1 + t + d}\left(1 - \frac{\alpha_1}{r - b}\,\frac{1 + d}{1 + t + d}\right) + \frac{\Delta\xi}{r - b}$$

Using (11.2-2′) this last expression can be manipulated into the form

$$\frac{\Delta(r - b)}{r - b} + \frac{\Delta(t + d)}{1 + t + d} = \frac{\Delta d}{d} - \frac{\Delta u}{u} + \frac{\alpha_1 \Delta u}{r - b} + \frac{\Delta\xi}{r - b} \tag{11.2-5}$$

where from (11.2-2)

$$(1 - u)\frac{\Delta t}{t} = (1 - u)\frac{\Delta d}{1 + d} - \frac{\Delta u}{u} \qquad (11.2\text{-}5')$$

is used.

Substituting (11.2-5) into (11.2-4), and from (11.2-3) we see that m_{t+1} is given by

$$\ln\frac{m_{t+1}}{m_t} = \Delta \ln \left.\frac{(1 + k)u}{1 + d}\right|_t^{t+1} - \alpha_1 \int \frac{\Delta u}{r - b} + \int \frac{\Delta \xi}{r - b}$$

where the integration is over one-quarter period from the tth to the $(t + 1)$st quarter. Thus we obtain

$$\ln\frac{m_{t+1}}{m_t} \cong \ln\left(\frac{\zeta_{t+1}}{\zeta_t}\right) - \alpha_1 \frac{u_{t+1} - u_t}{(r - b)_t} + \frac{\xi_{t+1} - \xi_t}{(r - b)_t} \qquad (11.2\text{-}6)$$

where $\zeta = (1 + k)u/(1 + d) = (1 + k)/(1 + t + d)$, and where expressions such as $\Delta u \cdot \Delta(r - b)$ are ignored as terms of a higher order of smallness.

From (11.2-6), we see that m_{t+1} is composed of two factors: one deterministic and known at the beginning of the $(t + 1)$th quarter and the other random and not known. From (11.2-6) we see that

$$m_{t+1} = F_{t+1}e^{\eta_{t+1}} \qquad (11.2\text{-}7)$$

where

$$F_{t+1} = m_t\zeta_{t+1} \exp\left[-\alpha_1(u_{t+1} - u_t)/(r - b)_t\right]/\zeta_t \qquad (11.2\text{-}8)$$

and where

$$\eta_{t+1} = (\xi_{t+1} - \xi_t)/(r - b)_t$$

In the above, we assume that k_{t+1}, t_{t+1} and d_{t+1}, hence ζ_{t+1} and u_{t+1}, are known at the beginning of the $(t + 1)$th quarter, whereas m_{t+1} is not known since η_{t+1} is not known. We thus express the multiplier for $(t + 1)$th quarter as the product of two factors, one is deterministic and known and the other is not, reflecting the fact that some part of the money multiplier is predictable or becomes known at the beginning of a quarter while the rest is the forecasting error term.

We denote by $\mathcal{I}_t$ the information set, known at the beginning of $(t + 1)$th quarter. The information state of the controller at the end of the

tth quarter and at the beginning of $(t+1)$th quarter is specified by

$$\mathcal{G}_t = \{\ldots, \eta_t, \ldots, (r-b)_t, \ldots, \Delta u_{t+1}, \ldots, \xi_t\}$$

In other words, at the beginning of the tth quarter, u_t, ξ_{t-1} and $(r-b)_{t-1}$ are known.

The random variable $\xi_{t+1} - \xi_t$ is observed at the end of the $(t+1)$st quarter but is unknown at the beginning of the $(t+1)$st quarter. We assume $\{\Delta\xi_t\}$ is a sequence of zero mean independent random variables. Typically ξ is close to zero in the economic application considered in this section. We have then (see Aitchison and Brown, 1957)

$$E(\eta_{t+1}|\mathcal{G}_t) = 0, \qquad E(\eta_{t+1}^2|\mathcal{G}_t) = \sigma_t^2$$

$$E(\eta_{t+1}|\mathcal{G}_{t-1}) = 0, \qquad \text{where } \sigma_t^2 = \sigma^2/(r-b)_t^2$$

$$E(\eta_{t+1}^2|\mathcal{G}_{t-1}) = \frac{\sigma^2}{(r-b)_{t-1}^2} E\{1/(1+\eta_t+\alpha_1\Delta u_t)^2|\mathcal{G}_{t-2}\}$$

$$\cong \frac{\sigma^2}{(r-b)_{t-1}^2} E\{1 - 2\eta_t - 2\alpha_1\Delta u_t + 3(\eta_t+\alpha_1\Delta u_t)^2|\mathcal{G}_{t-2}\} \qquad (11.2\text{-}9)$$

Under the assumption that σ^2 and σ_1^2 are small compared with 1, we have

$$E(e^{\eta_t}|\mathcal{G}_{t-1}) = e^{\sigma_t^2/2} \cong 1 + \sigma^2/2(r-b)_{t-1}^2$$

and

$$E(e^{\eta_t}|\mathcal{G}_{t-2}) = e^{\sigma_{t-1}^2/2} \cong 1 + \sigma^2/2(r-b)_{t-2}^2, \text{ etc.}$$

11.3 CONTROL OF A MACROECONOMIC MODEL

Having specified the linkage between M and the control variable B (or its rate of increase), we now need a set of equations connecting the money stock to real variables. Since the objective is not to construct a realistic macroeconomic model but rather to illustrate a control problem of dynamic systems with multiplicative noise disturbance, we choose one of the simplest possible transmission mechanisms of the monetary effect to the real sector. We assume the nominal GNP at the tth quarter is related to the stock of money at the beginning of the tth quarter M_t by

$$Y_t = V_t M_t \tag{11.3-1}$$

where the velocity V_t is assumed to be a martingale (see Aoki, 1967)

$$V_t = V_{t-1} + \epsilon_{t-1} \tag{11.3-2}$$

where ϵ_t is assumed to be a sequence of independent identically distributed random variables with mean zero and known variance σ_ϵ^2. A recent statistical test seems to indicate that (11.3-2) is not statistically inconsistent with the historically observed data (Gould and Nelson, 1974).

The instrument of the central bank is taken to be the rate of change of the net source base,

$$v_t = (B_{t+1} - B_t)/B_t \tag{11.3-3}$$

From (11.3-1)–(11.3-3), the difference equation governing Y_t is given by

$$Y_{t+1} = (Y_t + \epsilon_t M_t)\theta_t(1 + v_t) \tag{11.3-4}$$

where

$$\theta_t = m_{t+1}/m_t = f_t e^{\eta_{t+1}}$$

Note that Y_t and M_t are known at the end of the tth quarter.

We can either assume that u_t is set at the beginning of each quarter and the measurement on M_t is available each quarter, or u_t is chosen each week, say, and the money stock between quarters must be estimated. We assume the former and do not discuss estimating M between quarters here, although there is no theoretical difficulty in doing so.

Equation (11.3-4) is the dynamic equation of our model. The multiplicative disturbance is in θ_t as shown above. It appears that this type of equation has not been explicitly treated in the stochastic control literature. What is available in the literature is on the system

$$z_{t+1} = A_t z_t + m_t v_t$$

where v_t is the control.

When $\{m_t\}$ is taken to be a sequence of independent random vectors, the control problem is a special case considered by Joseph and Tou (1961),

Gunckel and Franklin (1963). Systems with imprecisely known constant vector gains are considered by Aoki (1967), Murphy (1968). The systems with dependent gains were considered by Bar-Shalom and Sivan (1969), Tse and Athans (1972) have considered a special case of (11.3-1) in which the gain vectors are generated by a known dynamic system with additive noise. Then they could reduce their problem to a stochastic control problem with an augmented state vector. In all cases, suboptimal controls using open-loop feedback control policies were investigated. Control problems with random gain may also be regarded as the ones with control dependent noise where randomness is introduced by an additive term. Kleinman (1969), MacLane (1971) and Wonham (1967) consider continuous-time stochastic control problems with state and control dependent plant noises and with exact state vector measurements. Alspach (1973) considered discrete-time systems with additive noise in the gain and noisy state measurements. In all these previous works, random elements in the gains are introduced as additive disturbances.

Here m_t contains a multiplicative random variable,

$$m_t = F_t e^{\eta_t}$$

with known conditional mean and variance. They depend endogenously on various ratios we introduced in connection with (11.2-1).

We now return to our model (11.3-4). The problem is to find a sequence of controls to minimize the expected value of a cost function J. For a special case of J, we derive the optimal closed loop control. In general, however, we cannot obtain optimal closed loop policies. Some approximations, such as the ones based on open-loop feedback control laws must be employed (Aoki 1967, Dreyfus 1964, Tse and Athans 1972). We assume that z_t and v_t are observed exactly.

11.3.1 Simple Optimization Problem

To illustrate potential effects of random multipliers, let us consider a control problem of keeping Y_t close to the desired level Y^* over the horizon of N, $N \geqslant 1$, quarters. As stated in the Appendix, during this period, reserve requirements, discount rate, etc., are assumed not to have changed. We use a quadratic loss function $J = \Sigma_{t=1}^{N}(Y_t - Y^*)^2$ for ease of presentation.

We illustrate the procedure first for $N = 1$ using (11.3-4). We show later that the monetary policy for $N > 1$ reduces to a sequence of the one-period optimal controls in which at each t, $E((Y_{t+1} - Y^*)^2 | \mathcal{G}_t)$ is minimized with respect to v_t. Let

$$\alpha_t = E(\theta_t|\mathcal{I}_t) = E(f_t e^{\eta_{t+1}}|\mathcal{I}_t) = f_t e^{\sigma_t^2/2}$$
$$\beta_t^2 = \text{var}(\theta_t|\mathcal{I}_t) = f_t^2 e^{\sigma_t^2}(e^{\sigma_t^2} - 1) \tag{11.3-5}$$

From (11.3-4), and (11.3-5)

$$E\left[(Y_{t+1} - Y^*)^2|\mathcal{I}_t\right] = \left[Y_t\alpha_t(1 + v_t) - Y^*\right]^2 + Y_t^2\beta_t^2(1 + v_t)^2 + \sigma_\epsilon^2 M_t^2(\alpha_t^2 + \beta_t^2)(1 + v_t)^2$$

Minimizing the above with respect to v_t, the optimal one-stage control is given by

$$1 + v_t^0 = \frac{\alpha_t}{\alpha_t^2 + \beta_t^2} \cdot \frac{Y_t Y^*}{Y_t^2 + \sigma_\epsilon^2 M_t^2} \tag{11.3-6}$$

This is the optimal rate of increase of the net source base.

The second factor in (11.3-6) is seen to be the control that would be optimal if the multiplier were a fixed constant, i.e., the rate of increase of the money stock is $\alpha_t^2/(\alpha_t^2 + \beta_t^2)$ of the one when the multiplier is a fixed constant.† This is to be expected.

The minimal one stage cost is given as

$$(Y^*)^2\left\{1 - \frac{\alpha_t^2}{\alpha_t^2 + \beta_t^2} \cdot \frac{Y_t^2}{Y_t^2 + \sigma_\epsilon^2 M_t^2}\right\}$$

When Y_t and M_t are expressed in terms of Y_{t-1}, M_{t-1}, θ_{t-1}, ϵ_{t-1} and v_{t-1}, and the expectation with respect to ϵ_{t-1} is taken, we see that

$$E\left[Y_t^2/(Y_t^2 + \sigma^2 M_t^2)\right] = Y_{t-1}^2/(Y_{t-1}^2 + \sigma^2 M_{t-1}^2)$$

and it is independent of v_{t-1}.

Thus, the one stage optimal policy is the optimal policy for the multistage problem.

†From (11.3-5), $\alpha_t/(\alpha_t^2 + \beta_t^2) = e^{-(3/2)\sigma_t^2}/f_{t+1}$.

11.4 DISCUSSION

The purpose of this chapter is to outline in as simple a setting as possible the way multiplicative stochastic disturbances may be integrated into the control theory of such dynamic systems. We have therefore avoided many real complications such as uncertain system parameters and the associated parameter estimates and parameter learning problems.

There, the conclusion of §11.3 is not to be taken literally. Full treatments of the control problems incorporating uncertain parameters and multiplicative disturbances are not available in the literature and must be developed utilizing many of the techniques discussed in this book. Some preliminary simulation results seem to validate the theoretical prediction: The larger α_t^2, the poorer the performance of optimal controls derived from the model with additive noises when the controls are applied to the system with multiplicative noises. See Aoki (1975b) for some simulation results.

12

Price and Quantity Adjustment Schemes

12.1 PRICE ADJUSTMENTS AND COBWEB THEOREM

We now turn our attention to price and quantity adjustment decisions of economic agents, such as firms, made mostly under imperfect information. Our purpose is not to give a comprehensive survey of results available on decision-making under imperfect information or on search behavior of economic agents in imperfect information situations. See, for example, Phelps (1970) or Rothschild (1973) for some recent references on these subjects.

Our purpose is, rather, to discuss selectively those types of price and quantity adjustment behavior of economic agents out of equilibrium that can usefully be discussed by system or control-theoretic techniques we have developed earlier in this book. We will emphasize the role of information structures or patterns in decision making. Different information set generally induces different behavior on economic agents. We will also examine suboptimal behavior and parameter-learning or parameter-adaptive aspects of their decision-making processes.

We begin by discussing the familiar cobweb theorem under a standard set of assumptions, primarily to underscore the unrealistic nature of these assumptions. Then, we reexamine several price adjustment schemes under imperfect information assumptions.

The cobweb theorem is about the successive price adjustments of a representative supplier (firm) of a homogeneous nonstorable good. The firm decides how much of the good to produce in response to the market clearing price observed in the previous periods. Assume that demand and supply schedules are stationary so that the same schedules apply period after period. Let us call one period a market day. Assume also that the production delay is one market day, so that the supply for the tth day S_t is the result of the supply decision taken at the $(t-1)$th day. We do not wish to complicate matters at this stage by including inventory. We therefore assume that the good is nonstorable. The price is then assumed to adjust in a day to clear the market.

12.1.1 A Deterministic Market

With price P_{t-1} clearing the market on the $(t-1)$th day, we assume that the supply decision is taken by the (representative) supplier as a function of P_{t-1}, according to his supply schedule $S(\cdot)$

$$S_t = S(P_{t-1})$$

We thus assume the adjustment is made on the supply side, i.e., supply follows demand. Since the good is not storable, the price that clears the market on the tth day is determined by the intersection of the demand schedule for the good with S_t, i.e.,

$$D(P_t) = S(P_{t-1}) \tag{12.1-1}$$

where $D(\cdot)$ is the demand schedule.

The firm then takes P_t as given on the tth day and decides to supply $S(P_t)$ for the $(t+1)$th day that will be cleared at P_{t+1} on the $(t+1)$th day where P_{t+1} is determined by (12.1-1) with t replaced by $t+1$ and so on.

Equation (12.1-1) provides the difference equation for the market clearing prices. For example, suppose for the time being that we know for the purpose of analysis (even though the firm may not share our knowledge) that the demand and supply schedules are given by

$$D(P) = a - bP, \qquad b \neq 0$$

and

$$S(P) = c + dP \tag{12.1-2}$$

Then from (12.1-1) the difference equation for the market clearing price becomes

$$a - bP_t = c + dP_{t-1}$$

or

$$P_t - P_e = \alpha(P_{t-1} - P_e) \tag{12.1-2$'$}$$

where†

$$\alpha = -d/b \qquad \text{and} \qquad P_e = \frac{a-c}{b+d}$$

Equation (12.1-2′) is a first-order difference equation for P_t. By solving it we see that the price adjusts according to

$$P_t = P_e + (P_0 - P_e)\alpha^t, \qquad t = 0, 1, \ldots \tag{12.1-3}$$

Thus, P_t converges to P_e if and only if $|d/b| < 1$, or $|d| < |b|$. The variable P_e is the equilibrium price. These facts are well-known. (See for example Allen, 1967, or Baumol, 1970).

For (12.1-1) or (12.1-3) to be the correct equations governing the market clearing prices, however, we must assume that the firm is completely ignorant of the demand schedule or the parameters in the demand schedule, and it acts as if it has no memory of the past prices, i.e., we assume that it does not know $D(\cdot)$ at all or that even if it were to know that $D(\cdot)$ is linear, it still would not know a and b, nor would it learn these parameters. The firm reacts passively to the market clearing prices and moves along its own supply schedule.

Suppose that the firm knows that $D(\cdot)$ is linear for the sake of the following discussion. If the firm has even a vague idea of the value of a and b, then since the firm has some perhaps more precise knowledge of the parameters of the supply schedule c and d, the firm can or tries to predict the equilibrium market clearing price P_e and avoid many days of disappointment. In case b and d are positive $\alpha < 0$, hence the sign of $P_t - P_e$ is the opposite of that of $P_{t-1} - P_e$. This means, of course, that the firm's selling price is alternately overshooting or undershooting P_e. Actually, when the schedules are known to be linear in the absence of any exogenous disturbances, the firm can obtain α and $(a - c)/b$ from

$$P_1 = \frac{a-c}{b} + \alpha P_0$$

$$P_2 - P_1 = \alpha(P_1 - P_0)$$

In other words, the firm can compute

$$\alpha = \frac{P_2 - P_1}{P_1 - P_0}$$

†More generally we obtain (12.1-2′) by assuming that $S(\cdot)$ and $D(\cdot)$ have continuous second-order derivatives and using the Taylor series expansion.

and

$$\frac{a-c}{b} = P_1 - \alpha P_0$$

Then the firm can determine P_e to be given by

$$P_e = (a-c)/b(1-\alpha) = \left(P_1^2 - P_0 P_2\right)/\left(P_2 - P_0\right)$$

Even when the demand schedule is not linear, so long as it is representable by a finite number of parameters (a and b in the case of linear demand schedule), the firm can find P_e after a few iterations. Therefore the infinite steps of price adjustments that accompany the cobweb theorem are not economically plausible in the perfect-information deterministic world. For clear discussions on what information is tacitly assumed in the textbook discussion of the supply and demand curve cross, see Haavelmo (1974).

We reexamine the price adjustment in detail later by spelling out precisely the nature of information available to the firm and the market signals it receives in the course of price adjustments. Analysis for deterministic models has been carried out in §7.1. We therefore admit exogenous random disturbances in the model we consider next.

12.1.2 Noisy Supply Schedule

Suppose now that the actual amount of the good supplied at the tth day is disturbed by an exogenous random variable ξ_t (weather, for example). The supply on the tth day is now given as the result of supply decision taken by the representative firm, perturbed by noise

$$S_t = S(P_{t-1}) + \xi_{t-1}$$

Assume, to simplify the subsequent analysis, that ξ's are independently and identically distributed random variables with $E(\xi_t) = 0$, and finite variance.

With the disturbances, the market-clearing price P_t is determined by

$$D(P_t) = S(P_{t-1}) + \xi_{t-1} \tag{12.1-4}$$

instead of (12.1-1).

This is a stochastic difference equation in P. With the linear demand and supply schedules given by (12.1-2), the sequence of prices is now governed by a stochastic linear difference equation. Now, even if the supplier knew the demand schedule exactly, he would not be able to predict P_t exactly nor reach P_e exactly due to the random disturbances. Let us suppose first that the firm merely reacts passively the market clearing prices which are the only signals the firm receives.

The stochastic difference equation is†

$$P_{t+1} = \beta + \alpha P_t - \zeta_t \tag{12.1-5}$$

where

$$\alpha = -d/b, \qquad \beta = (a-c)/b, \qquad \zeta_t = \xi_t/b$$

We first show that even if α and β are known, $P_t \nrightarrow P_e$, $t \to \infty$ where $P_e = (a-c)/(b+d)$. As we shall see later, P_e is the mean about which P_t will be distributed as $t \to \infty$.‡

With P_e assumed known, let

$$p_t = P_t - P_e, \qquad t = 0, 1, \ldots$$

be the deviation of the market clearing price on the tth day from P_e.

Then (12.1-5) becomes

$$p_{t+1} = \alpha p_t - \zeta_t \tag{12.1-6}$$

The mean prices are governed by

$$\bar{p}_{t+1} = \alpha \bar{p}_t$$

†By considering firms' subjective profit maximization behavior, we arrive at a different supply decision by this firm that in turn generates different price sequences. For example, suppose the firm wishes to maximize profit on the tth market day that is given by

$$\pi_t = P_t(Q_t)Q_t - c(Q_t)$$

where we regard the market clearing price to be a function of Q_t to be determined by $D(P_t) = Q_t + \xi_t$ and where $c(Q_t)$ is the cost of production. Assuming the existence of the unique inverse $D^{-1}(\cdot)$, we can write

$$P_t = D^{-1}(Q_t + \xi_t)$$

Then if the firm chooses Q_t that maximizes $E(\pi_t | \mathcal{I}_t)$ where $\mathcal{I}_t$ is the information set available to this firm, it gives rise to a sequence of the market clearing prices different from that governed by (12.1-5). See the next few pages as well as Chapters 10, 13 and Aoki (1974b, 1974d, 1975d).

‡Therefore it is no longer correct to refer to P_e as "the" equilibrium price.

or

$$\bar{p}_t = \alpha^t \bar{p}_0, \qquad t = 0, 1, \ldots$$

where

$$\bar{p}_t = E(p_t)$$

where the expectation $E(\cdot)$ is taken with respect to the ζ's and the variance of the prices are given by

$$V_{t+1} = \alpha^2 V_t + \sigma^2$$

where

$$V_t = \mathrm{var}(p_t), \qquad \sigma^2 = \mathrm{var}(\zeta_t), \qquad t = 0, 1, \ldots$$

Iterating this recursion equation for variance with respect to t, we see that

$$V_t = \alpha^{2t} V_0 + \sigma^2 \frac{1 - \alpha^{2t}}{1 - \alpha^2}$$

By the Chebyshev inequality, we see that for $|\alpha| < 1$

$$\text{Probability}\ (|p_t - \bar{p}_t| > \varepsilon) \leqslant \frac{V_t}{\varepsilon^2} \to \sigma^2/\varepsilon^2(1 - \alpha^2) \neq 0$$

as $t \to \infty$, showing that $p_t \nrightarrow 0$ in probability, or $P_t \nrightarrow P_e$ in probability.

If the ζ's are assumed to be Gaussian, then the distribution function of p_t converges to $N(0, \sigma^2/(1 - \alpha^2))$ as $t \to \infty$.

In other words, with the noisy supply schedule the limiting price P_∞ is normally distributed with mean P_e and variance $\sigma^2(1 - \alpha^2)$. Here, we have analyzed the adjustment of the market clearing prices when the supplies are distributed exogenously, assuming that the firm does not know the demand schedule or does not utilize whatever information it may have on the demand schedule, but merely reacts to the market clearing price on the previous day. The price dispersion persists due to the exogenous disturbances. In the above, the firm did not use P_e. If it knows P_e, it will disregard the price variability and maintain P_e since the resulting price variance $\sigma^2/(1 - \alpha^2)$ cannot be eliminated. If it does not know P_e (because it does not know a and b, say), then P_e can be estimated by averaging all past market clearing prices as we see shortly. Then the firm will behave passively initially but switch to P_e as soon as the firm has enough data to estimate P_e with "sufficient" accuracy.

To see this, compute the average price $\hat{P}_t$ from (12.1-5).
Since

$$P_s = \alpha^s P_0 + \frac{1-\alpha^s}{1-\alpha}\,\beta - \sum_{\tau=0}^{s-1} \alpha^{s-1-\tau}\zeta_\tau$$

$$\hat{P}_t = \frac{1}{t}\sum_{s=0}^{t-1} P_s$$

$$= \frac{1}{t}\,\frac{1-\alpha^t}{1-\alpha}\,P_0$$

$$+ \frac{1}{1-\alpha}\left[\left(1 - \frac{1-\alpha^t}{t(1-\alpha)}\right)\beta - \frac{1}{t}\sum_{\tau=0}^{t-2}\frac{1-\alpha^{t-\tau-1}}{1-\alpha}\,\zeta_\tau\right]$$

As $t \to \infty$, the first term vanishes and the second term approaches $\beta/(1-\alpha) = P_e$, if $|\alpha| < 1$. We show now that the third term goes to 0 in the mean square:

$$\operatorname{var}\left[\frac{1}{t}\sum_{\tau=0}^{t-2}\frac{1-\alpha^{t-\tau-1}}{1-\alpha}\,\zeta_\tau\right] = \frac{\sigma^2}{t^2}\sum_{\tau=0}^{t-2}\left(\frac{1-\alpha^{t-\tau-1}}{1-\alpha}\right)^2$$

$$\leqslant \frac{\sigma^2}{t(1-\alpha)^2}, \qquad \text{for } |\alpha| \leqslant 1,$$

$$\to 0, \qquad \text{as } t \to \infty$$

Therefore $\hat{P}_t$ converges to P_e in the mean square and hence in probability as $t \to \infty$.

The firm can therefore use $\hat{P}_t$ as an estimate of P_e, or α and β can be estimated directly by the least square method† to compute an estimate of P_e, by

$$\min_{\alpha,\beta}\sum_{s=0}^{t-1}(P_{s+1} - \alpha P_s - \beta)^2$$

that gives the estimates

$$\hat{\alpha} = \left[t\sum_{s=0}^{t-1}P_s P_{s+1} - \left(\sum_{s=0}^{t-1}P_s\right)\left(\sum_{s=0}^{t-1}P_{s+1}\right)\right]\Big/\Delta$$

†When the ζ's are Gaussian, the estimates are also the maximum-likehood estimates.

and

$$\hat{\beta} = \left[\left(\sum_{s=0}^{t-1} P_s^2\right)\left(\sum_{s=0}^{t-1} P_{s+1}\right) - \left(\sum_{s=0}^{t-1} P_s\right)\left(\sum_{s=0}^{t-1} P_s P_{s+1}\right)\right] \Big/ \Delta$$

where

$$\Delta = t\sum_{s=0}^{t-1} P_s^2 - \left(\sum_{s=0}^{t-1} P_s\right)^2$$

Here, we are primarily interested in pointing out the absurdity of the usual cobweb price adjustments and not so much with a number of ways we could introduce imperfect information price adjustment schemes or with technical discussions on how best to estimate the imprecisely known demand schedule. Our points are the following.

If α and β of (12.1-5) or α of (12.1-6) are known, then the firm is reasonably expected to predict P_{t+1} or p_{t+1} using (12.1-5) or (12.1-6). To be definite, let us use (12.1-6).

Let $\hat{p}_{t+1}$ be the firm's estimate of p_{t+1} made at the tth market day before the firm learns p_{t+1}. The variable $\hat{p}_{t+1}$ is some (measurable) function of the information held by the firm on the tth market day. It contains the *a priori* information (values of α, β and the noise distribution function) plus the record of past market clearing prices.

Let

$$e_{t+1} = p_{t+1} - \hat{p}_{t+1}$$

be the prediction error. Then $\hat{p}_{t+1}$ that minimizes the error covariance is then

$$\hat{p}_{t+1} = \alpha p_t$$

This line of reasoning gives rise to the so-called rational expectation procedure originally discussed by Muth (1961). We shall return to it shortly.

If the firm's *a priori* information does not include α and β, and the noises in (12.1-3) prevent the firm from learning them exactly as was the case in the deterministic cobweb situation, the firm may adjust prices in a number of ways. We must discuss the firm's learning behavior together with its price-setting behavior since the price adjustments designed to reduce the uncertainty of parameters most quickly are not necessarily most *profitable*. We must introduce the utility function for the firm and consider explicitly the firm's intertemporal utility or profit maximization problem to decide on the best price adjustment scheme. We shall then see that the firm must set prices to perform the dual task† of learning (imprecisely known parameters) and of maximizing the utility or profit, since these two objectives are often contradictory.‡ These topics will be discussed fully later.

12.1.3 Rational expectation

We first explain the approach of rational expectation, then show how unrealistic some of the underlying assumptions still are.

Suppose now that the firm displays a rudimentary intelligence and decides to predict the market clearing price on the next day and decides to supply $c + d\hat{P}_t$ that causes the amount given by

$$S_t = c + d\hat{P}_t + \xi_t$$

to be made available to the market on the tth day, where $\hat{P}_t$ is the price that the firm expects to prevail on the tth market period, given all the information available to the supplier at the $(t - 1)$th period, $\mathcal{I}_{t-1}$, i.e.,

$$\hat{P}_t = E(P_t | \mathcal{I}_{t-1})$$

where

$$\begin{aligned} \mathcal{I}_{t-1} &= \{\textit{a priori} \text{ information}, P_0, P_1, \ldots, P_{t-1}\} \\ &= \{\textit{a priori} \text{ information}, P_0, \xi_0, \ldots, \xi_{t-1}\} \end{aligned}$$

†These dual functions that the control variables (price may be considered as the control variable in the problem we are discussing) perform have been known in control literature since the publication of Fel'dbaum's paper (1960, 1961), Sworder (1966), Aoki (1967) and have been extended and elaborated since then (Tse *et al.* 1972, 1973).

‡As was recently pointed out by Rothschild (1973) perhaps one of the simplest examples to illustrate the dual functions that decision policies perform in imperfect information situations is the two-armed bandit problem: There are two slot-machines, Machine I with a known payoff probability s and Machine II an imprecisely known payoff probability r with a known *a priori* probability distribution of r. The objective is to maximize a discounted sum of total payoff. Clearly, if it is desired to learn r most quickly, then Machine II only will be played initially until it becomes reasonably clear that $s \geqslant r$. Clearly, however, this is not payoff maximizing.

Then $\{P_t\}$ is governed by

$$a - bP_t = c + d\hat{P}_t + \xi_t$$

The *a priori* information assumed to be available to the supplier in this instance includes such things as the knowledge of the equation for $\{P_t\}$, and the probability distribution of ξ's. Then, since P_e is known we have

$$p_t = \alpha\hat{p}_t - \zeta_t \qquad (12.1\text{-}7)$$

instead of (12.1-6) where $p_t = P_t - P_e$ as before. If we take $E(\cdot|\mathcal{I}_{t-1})$ of both sides, we see

$$(1 - \alpha)\hat{p}_t = 0$$

because $\hat{p}_t = \hat{P}_t - P_e$, or

$$\hat{p}_t = 0,$$

the expected price therefore equals the equilibrium price. Thus, the actual market clearing price at the tth market day is determined by the random variable ζ_t if $\alpha \neq 1$,

$$p_t = -\zeta_t \qquad (12.1\text{-}8)$$

Equation (12.1-8) is the result of several unrealistic assumptions about the behaviors of the firms. If several nonidentical firms, rather than the representative firm, supply the same good to the market and if their *a priori* information sets are incomplete and different (for example, each firm's subjective estimate of the demand schedule may be different), then there is no reason to believe that the estimated prices by the firms are identical and (12.1-8) does not hold. We examine this case later. Even if the representative firm knows P_e to be $(a - c)/(b + d)$ and we set $\hat{P}_t = P_e$ for $t = 0, 1, \ldots,$ if it does not know the noise variance it may take time before the firm is convinced that the actual price deviation from P_e that it observes is indeed statistically explainable as due solely to the exogenous noise, and not because $\hat{P}_t$ is not correctly set at P_e. In short, the rational expectation assumption is still unrealistic since it implies too much knowledge on the part of the firms. This point is later illustrated in greater detail.

12.1.4 Adaptive expectation

Since $E(p_t|\mathcal{I}_{t-1})$ may be difficult to compute exactly (e.g., due to the unknown demand function), suppose the firm revises the expected price $\hat{p}_t$

according to

$$\hat{p}_t = \hat{p}_{t-1} + \lambda_t(p_{t-1} - \hat{p}_{t-1}) \tag{12.1-9}$$

with some λ_t. Substituting (12.1-7) into (12.1-9) under the assumption that (12.1-7) is known to the firm, $\hat{p}_t$ is governed

$$\begin{aligned}\hat{p}_t &= \hat{p}_{t-1} + \lambda_t(\alpha - 1)\hat{p}_{t-1} - \lambda_t\zeta_{t-1} \\ &= [1 + \lambda_t(\alpha - 1)]\hat{p}_{t-1} - \lambda_t\zeta_{t-1}\end{aligned} \tag{12.1-10}$$

This way of generating the estimates is known as the adaptive expectation Nerlove (1972).† We can establish some probabilistic convergence behavior of $\hat{p}$'s generated by (12.1-10). For example, we can show that $\hat{p}_t \to 0$ in the mean square. Let

$$\sigma_t^2 = \mathrm{var}(\hat{p}_t)$$

Then it is governed by

$$\sigma_t^2 = [1 + \lambda_t(\alpha - 1)]^2\sigma_{t-1}^2 + \lambda_t^2\sigma^2$$

We choose λ_t to minimize σ_t^2, or choose λ_t to be given by

$$\lambda_t = -(\alpha - 1)\sigma_{t-1}^2/[(\alpha - 1)^2\sigma_{t-1}^2 + \sigma^2] \tag{12.1-11}$$

Then the minimal variance of $\hat{p}_t$ is given by

$$\begin{aligned}\min \sigma_t^2 &= \frac{\sigma^2\sigma_{t-1}^2}{(\alpha - 1)^2\sigma_{t-1}^2 + \sigma^2} \\ &= \frac{1}{1/\sigma_{t-1}^2 + (\alpha - 1)^2/\sigma^2}\end{aligned} \tag{12.1-12}$$

(12.1-10) becomes

$$\hat{p}_t = \frac{\sigma^2}{(\alpha - 1)^2\sigma_{t-1}^2 + \sigma^2}\hat{p}_{t-1} + \frac{(\alpha - 1)\sigma_{t-1}^2}{(\alpha - 1)^2\sigma_{t-1}^2 + \sigma^2}\zeta_{t-1} \tag{12.1-13}$$

From (12.1-12), with the adjustment parameter λ_t of (12.1-11), the variance

†Instead of $\{\lambda_t\}$, a same constant λ may be used for all $t = 0, 1, \ldots$.

satisfies the recursion equation

$$\begin{aligned} 1/\sigma_t^2 &= 1/\sigma_{t-1}^2 + (\alpha - 1)^2/\sigma^2 \\ &= 1/\sigma_0^2 + t(\alpha - 1)^2/\sigma^2 \\ &\to \infty, \qquad \text{as} \qquad t \to \infty \end{aligned}$$

namely

$$\sigma_t^2 \to 0, \qquad \text{as} \qquad t \to \infty \tag{12.1-14}$$

From (12.1-13) we establish

$$E(\hat{p}_t) = \frac{\sigma^2}{(\alpha - 1)^2\sigma_{t-1}^2 + \sigma^2} E(\hat{p}_{t-1})$$

Therefore iterating this, we obtain $E(\hat{p}_t) = \phi_t E(\hat{p}_0)$ where from (12.1-14) we see that $\phi_t = O(1/t)$ hence

$$E(\hat{p}_t) \to 0, \qquad \text{as} \qquad t \to \infty$$

or $\hat{p}_t$ converges to 0 in mean square. Thus from (12.1-7) and above

$$E(p_t) \to 0$$

and

$$\operatorname{var}(p_t) \to \sigma^2, \qquad \text{as} \qquad t \to \infty$$

If we drop the assumption that P_e is known, then (12.1-7) cannot be used to derive (12.1-10). We must work, for example, with

$$\Delta p_t = \alpha \Delta \hat{p}_t + \eta_t$$

where

$$\eta_t = -\zeta_t + \zeta_{t-1}$$

Before we leave this section, we should note that economic literature has very little to say about the interactions of firms on the market level, relying heavily on the assumption of representative firms. We will examine the questions of firms' interaction in § 12.2.

12.2 STOCHASTIC MARSHALLIAN QUANTITY ADJUSTMENT†

12.2.1 Case of Perishable Goods: Stochastic Difference Equation

We continue our discussion of the Marshallian quantity adjustment model of § 7.1. We make the model stochastic by introducing exogenous random noises that affect individual firms' decisions on the output rates.

We assume as in § 7.1 that there are n firms producing a homogeneous nonstorable good. The production delays of all the firms are the same, taken to be one market day and the total amount of the goods put on the market on the tth day is the total of individual firms' output rates

$$Q_t = \boldsymbol{e}^{\mathrm{t}}\boldsymbol{q}_t$$

where $\boldsymbol{q}_t = (q_{1t}, \ldots, q_{nt})^{\mathrm{t}}$ is the n-dimensional vector made up of the individual firms' output rates. Assume now that firms adjust output rates according to‡

$$\Delta q_{jt} = h_j(s_j(q_{jt}) - p_t^*) + \zeta_{jt}$$

where p_t^* is the market clearing price on the tth day and where

$$\Delta q_{jt} = q_{jt+1} - q_{jt}$$

where $s_j(\cdot)$ is firm j's supply price schedule, where ζ_{jt} is the exogenous disturbance, and where h_j, the adjustment step size, is the decision variable of firm j.

Suppose now that each firm's supply price schedule is linear.§ The quantity adjustment equation then becomes

$$\Delta q_{j,t} = h_j(\alpha_j + \beta_j q_{j,t} - p_t^*) + \zeta_{j,t}, \qquad j = 1, \ldots, n$$

Taking the difference of this equation for $t + 1$ and t we see that the quantity adjustments for all firms can be described by a stochastic linear

†See Aoki (1975d) for more detail on information processing that must be done by firms in the expected profit maximization under imperfect information.

‡As in §7.1, there are other ways for firms to incorporate their (subjective) estimates of the marginal rates of profit in adjusting output rates.

§This assumption is made primarily for ease of analysis and presentation. With nonlinear supply schedule, individual firms' local behavior will be analyzable assuming the supply schedules are linear (see § 2.6).

difference equation

$$\Delta \boldsymbol{q}_{t+1} = (I + HB)\Delta \boldsymbol{q}_t - \boldsymbol{h}\Delta p_t{}^* + \zeta_{t+1} - \zeta_t \tag{12.2-1}$$

where $H = \text{diag } (h_1, \ldots, h_n)$, $B = \text{diag } (\beta_1, \ldots, \beta_n)$, $\boldsymbol{h} = (h_1, \ldots, h_n)^{\text{t}}$ where $\zeta_t = (\zeta_{1t}, \ldots, \zeta_{nt})^{\text{t}}$. The change in the market clearing price is related to the change in the output rates by

$$\begin{aligned}\Delta p_t{}^* &= -\delta \Delta Q_t \\ &= -\delta \boldsymbol{e}^{\text{t}} \Delta \boldsymbol{q}_t\end{aligned}$$

as before.

Taking this relation into account, the stochastic difference equation governing the quantity adjustments is given by

$$\Delta \boldsymbol{q}_{t+1} = \Phi \Delta \boldsymbol{q}_t + \zeta_{t+1} - \zeta_t \tag{12.2-1$'$}$$

where

$$\Phi = I + HB + \delta \boldsymbol{h}\boldsymbol{e}^{\text{t}}$$

The third term in Φ represents the interaction or crosscoupling of different firms' output rates caused by the fact that all firms react to a common signal, the market clearing price.

12.2.2 Equilibrium Distribution

Proceeding as in § 7.1, we can compute the eigenvalues of Φ to be the roots of

$$\begin{aligned}0 &= |\lambda I - \Phi| \\ &= |(\lambda - 1)I - HB - \delta \boldsymbol{h}\boldsymbol{e}^{\text{t}}| \\ &= |(\lambda - 1)I - HB|\big(1 - \delta \boldsymbol{e}^{\text{t}}[(\lambda - 1)I - HB]^{-1}\boldsymbol{h}\big) \\ &= \Pi_{j=1}{}^n(\lambda - 1 - h_j\beta_j) \cdot \left\{1 - \delta \sum_{j=1}^n \frac{h_j}{\lambda - 1 - h_j\beta_j}\right\}\end{aligned}$$

With $h_j < 0$ and $|h_j\beta_j|$ small for all j, we see that n roots all have moduli less than one (see Appendix C).

Thus Φ in (12.2-1$'$) is an asymptotically stable matrix when h_j's are all suitably small negative numbers. Note that even a single firm could cause instability in the industry by using too large an adjustment step. We postpone detailed analysis of adjustments till the next section. We now examine first what happens to $\Delta \boldsymbol{q}_t$ when individual firms all maintain individual adjustment step sizes constant as $t \to \infty$. Let

$$z_t = E(\Delta \boldsymbol{q}_t) \quad \text{and} \quad \tilde{z}_t = \Delta \boldsymbol{q}_t - z_t$$

Then

$$z_{t+1} = \Phi z_t, \qquad z_0 = H(\boldsymbol{\alpha} + B\boldsymbol{q}_0) - \boldsymbol{h}p_0{}^*$$

where

$$\boldsymbol{\alpha} = (\alpha_1, \ldots, \alpha_n)^{\mathrm{t}}$$

From (12.2-1′), the difference equation for $\tilde{z}_t$ is

$$\tilde{z}_{t+1} = \Phi \tilde{z}_t + \zeta_{t+1} - \zeta_t \tag{12.2-2}$$

Denote the covariance matrix $\operatorname{cov}(\tilde{z}_t)$ by Z_t. From (12.2-2), then the covariance matrix evolves with time according to the difference equation

$$Z_{t+1} = \Phi Z_t \Phi^{\mathrm{t}} - \Phi\Sigma - \Sigma\Phi^{\mathrm{t}} + 2\Sigma, \qquad Z_0 = \Sigma_0$$

since

$$\begin{aligned} E\left[\tilde{z}_t(\zeta_{t+1} - \zeta_t)^{\mathrm{t}}\right] &= -E(\zeta_t \zeta_t^{\mathrm{t}}) \\ &= -\Sigma \end{aligned}$$

Since the eigenvalues of Φ are all less than one in modulus, $\tilde{z}_t \to 0$ as $t \to \infty$ and $Z_t \to Z^*$ where

$$Z^* = \Phi Z^* \Phi^{\mathrm{t}} - \Phi\Sigma - \Sigma\Phi^{\mathrm{t}} + 2\Sigma$$

or

$$\Phi^{-1}Z^* - Z^*\Phi^{\mathrm{t}} = -\Sigma - \Phi^{-1}\Sigma\Phi^{\mathrm{t}} + 2\Phi^{-1}\Sigma$$

This is a matrix equation for Z^* and has a solution if and only if eigenvalues of Φ^{-1} and Φ^{t} are not such that $\lambda_i + \mu_j = 0$ for any eigenvalues λ_i of Φ and μ_j of Φ^{-1}. (See Bellman 1960, p. 231.) The condition is satisfied since $|\lambda_i| < 1$ and $|\mu_j| > 1$ for all i and j. Thus, the distribution of

Δq_t will converge to that of $N(O, Z^*)$ if ζ's are normally distributed. Next, to find what happens to $\boldsymbol{q}_t$ as $t \to \infty$, solve (12.2-1) for $\Delta \boldsymbol{q}_t$

$$\Delta \boldsymbol{q}_t = \Phi^t \Delta \boldsymbol{q}_0 + \sum_{\tau=0}^{t-1} \Phi^{t-1-\tau}(\zeta_{\tau+1} - \zeta_\tau) \tag{12.2-3}$$

Then the output rate vector at the tth day is given as

$$\begin{aligned}
\boldsymbol{q}_t &= \boldsymbol{q}_0 + \Delta \boldsymbol{q}_0 + \cdots + \Delta \boldsymbol{q}_t \\
&= \boldsymbol{q}_0 + \sum_{s=0}^{t} \Phi^s \Delta \boldsymbol{q}_0 + \sum_{s=1}^{t} \sum_{\tau=0}^{s-1} \Phi^{s-1-\tau}(\zeta_{\tau+1} - \zeta_\tau) \\
&= \boldsymbol{q}_0 + (I - \Phi)^{-1}(I - \Phi^{t+1})\Delta \boldsymbol{q}_0 + \sum_{\tau=0}^{t-1} \sum_{u=\tau}^{t-1} \Phi^{u-\tau}(\zeta_{\tau+1} - \zeta_\tau) \\
&= \boldsymbol{q}_0 + (I - \Phi)^{-1}(I - \Phi^{t+1})\Delta \boldsymbol{q}_0 + \sum_{\tau=0}^{t-1} (I - \Phi)^{-1}(I - \Phi^{t-\tau})(\zeta_{\tau+1} - \zeta_\tau) \\
&= \boldsymbol{q}_0 + (I - \Phi)^{-1}(I - \Phi^{t+1})\Delta \boldsymbol{q}_0 + \boldsymbol{\eta}_t
\end{aligned} \tag{12.2-4}$$

where

$$\boldsymbol{\eta}_t = \sum_{\tau=1}^{t} \Phi^{t-\tau}\zeta_\tau - (I - \Phi)^{-1}(I - \Phi^t)\zeta_0 \tag{12.2-5}$$

If ζ's are normally distributed with $N(0, \Sigma)$, then so is $\boldsymbol{\eta}_t$ with $N(0, \Gamma_t)$ where

$$\Gamma_t = \sum_{\tau=1}^{t} \Phi^{t-\tau}\Sigma(\Phi^{\mathrm{t}})^{t-\tau} + (I - \Phi)^{-1}(I - \Phi^t)\Sigma(I - \Phi^t)^{\mathrm{t}}(I - \Phi^{\mathrm{t}})^{-1}$$

$$\to \Sigma + 2(I - \Phi)^{-1}\Sigma(I - \Phi^{\mathrm{t}})^{-1} \quad \text{as} \quad t \to \infty$$

Define this limit matrix as Γ_∞.

Therefore the output rate vector as $t \to \infty$ approaches $\boldsymbol{q}_\infty$ that is also normally distributed with mean $E[\boldsymbol{q}_0 + (I - \Phi)^{-1}\Delta \boldsymbol{q}_0] = \boldsymbol{q}_0 + (I - \Phi)^{-1}\cdot(H\boldsymbol{\alpha} + HB\boldsymbol{q}_0 - \boldsymbol{h}P_0^*)$, and variance $\Gamma_\infty + \Sigma$. The mean can be simplified to

$$\frac{(P_0^* + \delta Q_0)B^{-1}\boldsymbol{e}}{1 + \delta \boldsymbol{e}^{\mathrm{t}}B^{-1}\boldsymbol{e}} - \left(I - \delta \frac{B^{-1}\boldsymbol{e}\boldsymbol{e}^{\mathrm{t}}}{1 + \sigma \boldsymbol{e}^{\mathrm{t}}B^{-1}\boldsymbol{e}}\right)B^{-1}\boldsymbol{\alpha}$$

$$= -B^{-1}\boldsymbol{\alpha} + \frac{\gamma + \delta(\boldsymbol{e}^{\mathrm{t}}B^{-1}\boldsymbol{\alpha})}{1 + \gamma \boldsymbol{e}^{\mathrm{t}}B^{-1}\boldsymbol{e}} B^{-1}\boldsymbol{e}$$

where γ is the intercept parameter in the market clearing price schedule $p = \gamma - \delta Q$. Note that the mean as well as the variance is independent of the adjustment sizes used by the firms, provided they remain sufficiently small to maintain the asymptotic stability of Φ.

12.2.3 Firms' Output Rate Adjustment By Stochastic Approximation

We shall outline possible schemes for firms to select "suitable" output rate adjustment sizes h_j, $j = 1, \ldots, n$.

The market clearing prices are governed in general by the nth order linear stochastic difference equation with unknown coefficients and these coefficients are rather complicated functions of the parameters δ, the β's, the h's and the noise covariance Σ. Even with moderately large n, estimation of these parameters poses a difficult estimation problem for firms.

Unless we introduce some simplifying and *ad hoc* assumptions such as that all the firms' marginal supply price schedules are identical,† it is not realistic to assume that firms manage to obtain enough information on unknown parameters and consequently can predict future market clearing prices "optimally" so that firms can choose output rates to perform some intertemporal optimization of their utility functions.

Lacking sufficient information to carry out any sort of intertemporal optimization, firms must adjust their h's nevertheless. Here, we can borrow the techniques of direct optimization of functions of several variables developed in the discipline of nonlinear programming and their extensions to stochastic function maximization by algorithms analogous to the one proposed by Kiefer and Wolfowitz (1952) to seek the maximum of regression function (Kushner 1972, Wasan 1969, Elliott and Sworder 1970).

This is due to the fact that any firm's profit or loss over the past is a well-behaved function of the h's. We carry out our discussion for firm one. Suppose that all the firms have been in the same business from the same date and the industry is closed to new entry. We take this date as the market day referred to as number zero. On the tth market day, then, firm one knows how much profit and losses it has experienced on days

†If we assume that $h_i\beta_i = \text{const}$, $i = 1, \ldots, n$, then Δp_t^* is governed by the first-order difference equation $\Delta p_{t+1}^* = (1 + c)\Delta p_t^* + \epsilon_t$, where $c = h\beta - (\Sigma h_i)\delta$, $h\beta$ is the common value of $h_i\beta_i$, and where $\epsilon_t = \delta e'(\zeta_{t+1} - \zeta_t)$. On the other hand, if firm one assumes that the rest of the industry is represented by a representative firm, i.e., $\beta_2 = \cdots = \beta_n$, $h_2 = \cdots h_n$, then Δp_t^* can be shown to be governed by a third-order stochastic difference equation. See Aoki (1975d) for detailed analysis of the future price predictions.

0, 1, . . . , t. Denote the cumulative profit up to now by

$$L(\boldsymbol{h}) = \sum_{\tau=0}^{t} \left[p_\tau{}^* - s_1(q_{1\tau}) \right] q_{1\tau}$$

Note that $L(\cdot)$ is quadratic in $\boldsymbol{h}$, at least explicitly.

Suppose $\boldsymbol{h}^*$ is the optimal adjustment step size vector. Expand $L(\cdot)$ into Taylor series

$$L(\boldsymbol{h}) = L(\boldsymbol{h}^*) + \frac{1}{2}(\boldsymbol{h} - \boldsymbol{h}^*)^{\mathrm{t}} L_{hh}(\boldsymbol{h} - \boldsymbol{h}^*) + \cdots$$

Let $\boldsymbol{\rho}$ be a vector. Then

$$\begin{aligned} L(\boldsymbol{h} + \boldsymbol{\rho}) - L(\boldsymbol{h} - \boldsymbol{\rho}) &= \frac{1}{2}\Big[(\boldsymbol{h} - \boldsymbol{h}^* + \boldsymbol{\rho})^{\mathrm{t}} L_{hh}(\boldsymbol{h} - \boldsymbol{h}^* + \boldsymbol{\rho}) \\ &\quad - (\boldsymbol{h} - \boldsymbol{h}^* - \boldsymbol{\rho})^{\mathrm{t}} L_{hh}(\boldsymbol{h} - \boldsymbol{h}^* - \boldsymbol{\rho})\Big] + \cdots \\ &= \boldsymbol{\rho}^{\mathrm{t}}(L_{hh})(\boldsymbol{h} - \boldsymbol{h}^*) + \cdots \end{aligned}$$

This expression is correct up to the third order of ρ.

This relationship is the key relation† to allow us to show the convergence in probability and in mean square of h to h^*. The proof can be carried out by suitable extensions to the vector case of the stochastic adjustment methods due to Derman (1956), Kiefer-Wolfowtiz (1952), Dupač (1960), and Wasan (1969). This basic scheme has been recently extended by Kushner (1972).

In our problem we need to show that $h \to h^*$ in some probabilistic sense when individual h_j's are adjusted independently. This is analogous to the cyclic gradient method rather than the usual steepest ascent (descent) method or fictitious plays in games. See for example Brown (1951). The convergence analysis using fictitious play can also be carried out. See Aoki (1974d).

Firm one maximizes $L(h)$ by iterating on the adjustment step size using the data set covering the period 0, 1, . . . , t by‡

$$h_{1,m+1} = h_{1,m} + \frac{a_m}{c_m}\left[L(h_{1,m} + c_m) - L(h_{1,m} - c_m)\right]$$

†Other conditions on the boundedness of the second moments, etc., are all satisfied for our problem.

‡Assume other firms do not change their adjustment step sizes. Thus, this procedure is useful in establishing the existence of Nash equilibrium.

where $h_{1,m}$ is the mth iterate, a_m and c_m are such that

$$c_m \to 0, \quad \Sigma a_m = \infty, \quad \Sigma a_m c_m < \infty, \quad \Sigma a_m^2 / c_m^2 < \infty$$

$a_m = 1/m^{1-\epsilon}$, $c_m = 1/m^{1/2-\eta}$ for some small $\epsilon, \eta > 0$.

The limit h_1^* of these iterations is the adjustment step size firm one should have employed in the light of the firm's past loss and profit record. Firm one could periodically repeat these iterations, say every week, utilizing all the data accumulated up to that time. Then we have a sequence of such h^*'s, $h_{1,1}^*, h_{1,2}^*, \cdots$ where $h_{1,k}^*$ is the best h_1^* (in some probabilistic sense such as the mean sequence convergence) based on the data set of k weeks duration, say. The convergence of $h_{1,k}^*$ as $k \to \infty$ is examined elsewhere.†

12.3 PRICING POLICY OF A MARKETEER‡

In the previous two sections, we have analyzed quantity adjustment models in which prices adjust in one period to clear the markets. In these models, therefore, excess demands are zero at the end of each market day. Now we consider a price adjustment model in which an economic agent does not know the demand schedule facing him and adjusts prices depending on the excess demand. In other words, prices do not adjust fast enough to render the excess demands zero in one market day. Such a model is therefore appropriate for a market of nonperishable goods. We assume that an economic agent, called a marketeer, holds enough inventory of a homogeneous good to meet any demand for it. A trading specialist would be one example of such marketeers. Nonprice rationing behavior will not be discussed. We assume for the most part that the storage costs, reorder costs, etc., are negligible, ignoring the aspect of inventory entirely in this section. An example of nonnegligible storage cost will be discussed separately later in Chapter 13. In this section, we put emphasis of analysis on the learning processes of the marketeer of the demand schedule that faces him.

Consider a trading specialist dealing in an isolated market who trades in a single homogeneous good. As pointed out in Clower (1955, 1959a, 1959b), monopolistic or oligopolistic price adjustment can be formulated in an entirely analogous manner. The only technical difference is due to the fact that both the price and the output rate, rather than just the price, must be treated as decision variables.

†See Aoki (1974d).

‡This section is based on Aoki (1973, 1974b).

We consider a Marshallian "short period" market. At the beginning of each trading period (day, say), the specialist posts the price at which the good is to be traded. There are several possibilities. His selling and buying prices could differ by a fixed percentage, or he could face imprecisely known demand but known supply, and so on. In each case, the situation may be reduced to that where his subjective estimate of excess demand is given by $f(p; \theta)$ with appropriately chosen parameter θ in some known space Θ. Since he trades in disequilibrium, he is not certain what price p will clear the market. When he changes the price of the good, he has only his subjective estimate of the effects of the price change. Faced with this uncertainty, however, the trading specialist sets the price which, in his estimate, will maximize some chosen intertemporal criterion function.

The model to be discussed in this section could arise in several economically interesting contexts; for example, in considering the optimal price setting policy of a monopolist (Barro (1972),† or a trading specialist or inventory-holding middleman in some futures market.

The price he sets does not usually result in zero excess demand because (1) the agent's estimate of the parameter θ is different from the true parameter θ^*; and (2) there is an exogenous disturbance on the demand, i.e., there is a random variable ξ that represents effects or demands which are assumed by the agent to be nonsystematic. Various anticipated or systematic trends such as price expectation on the part of the buyers could be modeled.† The probability distribution of the random variable ξ is assumed to be known to the agent. The type of consideration to be presented below can easily be extended to the case where the distribution is known imperfectly; for example, up to certain parameter values specifying the distributions uniquely. This added generality is not included here since it represents a straightforward extension.

12.3.1 Model

We assume, therefore, the trading specialist's criterion function over the next T periods is a function of x_t, $t = 0, \ldots, T$ where x_t is the excess demand at time t

$$x_t = d_t - s_t \tag{12.3-1}$$

†Barro's approach may be considered a special case of this section where θ is a singleton. We do not consider price adjustment cost, however, in the criterion function of the specialist. This is done for the sake of simplicity of presentation.

†To consider these, p in $f(p; \theta)$ must be replaced by the history of past prices. We do not consider this case explicitly here.

with d_t and s_t being the market demand and supply at the end of the tth market day.

We assume the market excess demand function $x_t(p)$ is to be given parametrically as

$$x_t(p) = f(p; \theta^*) + \xi_t, \qquad \theta^* \in \Theta \tag{12.3-2}$$

where θ^* is the parameter vector that specifies a particular excess demand curve out of a family of such curves, and where $\{\xi_t\}$ is taken to be a sequence of independent, identically distributed random variables† with

$$E\xi_t = 0, \qquad \text{all } t \geqslant 0$$

$$\operatorname{var} \xi_t = \sigma^2 < \infty, \qquad \text{all } t \geqslant 0$$

We assume σ is known.

One example discussed later considers a linear excess demand function

$$x_t(p_t; \alpha, \beta) = -\alpha p_t + \beta + \xi_t, \qquad \alpha > 0, \beta > 0 \tag{12.3-3}$$

In this case the unknown parameters are α and β.‡ It is assumed that the trading specialist has an *a priori* probability density function for θ, $p_0(\theta)$.

We examine an optimal price setting policy by formulating this problem as a parameter adaptive (or a parameter learning) optimal control problem, since he learns about the unknown parameters that specify the excess demand function. We then apply the dual control theory to derive the equation for the optimal price policy (Aoki 1973).

This model represents one possible mathematical formalization of price setting mechanisms discussed by Clower who emphasized the subjective nature of economic agents' decision making processes (Arrow 1959, Clower 1955, 1959a, b).

12.3.2 Pricing Policies

ONE PERIOD OPTIMIZATION Suppose the marketeer takes the objective function to be minimized at the beginning of the tth day as

$$J_t = \phi_t(x_t) \tag{12.3-4}$$

†Serially correlated ξ's can be handled with no conceptual or technical difficulties. The independence assumption is chosen for the sake of simplicity.

‡Theoretically, there is no difficulty in increasing the number of unknown parameters more to include σ, say. See § III.2 of (Aoki 1967). The computation becomes more involved, of course.

where

$$\phi_t(x_t) \geqslant 0$$

To be specific we assume

$$\phi_t(x_t) = x_t^2 \tag{12.3-5}$$

In other words, the marketeer has a myopic objective of achieving $x_t = 0$.

The problem may now be stated as: Determine the sequence of prices $p_0, p_1, \ldots, p_T$ such that the expectations

$$E[J_t | \mathcal{I}_t], \qquad t = 0, 1, \ldots, T$$

are minimized, where

$$\mathcal{I}_t = \{x_{t-1}, p_{t-1}, \mathcal{I}_{t-1}\}, \qquad \mathcal{I}_0 = \{\textit{a priori} \text{ knowledge of } \theta\}$$

Denote by $g(\theta | \mathcal{I}_t)$ the *a posteriori* probability density function of the uncertain parameter vector θ at the beginning of the tth market day before the marketeer sets his price p_t for the tth day. The *a priori* density function $g_0(\theta) = g(\theta | \mathcal{I}_0)$ is also assumed given.

Then from (12.3-4), p_t is chosen to minimize

$$E(x_t^2 | \mathcal{I}_t)$$

From (12.3-3), this expected value is expressible as

$$E(x_t^2 | \mathcal{I}_t) = (\boldsymbol{\theta}_t^{\mathrm{t}} \tilde{\boldsymbol{p}}_t)^2 + \sigma^2 + \tilde{\boldsymbol{p}}_t^{\mathrm{t}} \Lambda_t \tilde{\boldsymbol{p}}_t \tag{12.3-6}$$

where

$$\tilde{\boldsymbol{p}}_t = \begin{pmatrix} p_t \\ 1 \end{pmatrix}$$

$$\boldsymbol{\theta}_t = E(\boldsymbol{\theta} | \mathcal{I}_t), \qquad \alpha_t = E(\alpha | \mathcal{I}_t), \qquad \beta_t = E(\beta | \mathcal{I}_t)$$

$$\Lambda_t = \operatorname{cov}(\boldsymbol{\theta} | \mathcal{I}_t)$$

These quantities are computed in the Appendix 12-A. Minimizing (12.3-6) with respect to $\tilde{\boldsymbol{p}}_t$, we obtain

$$p_t^* = (\alpha_t \beta_t - \Lambda_t^{12}) / (\alpha_t^2 + \Lambda_t^{11}), \qquad t = 0, 1, \ldots \tag{12.3-7}$$

where

$$\Lambda_t = \begin{pmatrix} \Lambda_t^{11} & \Lambda_t^{12} \\ \Lambda_t^{12} & \Lambda_t^{22} \end{pmatrix}$$

Denote the minimum of (12.3-6) by γ_t^*,

$$\gamma_t^* = \min E(x_t^2 | \mathcal{G}_t)$$

where

$$\gamma_t^* = \Lambda_t^{22} + \sigma^2 + \frac{\beta_t(\beta_t\Lambda_t^{11} + \alpha_t\Lambda_t^{12}) + \Lambda_t^{12}(\alpha_t\beta_t - \Lambda_t^{12})}{\alpha_t^2 + \Lambda_t^{11}}$$

Another possible type of myopic behavior for the marketeer makes him set p_t at the beginning of the tth day so that he expects it to clear the market, i.e., p_t which makes the expected excess demand zero. This pricing policy is an example of the certainty equivalence principle applied to pricing, Simon (1956) and Theil (1957). From (12.3-3)

$$\begin{aligned} 0 &= E(x_t | \mathcal{G}_t) \\ &= -\alpha_t p_t + \beta_t \end{aligned}$$

or the certainty-equivalent price is given by

$$p_t = \beta_t / \alpha_t \tag{12.3-8}$$

One-period pricing rules such as (12.3-7) or (12.3-8) imply that prices obey a certain recursion equation since the parameter estimates $\boldsymbol{\theta}_t$ and the associated error covariances are being updated by Bayes' rule. We return to this topic after we mention pricings involving optimization of criteria over several periods.

MANY PERIOD OPTIMIZATION If the marketeer chooses (12.3-9) instead of (12.3-5),

$$J_t = \sum_{s=t}^{T} x_s^2 \tag{12.3-9}$$

as the objective function, we can give a recursive formula for solving optimal pricing policy. The variable T is the last day in his planning horizon. Unfortunately, we cannot give an analytical explicit solution. We can see this as follows.

On the last day, the past optimal prices $p_0, \ldots, p_{T-1}$ have been determined. The optimal price on the Tth day $p_T{}^*$ is obtained from

$$\min E\left(x_T{}^2 | \mathcal{G}_T\right)$$

The optimal price is given by (12.3-7) with T replacing t. The optimal price at the $(T-1)$th day is obtained from

$$\min_{p_{T-1}} E\left[x_{T-1}{}^2 + \gamma_T{}^* | \mathcal{G}_{T-1}\right]$$

As is often the case in problems of this type, we cannot evaluate $E(\gamma_T{}^* | \mathcal{G}_{T-1})$ in an analytically closed form.

Let

$$\gamma_{T-1}{}^* = \min_{p_{T-1}} \left[x_{T-1}{}^2 + \gamma_T{}^* | \mathcal{G}_{T-1}\right]$$

Then $p_{T-2}{}^*$ is obtained from

$$\min_{p_{T-2}} \left[x_{T-2}{}^2 + \gamma_{T-1}{}^* | \mathcal{G}_{T-2}\right]$$

and so on.

Any other multiperiod objective function such as

$$J_t = \psi(S_{T-1}) + \lambda \sum_{s=t}^{T} \phi_s(x_s)$$

also encounters similar difficulties, where S_{T-1} is the stock of the good at the end of the last day in the planning horizon.

We can discuss an approximately optimal policy using the so-called open-loop feedback control pricing policy in §10.2.2 (also see Aoki 1973).

12.3.3 Recursive Pricing Policies

By Bayes' rule, we know how the marketeer's estimate of the parameter and its error covariance matrix change with past prices he sets. Suppose, for simplicity of presentation, that the marketeer is following (12.3-8), i.e., the marketeer always sets the price to clear the expected excess demand.

Then p_{t+1} is related to p_t by

$$\begin{aligned} p_{t+1} &= \alpha_{t+1}/\beta_{t+1} \\ &= (\alpha_t - \delta_2)/(\beta_t + \delta_1) \\ &\cong p_t + k_t x_t \end{aligned} \qquad (12.3\text{-}10)$$

where δ_1 and δ_2 and k_t are functions of past prices and of the elements of the error-covariance matrix (see Appendix 12-A for the expression). Equation (12.3-10) is a recursive price adjustment rule. Many pricing policies turn out to have this kind of recursive structure.

For example, when k_t is a function of time only, for example $k_t = c/t$ for some c then (12.3-10) is a case of stochastic approximation.

Another useful interpretation of k_t when it is a function of t (and possibly of some *a priori* information) is to regard (12.3-10) as resulting from an open-loop or open-loop feedback approximation of some intertemporal optimization problems. Similarly, with k_t dependent on t and the histories of prices and excess demands, (12.3-10) may be regarded as an approximate optimal price policy for some intertemporal optimization problems.

PRICE ADJUSTMENTS BY STOCHASTIC APPROXIMATION When prices are adjusted by

$$p_{t+1} = p_t + a_t x_t \tag{12.3-11}$$

with $a_t = c/t$, where $x_t = -\alpha p_t + \beta + \xi_t$ we have a price adjustment scheme of the Robbins-Monro stochastic approximation. We assume as before that $\{\xi_t\}$ is a sequence of independently and identically distributed random variables with mean zero and finite variances σ^2.

Chung (1954) has shown that the prices generated by (12.3-11) converge in probability to β/α, i.e., to the correct market clearing price when no noise is present. See Appendix 12-B at the end of this chapter for the proof of this fact.

We can also show that p_t generated by (12.3-11) converges in mean square. Define the variance

$$v_t = E(p_t - \bar{p}_t)^2$$

where $\bar{p}_t = E(p_t)$. We use the symbol $\sim$ to indicate the order of magnitude relation.

Hodges and Lehmann (1956) have shown that with the adjustment parameter $a_t = c/t$, we have the order of magnitude relations for the means and variances of prices generated by (12.3-11)

$$E(p_{t+1}) \sim E(p_1)/t^{\alpha c} + c(1 - t^{-\alpha c})\beta/\alpha$$

$$v_{t+1} \sim v_1/t^{2\alpha c} + \sigma^2/\alpha^2 t \cdot (\alpha c)^2/(2c\alpha - 1) \tag{12.3-12}$$

with any constant c such that $2\alpha c > 1$.

Equation (12.3-12) remains valid for any other choice of a_t such that $ta_t \to c$, $2\alpha c > 1$, as $t \to \infty$ (see Appendix 12-C for a proof).

We can establish the convergence with probability one for the price adjustment equation (12.3-11) for any a_t such that $ta_t \to 1/\alpha$. The average price

$$\hat{p}_t = \frac{1}{t} \sum_{s=1}^{t} p_s$$

also converges with probability 1 (see Appendix 12-D for a proof).

We have discussed two examples of pricing policies which may be considered as applications of stochastic approximation: one is the Kiefer-Wolfowitz type in which a subjectively conceived profit function is maximized; the other is the Robbins-Monro type in which the price which clears the market with unknown excess demands is generated.

The stochastic approximation adjustment schemes do not incorporate learning in adjustment behavior since the sequences of adjustment constants are chosen *a priori*. More generally, they can be made to depend on previous experiences of economic agents.

APPENDIX 12-A: CALCULATIONS OF PRICES

The marketeer's subjective knowledge on $\boldsymbol{\theta}$ at time t is embodied in his posterior probability density function $p(\boldsymbol{\theta} | \mathcal{I}_t)$.

It is computed by the Bayes rule recursively from $p_0(\boldsymbol{\theta})$ by

$$p(\boldsymbol{\theta} | \mathcal{I}_{t+1}) = \frac{p(\boldsymbol{\theta} | \mathcal{I}_t) p(x_t | \mathcal{I}_t, \boldsymbol{\theta}, p_t)}{p(x_t | \mathcal{I}_t, p_t)} = \frac{p(\boldsymbol{\theta} | \mathcal{I}_t) p(x_t | \boldsymbol{\theta}, p_t)}{p(x_t | \mathcal{I}_t, p_t)}$$

where

$$p(\boldsymbol{\theta} | \mathcal{I}_0) = \frac{p(x_0 | \boldsymbol{\theta}, p_0) p_0(\boldsymbol{\theta})}{\int_{\theta} p(x_0 | \boldsymbol{\theta}, p_0) p_0(\boldsymbol{\theta}) \, d\boldsymbol{\theta}}$$

where we compute $p(x_t | \boldsymbol{\theta}, p_t)$ from our knowledge of the probability density function for the noise ξ_t. For example, when ξ_t is Gaussian, with mean 0 and standard deviation σ, then we have

$$p(x_t | \boldsymbol{\theta}, p_t) = \frac{1}{\sqrt{2\pi}\ \sigma} \exp - \frac{1}{2\sigma^2} (x_t - f(\boldsymbol{\theta}, p_t))^2$$

From the above, it is straightforward to see that

$$\hat{\boldsymbol{\theta}}_{t+1} = (I - K_{t+1})\left(\hat{\boldsymbol{\theta}}_t + \frac{\Lambda_t}{\sigma^2}\begin{pmatrix} p_t \\ 1 \end{pmatrix} x_t\right) \tag{12.A-1}$$

where

$$\Lambda_t = \text{cov}(\boldsymbol{\theta} \mid \mathcal{G}_t)$$

$$K_{t+1} = \Lambda_t \tilde{\boldsymbol{p}}_t \tilde{\boldsymbol{p}}_t^{\mathrm{t}} / (\sigma^2 + \tilde{\boldsymbol{p}}_t^{\mathrm{t}} \Lambda_t \tilde{\boldsymbol{p}}_t)$$

The covariance matrix satisfies

$$\Lambda_t^{-1} = \Lambda_{t-1}^{-1} + \frac{1}{\sigma^2}\begin{pmatrix} p_{t-1} \\ 1 \end{pmatrix}(p_{t-1} \quad 1) \tag{12.A-2}$$

Denote the elements of Λ_t by

$$\Lambda_t / \sigma^2 = \begin{pmatrix} \lambda_{1t} & \lambda_{2t} \\ \lambda_{2t} & \lambda_{3t} \end{pmatrix}$$

Let $\hat{\boldsymbol{\theta}}_t = \begin{pmatrix} -\alpha_t \\ \beta_t \end{pmatrix}$ and write (12-A.1) in terms of components as

$$\begin{aligned} \alpha_{t+1} &= \alpha_t - (\lambda_{1t} p_t + \lambda_{2t}) \frac{x_t - \hat{x}_t}{1 + \tilde{\boldsymbol{p}}_t^{\mathrm{t}} \Lambda_t \tilde{\boldsymbol{p}}_t / \sigma^2} \\ \beta_{t+1} &= \beta_t + (\lambda_{2t} p_t + \lambda_{3t}) \frac{x_t - \hat{x}_t}{1 + \tilde{\boldsymbol{p}}_t^{\mathrm{t}} \Lambda_t \tilde{\boldsymbol{p}}_t / \sigma^2} \end{aligned} \tag{12.A-3}$$

where

$$\hat{x}_t = \tilde{\boldsymbol{p}}_t^{\mathrm{t}} \hat{\theta}_t = -\alpha_t p_t + \beta_t$$

This term is zero for the pricing policy $p_t = \beta_t / \alpha_t$, but nonzero for other policies.

We have computed from (12.A-2) that

$$\sigma^2 \Lambda_t^{-1} = \begin{bmatrix} \lambda_1 + \sum_{s=0}^{t-1} p_s^2 & \lambda_2 + \sum_{s=0}^{t-1} p_s \\ \lambda_2 + \sum_{s=0}^{t-1} p_s & \lambda_3 + t \end{bmatrix}$$

where

$$\sigma^2 \Lambda_0^{-1} = \begin{pmatrix} \lambda_1 & \lambda_2 \\ \lambda_2 & \lambda_3 \end{pmatrix}$$

There is a very close and interesting relation with the so-called input-signal synthesis problem of control theory. The problem is to design input signals to excite the dynamic systems so as to minimize some measure of estimation error. While this problem makes sense in a control context, it is not too appropriate in an economic context, since there is a real cost and information (search) cost associated with changing price in an economic context. See, for example, Alchian (1970) on the search cost associated with changing price. We do not explore this aspect.

Inverting the above matrix we obtain

$$\frac{\Lambda_t}{\sigma^2} = \frac{1}{\Delta} \begin{pmatrix} \lambda_3 + t & -(\lambda_2 + \Sigma p_s) \\ -(\lambda_2 + \Sigma p_s) & \lambda_1 + \Sigma p_s^2 \end{pmatrix}$$

with

$$\begin{aligned} \Delta &= (\lambda_1 + \Sigma p_s^2)(\lambda_3 + t) - (\lambda_2 + \Sigma p)^2 \\ &= (\lambda_1 \lambda_3 - \lambda_2^2) + \lambda_1 t + \lambda_3 \Sigma p_s^2 - 2\lambda_2 \Sigma p_s \\ &\quad + t\Sigma p_s^2 - (\Sigma p_s)^2 \end{aligned}$$

For simplicity, assume $\lambda_2 = 0$.† Then

$$\lambda_{1t} = \frac{1}{\Delta}(\lambda_3 + t)$$

$$\lambda_{2t} = -\frac{1}{\Delta} \sum_{s=0}^{t-1} p_s$$

$$\lambda_{3t} = \frac{1}{\Delta}\left(\lambda_1 + \sum_{s=0}^{t-1} p_s^2\right)$$

with

$$\Delta = \left(\lambda_1 + \sum_s p_s^2\right)(\lambda_3 + t) - \left(\sum_s p_s\right)^2$$

---

†It is reasonable to assume that the marketeer's *a priori* knowledge of the slope of the excess demand curve and the point of intercept are uncorrelated.

Define

$$\bar{p}_t = \frac{1}{t}\sum_{s=0}^{t-1} p_s: \qquad \text{average price over } [0, t-1]$$

and

$$s_t^2 = \frac{1}{t}\sum_{s=0}^{t-1} (p_s - \bar{p}_t)^2: \qquad \text{sample variance over } [0, t-1]$$

Using these quantities, we can express the elements of Λ_t/σ^2 as

$$\begin{aligned} \lambda_{1t} &= (\lambda_3 + t)/\Delta \\ \lambda_{2t} &= -t\bar{p}_t/\Delta \\ \lambda_{3t} &= \left[\lambda_1 + t(\bar{p}_t^2 + s_t^2)\right]/\Delta \end{aligned} \qquad (12.\text{A-}4)$$

with

$$\Delta = t^2 s_t^2 + t\left[\lambda_1 + \lambda_3(\bar{p}_t^2 + s_t^2)\right] + \lambda_1\lambda_3$$

Hence for large t with $s_t^2 > 0$

$$\begin{aligned} \lambda_{1t} &= \frac{1}{ts_t^2} + o(1/ts_t^2) \\ \lambda_{2t} &= -\frac{\bar{p}_t}{ts_t^2} + o(1/ts_t^2) \\ \lambda_{3t} &= \frac{\bar{p}_t^2 + s_t^2}{ts_t^2} + o(1/ts_t^2) \end{aligned} \qquad (12.\text{A-}4')$$

For further details of these relations, see (Aoki, 1974b).

Unless p_s = const, $s = 0, 1, \ldots, t-1$, we have $s_t^2 > 0$. We also assume that at most $\bar{p}_t^2/s_t^2 = o(1)$.

From the consideration of information search cost, it is reasonable to assume that p's will not be violently changing for large t. Then $\bar{p}_2$ will be nearly a constant and s_t^2 will only be growing slowly for large t.

We consider the case where $s_t^2 > 0$, and evaluate (12.A-2). Expressions similar to those below can be easily obtained for p_s = const. From

(12.A-4),

$$\lambda_{1t}p_t + \lambda_{2t} = \{(\lambda_3 + t)p_t - t\bar{p}_t\}/\Delta$$
$$= [\lambda_3 + t(p_t - \bar{p}_t)]/\Delta$$
$$\lambda_{2t}p_t + \lambda_{3t} = [\lambda_1 + ts_t^2 - t\bar{p}_t(p_t - \bar{p}_t)]/\Delta$$

We have also

$$\tilde{\boldsymbol{p}}_t{}'\Lambda_t\tilde{\boldsymbol{p}}_t/\sigma^2 = [\lambda_1 + \lambda_3 p_t^2 + t(p_t - \bar{p}_t)^2 + ts_t^2]/\Delta$$

Therefore, from the above and (12.A-3),

$$\begin{aligned} \alpha_{t+1} &= \alpha_t - \delta_2 \\ \beta_{t+1} &= \beta_t + \delta_1 \end{aligned} \tag{12.A-5}$$

where

δ_2

$$= \frac{[\lambda_3 + t(p_t - \bar{p}_t)](x_t - \hat{x}_t)}{t(t+1)s_t^2 + t[(p_t - \bar{p}_t)^2 + \lambda_1 + \lambda_3(\bar{p}_t^2 + s_t^2)] + \lambda_1(1+\lambda_3) + \lambda_3 p_t^2}$$

$$= \frac{[(p_t - \bar{p}_t) + \lambda_3/t](x_t - \hat{x}_t)}{(t+1)s_t^2 + (p_t - \bar{p}_t)^2 + \lambda_1 + \lambda_3(\bar{p}_t^2 + s_t^2) + \lambda_1(1+\lambda_3)/t + \lambda_3 p_t^2/t}$$

$$= \frac{(p_t - \bar{p}_t)}{(t+1)s_t^2}(x_t - \hat{x}_t) + o(1/ts_t^2) \tag{12.A-6}$$

$$\delta_1 = \frac{s_t^2 - \bar{p}_t(p_t - \bar{p}_t)}{(t+1)s_t^2}(x_t - \hat{x}_t) + o(1/ts_t^2)$$

Therefore,

$$\begin{aligned} (p_{t+1} - p_t)/p_t &= \frac{\delta_1}{\beta_t} + \frac{\delta_2}{\alpha_t} \\ &= k_t(x_t - \hat{x}_t) \end{aligned} \tag{12.A-7}$$

where

$$k_t = \frac{1}{(t+1)\beta_t} + \frac{1}{(t+1)\beta_t}\,\frac{(p_t - \bar{p}_t)^2}{s_t^2} = O\left(\frac{1}{t}\right) \tag{12.A-8}$$

APPENDIX 12-B: CONVERGENCE IN PROBABILITY

We establish convergence in probability of the sequence of prices generated by $p_{t+1} = p_t + a_t x_t$ to β/α when x_t is linear in p, i.e., $x_t = -\alpha p_t + \beta + \xi_t$. Here α and β are taken to be unknown constants, not random variables. The linear case has been solved by Chung (1954). We follow his proof.

Assume as before that the excess demand $x(p)$ at time t, for price p, is given by

$$x_t = x(p_t) = -\alpha p_t + \beta + \xi_t, \qquad t = 0, 1, \ldots$$

where ξ's are assumed to have mean zero and be independently and identically distributed.† We assume the commodity traded is not a free good; hence $\beta, \alpha > 0$.

Let $F(\cdot)$ be the distribution function for ξ with $\phi(\cdot)$ being its characteristic equation. Then

$$\begin{aligned} E\left(e^{iux_t} \mid p_t\right) &= \int e^{iux_t}\, dF(x_t \mid p_t) \\ &= e^{iu(-\alpha p_t + \beta)} \int e^{iu\xi_t}\, dF(\xi_t) \\ &= e^{iu(-\alpha p_t + \beta)} \phi(u) \end{aligned}$$

Let the characteristic function of p_t by $\phi_t(u)$, $t = 1, \ldots$. From $p_{t+1} = p_t + a_t x_t$

$$\begin{aligned} \phi_{t+1}(u) &= E\left(e^{iup_{t+1}}\right) = E\left(e^{iu(p_t + a_t x_t)}\right) \\ &= E\left\{e^{iup_t} E\left(e^{iua_t x_t} \mid p_t\right)\right\} \\ &= E\left\{e^{iup_t} e^{iua_t(-\alpha p_t + \beta)} \phi(a_t u)\right\} \\ &= e^{iua_t\beta} E\left\{e^{iu(1-\alpha a_t)p_t}\right\} \cdot \phi(a_t u) \\ &= e^{iua_t\beta} \phi_t((1 - \alpha a_t)u) \phi(a_t u) \end{aligned}$$

†The assumption of the identical distribution is not essential.

By recursion

$$\phi_{t+1}(u) = \exp\left[\frac{iu\beta}{\alpha}\left\{1 - \prod_{s=1}^{t}(1 - \alpha a_s)\right\}\right] \cdot \times \prod_{s=1}^{t} \phi(-(1 - \alpha a_t) \cdots (1 - \alpha a_{s+1}) a_s u) \cdot \phi\left(\prod_{s=1}^{t}(1 - \alpha a_s) u\right) \tag{12.B-1}$$

Choose the time-varying constant (gain) of the stochastic approximation by

$$a_s = \frac{1}{s\alpha} \tag{12.B-2}$$

Then in (12.B-1) we have

$$\prod_{s=1}^{t}(1 - \alpha a_s) = 0$$

$$(1 - \alpha a_t) \ldots (1 - \alpha a_{s+1}) a_s = \frac{1}{t\alpha}$$

Then (12.B-1) becomes

$$\phi_{t+1}(u) = \exp\left(\frac{iu\beta}{\alpha}\right) \cdot \prod_{s=1}^{t} \phi\left(-\frac{u}{t\alpha}\right)$$

$$= \exp\frac{iu\beta}{\alpha}\left[\phi\left(-\frac{u}{t\alpha}\right)\right]^t \tag{12.B-3}$$

This equation determines the distribution of p_s, $s \geqslant 2$.

The characteristic function of

$$-\sum_{s=1}^{t} \frac{\xi_s}{t\alpha}$$

is exactly $[\phi(-u/\alpha t)]^t$.

Since $\phi(u) = E(e^{i\xi u}) = 1 + O(u)$, we see that

$$\left[\phi\left(-\frac{u}{\alpha t}\right)\right]^2 = \left[1 + O\left(-\frac{u}{\alpha t}\right)\right]^t$$

$$\to 1, \qquad \text{as} \quad t \to \infty$$

Thus

$$\lim_{t\to\infty} \phi_{t+1}(u) = \exp \frac{iu\beta}{\alpha}$$

This is a special case of Khintchine's theorem. See, for example, Cramér (1946) for the statement and proof of this theorem.

This is the characteristic equation of the probability distribution with the probability mass one at $p = \beta/\alpha$, i.e.,

$$p_t \to \beta/\alpha \text{ in probability.}$$

In the above, the choice of a_t as in (12.B-2) depends on the unknown α. With a minimal amount of modifications, a_s of (12.B-2) can be replaced by $a_s = 1/s\alpha_s$ such that $s\alpha_s \to 1/\alpha$ as $s \to \infty$.

APPENDIX 12-C: CONVERGENCE IN MEAN SQUARE

We derive the recursive equation for $\text{var}(p_t)$ and investigate its asymptotic behavior, following the treatment by Hodges and Lehmann (1956).

Consider a price adjustment equation given by

$$p_{t+1} = p_t + a_t x_t \tag{12.C-1}$$

where

$$x_t = -\alpha p_t + \beta + \xi_t \tag{12.C-2}$$

The random variables ξ's are assumed to be independent with mean 0 and variance σ^2. The parameters α and β have unknown values. We assume that $\alpha > 0$ and $\beta > 0$.

From (12.C-1) and (12.C-2) we have

$$p_{t+1} = (1 - \alpha a_t) p_t + a_t \beta + a_t \xi_t$$

Taking the expectation of both sides,

$$E(p_{t+1}) = (1 - \alpha a_t) E(p_t) + a_t \beta \tag{12.C-3}$$

From it, we obtain

$$E(p_{t+1}) = \phi_{t+1,1} E(p_1) + \frac{\beta}{\alpha} \sum_{s=1}^{t} \phi_{t+1,s+1} a_s \alpha$$

where

$$\phi_{t+1,s} = (1 - \alpha a_t)\phi_{t,s}, \qquad s = 1, \ldots, t$$

$$\phi_{t+1,t+1} = 1$$

From (12.C-1) ~ (12.C-3)

$$p_{t+1} - E(p_{t+1}) = (1 - \alpha a_t)(p_t - E(p_t)) + a_t \xi_t$$

Let v_t be the variance of p_t,

$$v_t = E(p_t - E(p_t))^2$$

Then it is governed by

$$v_{t+1} = (1 - \alpha a_t)^2 v_t + a_t^2 \sigma^2 \tag{12.C-4}$$

since p_t and ξ_t are independent.

Iterating (12.C-4) we obtain

$$v_{t+1} = \psi_{t+1,1} v_1 + \left(\frac{\sigma}{\alpha}\right)^2 \sum_{s=1}^{t} \psi_{t+1,s+1}(a_s \alpha)^2 \tag{12.C-5}$$

where

$$\psi_{t+1,s} = (1 - \alpha a_t)^2 \psi_{t,s}, \qquad s = 1, \ldots, t$$

$$\psi_{t+1,t+1} = 1$$

Now suppose a_s, the adjustment parameter (gain parameter of the differential correction term) is taken to be

$$a_s = c/s, \qquad s = 1, 2, \ldots, \qquad c > 0$$

then in (12.C-5), asymptotically

$$\psi_{t+1,1} = \prod_{s=1}^{t} \left(1 - \frac{c\alpha}{s}\right)^2 \sim \frac{1}{t^{2c\alpha}} \tag{12.C-6}$$

Here, ~ is used to indicate the order of the asymptotic behavior. In

(12.C-5), we also have asymptotically

$$\sum_{s=1}^{t} \psi_{t+1,\,s+1}(a_s\alpha)^2 \sim \sum_{s=1}^{t} \left(\frac{c\alpha}{s} \right)^2 t^{-2c\alpha} s^{2c\alpha}$$

$$= \frac{(c\alpha)^2}{t^{2c\alpha}} \sum_{s=1}^{t} s^{2c\alpha-2} \tag{12.C-7}$$

We are mostly interested in the case

$$c\alpha > 1/2$$

or in particular, $c\alpha = 1$. The following analysis is still valid for a_t such that $ta_t \to 1/\alpha$ as $t \to \infty$. For $c\alpha > 1/2$, we have

$$\sum_{s=1}^{t} \psi_{t+1,\,s+1}(a_s\alpha)^2 \sim \frac{(c\alpha)^2}{t(2c\alpha - 1)} \tag{12.C-8}$$

From (12.C-4) ~ (12.C-8), we have established the asymptotic behavior of v_t

$$v_{t+1} \sim \frac{v_1}{t^{2c\alpha}} + \frac{c^2\sigma^2}{2c\alpha - 1} \frac{1}{t}, \qquad \text{if} \qquad 2c\alpha > 1 \tag{12.C-9}$$

By similar arguments, we can establish asymptotic behavior of $E(p_t)$.
For $c\alpha = 1$, we have from (12.C-3)

$$\psi_{t+1,\,1} \sim 1/t$$

$$\sum_{s=1} \psi_{t+1,\,s+1} a_s \alpha \sim 1$$

Thus

$$E(p_{t+1}) \sim E(p_1)/t + \beta/\alpha, \qquad \text{if } c\alpha = 1 \tag{12.C-10}$$

Thus for the choice of $a_s = 1/s\alpha$

$$E(p_{t+1}) \sim E(p_1)/t + \beta/\alpha$$

$$\operatorname{var}(p_{t+1}) \sim \operatorname{var}(p_1)/t^2 + \sigma^2/\alpha^2 t \tag{12.C-11}$$

APPENDIX 12-D: CONVERGENCE WITH PROBABILITY ONE

With $a_t = 1/\alpha t$, p_t generated by (12.3-11) can be written as

$$p_{t+1} = \beta/\alpha + \eta_{t+1}$$

where

$$\eta_{t+1} = \frac{1}{t\alpha} \sum_{s=1}^{t} \xi_s$$

Define $\zeta_t = \Sigma_{s=1}^{t} \xi_s / s$, $t = 1, \ldots$. It is easy to verify that $\{\zeta_t\}$ is a martingale and $\sup_t E|\zeta_t| < \infty$. Thus, ζ_t converges to a finite limit with probability 1 (Chung, 1968). By the Kronecker's lemma (Chung 1968), the convergence of ζ_t (with probability 1) implies that $\eta_{t+1} \to 0$ (with probability 1), as $t \to \infty$.

The variable $\hat{p}_t$ is given as $\hat{p}_t = \beta/\alpha + \Sigma_{s=1}^{t} \eta_s / t$. Since $\sup_t E(\Sigma_{u=1}^{t} \eta_u / u)^2 < \infty$, $\Sigma_{u=1}^{t} \eta_u / u$ converges to a finite limit with probability 1 by Corollary 1 of Lemma 1, Kushner (1972). Thus by Kronecker's lemma $\Sigma_{s=1}^{t} \eta_s / t \to 0$, with probability 1, hence $\hat{p}_t \to \beta/\alpha$ with probability 1.

13

Price Adjustment of a Middleman with Inventory†

13.1 INTRODUCTION

Except for a brief discussion in §12.3, inventory has not been incorporated into our previous discussion of price adjustment consideration.

In an economy of uncertain and imperfect information, ex-ante plans of households and firms will not generally be coordinated. What they plan is not what they end up doing. In such an economy, inventory will be expected to play an important role as a buffer against the consequences of mismatched plans. Inventory holding behavior of a storekeeper or of a middleman in a world of uncertain and incomplete information has not been analyzed to any extent (see, however, Phelps *et al.* 1970, or Rothschild 1973).

We consider a model with a middleman who holds inventory and analyze his pricing policies under imperfect information assumption. We will assume that he has one important piece of general knowledge about his market, namely, that he is surrounded by numerous competitors selling the same goods as he does or close substitutes for it. What he does not know is what prices they are charging.‡ Consequently, he has to decide on his own pricing over time on the basis of what he can observe directly, i.e., his own sales. We examine economic implications of the model with focus on his price adjustment behavior. We carry out our analysis to the first order in price differentials. In this model, decisions are based not on the traditional price signals but rather on quantity signals.

†Some results in this chapter were reported as "A Pricing Policy of a Middleman Under Imperfect Information" at the 49th Annual Conference of the Western Economic Association, Las Vegas, June 1974.

‡There are other models in which agents must set prices without exact knowledge of the 'market average' price. See for example (Lucas 1973a) or (Mortensen 1974).

13.2 MODEL

Consider a middleman in a market with many other middlemen who all sell a storable consumption good.† A customer is indifferent from whom the customer purchases the goods, if the price is the same. We assume gradual customer responses to price changes and price differentials, i.e., customers respond to price changes with gradual learning in a manner to be made precise later. It is assumed that the middleman faces a stochastic demand, that he sets his price p and must decide either to maintain or how much to change p in the face of fluctuating demands. The signal he observes mixes two kinds of messages, one being the feedback from his setting of price and the other originating from exogenous sources (random shifts of product demand, for example) and occurring independently of his decision. This complicates his decision process either to confirm or to refute some or all of his subjectively held beliefs about the market he is in and the demand condition he faces. For example, an observed decline in his own sales over a succession of 'weeks' might mean either of two things. It could mean that competitiors have lowered the prices they charge so that his own price is now out of line with prices prevailing elsewhere in the market; if this were the case, he would be threatened by a continuing loss of customers to other sellers as consumers continue to learn about the prevailing price distribution.‡ On the other hand, it could mean that the market demand has declined generally with all sellers losing sales in roughly the same proportion—a less alarming situation.

To focus our attention on the stochastic demand side, assume that the middleman can get instant delivery of any amount of the good, with a constant setup cost δ and that he always replenishes his stock to the level of Q as soon as the stock is depleted. In other words, he follows (s, S) policy where $s = 0$, $S = Q$, where s may be taken to be zero due to the instantaneous delivery assumption (Scarf *et al.* 1963). Given fluctuating demands, he will attribute the demand variation in part to his price not in line with the average market price and in part to the exogenous change in economic conditions. How this is done is outlined later in this chapter. The middleman would generally change both Q and the price p. Here, we focus attention to the pricing scheme while holding Q fixed. A storage cost (rate) proportional to the level of inventory is charged at each instant of time. Customer arrivals are modeled as a renewal process. See Cox (1962), for example, for description of the renewal processes. To be specific, we take it

†With slight modifications of terminology, the model is also applicable to a good with a slight product differentiation.

‡See Aoki (1975c) on the customer flow rate.

to be a Poisson process with an arrival rate ρ that may depend on the price the middleman sets, and any exogenous variable θ indicating the "state" of the economy.†

It is important to realize that this "state of the world" parameter θ is not directly observable. It is revealed indirectly through the sales. Suppressing this dependence for the moment to lighten notation, we therefore assume that k customers arrive in an interval of length t with probability

$$P(N_t = k) = \frac{(\rho t)^k}{k!} e^{-\rho t}$$

where N_t is the number of customers who arrived in a time interval of length t if p is held constant over the interval. The probability of more than one customer arriving in a small time interval Δt is of the order $o(\Delta t)$, and can be neglected.

Once a customer arrives at a store, he purchases a random quantity $q \geqslant 0$ of the good. We assume q to be a continuous variable.‡ Let $G(q;\theta, p)$ be its probability distribution function with the probability density function $g(q)$ that is assumed to exist. (We will drop θ and p from the probability density to simplify notation.)

One example of $G(\cdot)$ is the Γ-distribution. Its probability density function is

$$g(q) = \frac{\mu(\mu q)^{\alpha-1} e^{-\mu q}}{\Gamma(\alpha)} \tag{13.2-1}$$

where α and μ are the parameters of the distribution. The variable α is known as the shape parameter; μ is called the scale parameter. They may also depend on p and θ. Their dependence on p and θ will be made explicit later. The mean is given by α/μ and the variance is α/μ^2.

The middleman is assumed to be rational in that he sets p so as to maximize "average" profit over some long time interval T.

The average profit rate π is taken to be given by

$$\pi = r - c \tag{13.2-2}$$

†A somewhat simpler model with a uniform arrival rate is used later in the chapter.

‡An alternate possibility is to assume that a customer has some nonzero probability of purchasing a unit of the good. Such a model may be appropriate when the good in question is a durable or capital good. The kinds of goods that fit our model would be more in the nature of nonperishable consumer goods; cans of soups, for example.

where r is the average revenue rate from sales and c is the average cost rate composed of setup cost and the storage cost. Explicit expressions will be given later. We discuss in the next section how r and c can be computed (as a function of the variable p, among others).

The most difficult question that the middleman faces in the light of random customer arrivals and varying amounts of purchases made by individual customers is a sequential statistical decision problem: when and how does he conclude that his price is out of line with the price prevailing in the market (assume, for example, that he is a new entrant to the market and that he does not know the prevailing price exactly, and/or the price information can be gathered only with cost and gradually over time), and when and how does he decide that a changing trend in the amounts sold is due to changes in the demand for the goods due to shift in preferences and so on.

At a first approximation to this problem, suppose that an exogenous change in the demand (shift in the preference) implies that every customer's purchase is changed by $\theta\%$, where θ is unknown, while if his price is much lower than the prevailing price many customers who normally have a large demand for the good will buy an abnormally large quantity and if his price is much higher, then many customers will buy considerably less.† Price differences do not much affect the customers with small demand for the good. We formalize these assumptions as follows.

Suppose that when there is an exogenous shift in the demand schedule then each customer buys $y = (1 + \theta)x$ instead of x. The $\theta = 0$ corresponds to the situation before the shift. The amount of purchase now is governed by the probability density function

$$g(y) = \nu \frac{(\nu y)^{\alpha - 1} e^{-\nu y}}{\Gamma(\alpha)}$$

where

$$\nu = \frac{\mu}{1 + \theta}$$

That is, the scale parameter μ is changed to $\mu/1 + \theta$, while the shape parameter α remains the same. In other words, the shift in demand is detected as a shift in the mean with no change in coefficient of variation defined as

$$C^2(x) = \frac{\text{var}(x)}{E(x)^2}$$

†This interpretation has been suggested by Leijonhufvud [Private Communication].

since the mean changes by $(1 + \theta)$ from α/μ to $(\alpha/\mu)(1 + \theta)$ but the standard deviation also increases by the same proportion from $\sqrt{\alpha}/\mu$ to $\sqrt{\alpha}(1 + \theta)/\mu$.

Suppose now the price the middleman charges is low compared with the price prevailing in the market $p < \bar{p}$, where $\bar{p}$ is the price that is prevailing elsewhere in the market. Then those who have larger demands (and those whose budget allows larger purchases) will buy larger amounts to stock up. If $p > \bar{p}$, however, those who would be buyers of large amounts curtail their purchases more sharply than those with smaller demands. Suppose that we interpret this possibility as the shift of the parameters of the probability distribution. Keeping our example of cans of soups in mind, we assume that

$$q(p) \cong q(\bar{p}) + \epsilon(p - \bar{p})^{\zeta} \qquad (13.2\text{-}3)$$

where

$$\epsilon(0) = 0$$

and where

$$\epsilon(p - \bar{p}) \text{ is small for small } |p - \bar{p}|$$

(see later pages in this section for details).

Such a form may be illustrated by considering a customer whose utility function is given by

$$U = (q_1^{\gamma} + q_2^{\gamma})q_3^{\beta} \qquad 0 < \gamma, \beta < 1$$

where good 1 and good 2 are substitutes. The total number of goods need not be three. This is chosen solely for ease of illustration.

The middleman maximizes U subject to his budget constraint $I = p_1q_1 + p_2q_2 + p_3q_3$. The result of the maximization is

$$q_1 = \text{const} \cdot p_1^{-1}\{1 + (p_2/p_1)^{\gamma/(\gamma-1)}\}^{-1}$$

In other words, if the price difference $p_2 - p_1$ between the substitutes is small, then

$$q_2 \cong q_1\{1 + (p_2 - p_1)/(\gamma - 1)p_1\}$$

$$= q_1 + \epsilon(p_2 - p_1)q_1^2 + o(\epsilon)$$

where

$$\epsilon(p_2 - p_1) \cong -2(\gamma + \beta)(p_2 - p_1)/(1 - \gamma)\gamma I$$

This is a special case of (13.2-3) with $\zeta = 2$. Equation (13.2-3) leads to the expression of the mean and the variance

$$m(p) = m(\bar{p}) + \epsilon(p - \bar{p})K \tag{13.2-4}$$

where

$$K = E(q^{\zeta}) = \Gamma(\alpha + \zeta)/\Gamma(\alpha)\,\mu\zeta$$

and

$$V(p) = V(\bar{p}) + 2\epsilon K\zeta/\mu \tag{13.2-5}$$

When $\zeta = 2$, $K = \alpha(\alpha + 1)/\mu^2$.

From these, combining the effects of exogenous shifts, we postulate that the parameters of the distribution can be approximated near $\bar{p}$ by

$$\begin{aligned}
\alpha(p) &= \alpha(\bar{p})\left(1 + \frac{\epsilon(p - \bar{p})K}{m(\bar{p})}\right)^2\left(1 + \frac{2\epsilon(p - \bar{p})K\zeta}{m(\bar{p})}\right)^{-1} \\
\mu(p) &= \mu(\bar{p})\left(1 + \frac{\epsilon(p - \bar{p})K}{m(\bar{p})}\right)\left(1 + \frac{2\epsilon(p - \bar{p})K\zeta}{m(\bar{p})}\right)^{-1}(1 + \theta)^{-1}
\end{aligned} \tag{13.2-6}$$

or

$$m(p) = (m(\bar{p}) + \epsilon(p - \bar{p})K)(1 + \theta)$$

$$V(p) = \{V(\bar{p}) + 2\epsilon(p - \bar{p})K\zeta/\mu\}(1 + \theta)^2$$

Redefine $\epsilon(p - \bar{p})$ to be $K\epsilon(p - \bar{p})/m(\bar{p})$. Then

$$\begin{aligned}
\alpha(p) &\cong \alpha(\bar{p})(1 + \epsilon(p - \bar{p}))^2(1 + 2\zeta\epsilon(p - \bar{p}))^{-1} \\
\mu(p) &\cong \mu(\bar{p})(1 + \epsilon(p - \bar{p}))(1 + 2\zeta\epsilon(p - \bar{p}))^{-1}(1 + \theta)^{-1} \\
m(p) &\cong m(\bar{p})(1 + \epsilon(p - \bar{p}))(1 + \theta)^{-1}
\end{aligned} \tag{13.2-7}$$

where

$$\epsilon(p - \hat{p}) < 0, \quad \text{for } p - \hat{p} > 0 \qquad \text{and} \qquad \epsilon(p - \hat{p}) > 0, \quad \text{for } p - \hat{p} < 0$$

We assume this to be valid in the short run at least.

We could adopt a basic time unit such as a week to compute various relevant averages and use the stock at the beginning of the week as the state variable, treating the customer arrivals as a modified renewal process (Cox, 1962, page 28). That is, customer arrivals are modeled as an ordinary renewal process except for the first customer arrival that is measured from the beginning of the week. Thus, the first customer's arrival is governed by a different probability distribution function.

Instead, we use the time interval between reorders T_Q to compute the averages. For convenience, we assume that Q is so much larger than the average order of any one customer that we may say $\alpha/\mu \ll Q$. We take the origin of time to be the time at which the stock has been replenished to the level Q, i.e., the time at which a customer arrives whose order depletes the remaining stock and makes reorders necessary. When a customer's order is bigger than the existing stock, his order is partially filled by exhausting the existing stock, and the middleman orders enough goods to fill the remainder of the customer's order and to bring his stock level back to Q, so that the process is truly a renewal process.

Strictly speaking, his reorder cost then is a random variable since reorder is Q plus some small quantity depending on the size of the last customer's order and the stock level at that time. We ignore this since Q is assumed to be sufficiently large. Alternately, we may assume that the last customer before reorder is rationed and his order is partially filled by depleting the stock and he does not get any more. We adopt this latter assumption again because of large Q. The occasional rationing of one customer per reorder cycle will not affect the results of the analysis seriously. The period between two successive reorders is referred to as one cycle. The sales commission is proportional to Q/T_Q.

13.3 ANALYSIS

13.3.1 Expected Reorder Time Interval $E(T_Q)$

Let t_i be the interarrival time between the $(i-1)$th and the ith customer. (We count the customer who triggers the reorder in the old cycle but not in the new cycle.) From the assumption of Poisson arrivals, we know the probability density of t_i to be $f(x) = \rho e^{-\rho x}$ where ρ is the arrival rate parameter (the exponential distribution probability density function). Denote the number of customers in one cycle of duration T_Q by N_Q rather than N_{T_Q}. We have

$$T_Q = \sum_{i=1}^{N_Q} t_i$$

Denote the quantity purchased by the ith customer by q_i. Then

$$\sum_{i=1}^{N_Q-1} q_i < Q \leqslant \sum_{i=1}^{N_Q} q_i$$

Note the equivalences of the probabilistic events

$$(T_Q > t) \Leftrightarrow \left(\sum_{i=1}^{N_t} q_i < Q \right)$$

and

$$(N_Q = r) \Leftrightarrow \left(\sum_{i=1}^{r-1} q_i < Q \leqslant \sum_{i=1}^{r} q_i \right)$$

Making use of these equivalences, we can compute $E(T_Q)$ that is given by

$$E(T_Q) = \frac{Q}{\rho}\left[\frac{1}{m} + \frac{1}{Q}\,\frac{\sigma^2 + m^2}{2m^2} + o(1/Q) \right] \tag{13.3-1}$$

See Appendix 13.A for the details of the derivation of this expression.

The variance of T_Q is also easily computed and is given by

$$\operatorname{var}(T_Q) = E(T_Q{}^2) - E(T_Q)^2$$

$$= \frac{Q}{\rho^2 m}\left(1 + \frac{\sigma^2}{m^2}\right)\left\{1 + \frac{7m^2 - \sigma^2}{12mQ} + o(1/Q)\right\} \tag{13.3-2}$$

where $m = \alpha/\mu$ and $\sigma^2 = \alpha/\mu^2$. The details of calculation are also carried out in Appendix 13-A.

The middleman therefore knows the expected time between reorders and its variance $E(T_Q)$ and Var (T_Q) based on his subjective estimates of α and μ. If his observed sample mean and sample variance of T_Q deviate sufficiently from his beliefs, then that is a signal to him to revise his estimates of the parameters such as α, μ, ρ and/or θ.

13.3.2 Storage Cost

During one cycle, assume that the storage cost is proportional to the amount stored, i.e., to the time integral of the stock he holds. It is given as

$$Qt_1 + (Q - q_1)t_2 + \cdots + \left(Q - \sum_1^{r-1} q_i\right)t_r$$

$$= T_Q\left[Q - q_1t_2/T_Q - (q_1 + q_2)t_3/T_Q - \cdots - \left(\sum_1^{r-1} q_i\right)t_r/T_Q\right]$$

for $N_Q = r$.

Taking the average over q_i's, this expression becomes approximately† equal to

$$T_Q[Q - m\{t_2 + 2t_3 + \cdots + (r-1)t_r\}/T_Q], \qquad \text{where } m = \alpha/\mu$$

The random variable average stock level S is then such that

$$E(S|N_Q = r) \cong Q - mE[t_2/T_Q + 2t_3/T_Q + \cdots + (r-1)t_r/T_Q] \tag{13.3-3}$$

where the expectation is with respect to q's and t_i/T_Q, $i = 2, 3, \ldots, n$. We carry out the expectation operations in Appendix 13-B.

The average stock level is given by

$$E(S) = E(E(S|N_Q = r)) \cong Q - mE\left(\frac{r-1}{2}\right)$$

Use

$$E(r) \cong \frac{\mu Q}{\alpha} + \frac{\alpha + 1}{2\alpha}$$

to obtain

$$E(S) \cong Q - \frac{m}{2}\left[\frac{\mu Q}{\alpha} + \frac{\alpha + 1}{2\alpha} - 1\right]$$

$$\cong \frac{Q}{2} + \frac{\alpha - 1}{4\mu} \tag{13.3-4}$$

---

†The approximation consists in treating $E(q_1|q_1 < Q) \cong E(q_1)$, $E(q_2|q_1 + q_2 < Q) \cong E(q_2)$, etc.

Alternately, the average storage cost may be computed as the limit as $N \to \infty$ of

$$I_N = \sum_{i=1}^{N} V_i \Big/ \sum_{i=1}^{N} T_i$$

where N is the number of reorder cycles, where T_i and V_i are the duration and the time integral of the inventory of the ith cycle respectively. Note that T_i are independently and identically distributed (iid). The random variable V_i is also iid

$$E(S) = \overline{V}/\overline{T} \cong \frac{Q}{2} + \frac{\alpha - 3}{4\mu} + \frac{(\alpha + 1)(\alpha - 7)}{24\mu^2} \frac{1}{Q} + o\left(\frac{1}{Q}\right) \quad (13.3\text{-}5)$$

Comparing (13.3-4) with (13.3-5), we see that these expressions differ by $o(1)$ since they emply different approximations. It is reassuring that they agree in the leading term. We denote $E(S)$ as

$$E(S) \cong \frac{Q}{2}\left(1 + \frac{\alpha - k}{4\mu Q} + o\left(\frac{1}{Q}\right)\right) \quad (13.3\text{-}6)$$

where $k = O(1)$.

13.3.3 Optimal Prices

The average profit rate then is given by

$$\pi = (p - c)Q/\overline{T}_Q - \delta/\overline{T}_Q - dE(S) \quad (13.3\text{-}7)$$

since

$$Q/\overline{T}_Q = \lim_{N \to \infty} \frac{Q}{\frac{1}{N}\sum_{1}^{N_T} Q^i} \text{ almost surely} \quad (13.3\text{-}8)$$

by the Kolmogorov's strong law of large numbers where δ is the cost for reorder, and where d is the cost of storage, p is chosen to maximize π.

Substitute (13.3-1), (13.3-6) and (13.3-8) into (13.3-7) to obtain the expression for the average profit rate π, when p is charged,

$$\pi = \frac{\rho\alpha}{\mu} \frac{p - c - \delta/Q}{1 + [(\alpha + 1)/2\mu Q]}\left(1 + \frac{\alpha + 1}{\mu Q}\right) - d\left(\frac{Q}{2} + s\right) + o\left(\frac{1}{Q}\right) \quad (13.3\text{-}9)$$

where $s = (\alpha - k)/4\mu$ and $k = O(1)$.

Ignoring terms of the order $o(1/Q)$, (13.3-9) can be rewritten as

$$\pi = Ap - (c + \delta/Q)A - d(Q/2 + s) \qquad (13.3\text{-}10)$$

where

$$A = \rho m: \begin{cases}\text{average quantity sold per} \\ \text{period}\end{cases}$$

$$(c + \delta/Q)A: \begin{cases}\text{cost of reordering and} \\ \text{purchasing quantity } A \text{ per} \\ \text{period, since A/Q is the} \\ \text{average number of reorders} \\ \text{per period}\end{cases}$$

$$d\left(\frac{Q}{2} + s\right): \{\text{cost of storage}$$

In the above, $dQ/2$ is the cost of storage if the quantity Q is sold at a constant rate. So long as Q is fixed, it is a fixed cost independent of prices. There is a correction term ds to the average cost of storage due to the fact that customers do not buy same quantities and only a finite number of them arrive per period. The variable s is not zero, even if the same number of customers are assumed to arrive at the store at a uniform rate, so long as they do not buy the same quantity. Thus, $s \neq 0$ reflects in an essential way the random purchase assumption of this model.

From (13.3-10), the first-order condition for the maximization of the average profit is given by an implicit function for p^*

$$p^* = \{c + \delta/Q + dfs/A\}/(1 - 1/e)$$

where

$$e = -\partial \ln A/\partial \ln p$$

and where

$$f = -\frac{1}{e}\,\partial \ln s/\partial \ln p: \begin{cases}\text{priced-induced ratio of \%} \\ \text{change in } s \text{ over \% change in } A\end{cases}$$

Substituting this price into (13.3-10), we obtain the expression for the maximum average profit

$$\pi^* = -dQ/2 + \{(c + \delta/Q)A/(e - 1)\} + ds\left(\frac{fe}{e - 1} - 1\right)$$

An exogenous shift θ changes the last two terms of the above, i.e., the part of the profit dependent on p changes to

$$\{(c + \delta/Q)A/(e-1) + ds(f-1)\}(1+\theta)$$

Thus, if the middleman has chosen Q at which π^* is zero, then $\theta > 0$ now makes his average profit positive and $\theta < 0$ would make it negative.

In evaluating e and f, we need price elasticities of α, μ, s and ρ. At the end of §13.2 we posited (13.2-4). From these relations we can obtain the needed price elasticities of them.

We assume $\rho' = 0$. In other words, the knowledge that the middleman's price is different from the price prevailing in the market does not spread instantaneously, hence does not affect in the short run the rate at which the customers come to him. See Aoki (1975c) for the case $\rho' \neq 0$ The actual calculation of $\partial\pi/\partial p$ is carried out in Appendix 13-D.

When the middleman is at equilibrium with the 'prevailing' market price, his price is given by (13.D-3) of Appendix 13-D or

$$\bar{p} = c + \frac{\delta}{Q} + \frac{d}{4\delta}\left(1 - \frac{(2\delta-1)k}{\bar{\alpha}}\right) - \frac{1}{\epsilon'(0)} \qquad (13.3\text{-}11)$$

where $\bar{\alpha} = \alpha(\bar{p})$.

Proposition *Suppose $\hat{p}$ is optimal and satisfies* (13.3-11) *and suppose $p \neq \hat{p}$. Then the middleman following the price equation*

$$p = c + \delta/Q - \frac{1}{\epsilon'(0)} + \frac{d}{4\rho}\left\{1 - \frac{k(2\zeta-1)}{\alpha(p)}\right\}$$

can eventually restore the price differential to 0, *if*

$$\left|\frac{d(2\zeta-1)(\zeta-1)k\epsilon'(0)}{2\rho\alpha(\hat{p})}\right| < 1 \qquad (13.3\text{-}12)$$

Proof Since $\hat{p}$ is unknown to the middleman, he sets his price to satisfy

$$p = c + \delta/Q - \frac{1}{\epsilon'(0)} + \frac{d}{4\rho}\left[1 - \frac{k(2\zeta-1)}{\alpha(p)}\right]$$

This is an implicit equation and can be solved by a recursive equation

$$p_{t+1} = c + \delta/Q - \frac{1}{\epsilon'(0)} + \frac{d}{4\rho}\left[1 - \frac{k(2\zeta-1)}{\alpha(p_t)}\right]$$

From (13.2-6), note that

$$\alpha(p_t) \cong \alpha(\hat{p})\{1 - 2(\zeta - 1)\epsilon'(0)(p_t - \hat{p})\}$$

Substituting this into the above, we obtain

$$p_{t+1} = c + \frac{\delta}{Q} - \frac{1}{\epsilon'(0)} + \frac{d}{4\rho}\left\{1 - \frac{k(2\zeta - 1)}{\alpha(\hat{p})} - \frac{2k(2\zeta - 1)(\zeta - 1)}{\alpha(\hat{p})}\epsilon'(0)(p_t - \hat{p})\right\}$$

or

$$p_{t+1} - p_t = -\lambda(p_t - p_{t-1})$$

where

$$|\lambda| = \left|\frac{d(2\zeta - 1)(\zeta - 1)d}{2\zeta\alpha(\hat{p})}\epsilon'(0)\right|$$

We see that $\lim_{t\to\infty} p_t = \hat{p}$ if $|\lambda| < 1$.

The condition of convergence (13.3-12), rewritten as

$$|k|\ |\epsilon'(0)| < \frac{2\rho\alpha(\hat{p})}{d(2\zeta - 1)(\zeta - 1)}$$

shows that the price sensitive behavior of the quantity purchased cannot be too great, otherwise the middleman tends to overcorrect. It also shows that the higher the storage cost the more unstable the price correction behavior is likely to be for the same $|\epsilon'(0)|$. In other words, the higher d, the smaller $|\epsilon'(0)|$ must be for this price adjustment method to be stable. The more frequently customers arrive, the higher $\epsilon'(0)$ could be.

In the special case of $\zeta = 2$, (13.3-12) becomes (recalling that $K\epsilon/m$ is renamed as ϵ in (13.2-7))

$$|k\epsilon'(0)| < \frac{2\rho}{3d}\frac{\alpha\mu}{(\alpha + 1)} < \frac{2\rho}{3d}\frac{m(\hat{p})}{V(\hat{p})} \tag{13.3-13}$$

showing that a smaller value of the ratio of m/V requires a smaller value of $|\epsilon'(0)|$ for stable adjustments. In other words, as variability of an individual's purchase become larger, the less sensitive the quantity purchased must be to price differential for the middleman to adjust his prices successfully.

CHANGE IN θ When the price difference is zero, the middleman's price is given by (13.3-5). In (13.3-5), none of the terms will change when general shifts in demand occur since we assume that ρ is not affected by θ in this model in the short run.

Thus we have the following proposition.

Proposition *The optimal price of the middleman is the same for $\theta \neq 0$ in the short run. This may be seen also directly from the expression for the long run profit* (13.3-9). *In* (13.3-9), *the portion of π that is price sensitive* (*i.e.*, *all terms except* $-dQ/2$) *is affected equally by* $(1 + \theta)$.

13.3.4 Parameter Estimates and Their Revisions

Obviously the price that the middleman sets is strongly dependent on his beliefs about the customer arrival rate, the parameters of the demand distribution and shift parameter of the demand schedule. Because the parameters are not known perfectly and are subject to change without his knowledge, the middleman must estimate them and be prepared to revise his beliefs in the face of contradiction with observed data.

In line with the market behavior already discussed, the middleman can monitor three sets of parameter combinations in order to determine whether or not a shift in the demand schedule has occurred or his price is out of line with the price prevailing elsewhere in the market.

Case 1. If the mean α/μ of the demand distribution changes and α does not, then θ (the demand shift parameter) has changed.

Case 2. If α or the coefficient of variation changes, then the asking price is not equal to the market price.

Case 3. If both the mean and the coefficient of variation change, then θ has changed and the price is not equal to market price.

It would place an unreasonable computational burden on the middleman to require him to update his parameter estimates after each customer. Even if he did update them that often it is highly unlikely that he would want to vary his prices continually. We thus assume that all the observed data is pooled until a new stock shipment is made (at time T) and the estimation and parameter revisions are carried out.† At time T it is assumed that the middleman knows how many customers N have come in

†When $\theta \neq 0$ or $p \neq \bar{p}$ are detected, old data may no longer be relevant in the changed environment. It may be best to discard the old data and restart pooling of data anew. A number of techniques is available in statistical literature for testing these hypotheses such as the likelihood ratio test and so forth. Therefore we merely suggest that these tests can be carried out, without further elaborations.

and how much, q_i, $i = 1, \ldots, N$, they purchased during the period for which the estimation is to be carried out (begin at time 0, say). Then it can be shown (see Appendix 13-E) that the maximum likelihood estimates of the various parameters of the interest are given by

$$\hat{\rho} = \frac{N}{T}$$

$$\hat{\mu} = \frac{\sum_{i=1}^{N} q_i}{\hat{\alpha} N}$$

$$\hat{\alpha} = \frac{1}{2N\left(\ln \Sigma q_i - \frac{1}{N} \ln \Sigma q_i\right)}$$

$$\hat{m} = \left(\frac{\hat{\alpha}}{\mu}\right) = \frac{1}{N} \sum_{i=1}^{N} q_i$$

$$\hat{\sigma} = \left(\frac{\hat{\alpha}^{1/2}}{\mu}\right) = \frac{\frac{1}{N} \Sigma q_i}{\hat{\alpha}^{1/2}}$$

The asymptotic distribution for these maximum likelihood estimates are known to be normal (see Wilks, 1962, page 360) and given by

$$\hat{\rho} \sim N\left(\rho, \frac{\rho}{T}\right)$$

$$\hat{\mu} \sim N\left[\mu, \frac{\mu^2}{N\left(\alpha - \frac{1}{\psi'}(\alpha)\right)}\right]$$

where $\psi'(\alpha)$ is the derivative of the digamma function $\psi(\alpha)$ (Cox and Lewis 1966)

$$\hat{\alpha} \sim N\left[\alpha, \frac{1}{N\left(\psi'(\alpha) - \frac{1}{\alpha}\right)}\right]$$

$$\hat{m} = \left(\frac{\hat{\alpha}}{\mu}\right) \sim N\left(m, \frac{m^2}{N\alpha}\right)$$

$$\hat{\sigma} = \left(\frac{\alpha^{1/2}}{\mu}\right) \sim N\left(\sigma, \frac{\sigma^2}{\alpha} \frac{\psi'(\alpha) - (3/4)\alpha}{\psi'(\alpha) - 1/\alpha}\right)$$

A simple test that the middleman can perform to determine whether or not his beliefs are correct is simply to monitor the maximum likelihood estimate of each parameter and to reject this belief in any parameter estimate that escapes from a band of width three standard deviations (given in the previous asymptotic normal distributions) on each side of his current belief in the parameter value. For example, if ρ_0 is the current value the middleman holds for ρ then he rejects that belief if $\hat{\rho} > \rho_0 + (3\sqrt{\rho_0}/\sqrt{T})$ or if $\hat{\rho} < \rho_0 - (3\sqrt{\rho_0}/\sqrt{T})$ at estimation time T. If a parameter value is rejected then the current maximum likelihood estimate of that parameter is assumed to be the correct value. This new parameter value is then tested in the manner just described. See Appendix 13.D for the details.†

Revision of Parameter Estimates and Price Changes As we indicated in §13.1, the most difficult estimation problem that faces the middleman is when to decide that his subjective estimates of the parameter values need revision and when to decide an exogenous shift in the demand has occurred.

It should be clear from the type of assumptions made in the previous analysis that the final results are applicable only if the middleman has a fairly good idea of the prevailing market price. If his price is very different from the prevailing market price, the customer arrival rate will be significantly affected and our assumption that ρ is independent of p will not hold. The results presented in this section improve the middleman's ability to increase his knowledge of his economic environment and to adjust his asking price.

At the time the middleman receives a new shipment of Q units of the commodity, he updates his estimates of the customer demand and arrival parameters using the maximum likelihood estimates described in §13.3.4. If these estimates do not reject (using the tests described in §13.3.4) his previous belief in the parameter values then he holds his price constant. If, on the other hand, the scale parameter, μ, of the demand changes while the coefficient of variation remains constant then a change in the demand for the good has occurred in the state of the economy. If μ is the previous estimate and ν is the new estimate of the scale parameter then the percentage change in the economy is given by $\theta = [(\mu/\nu) - 1)] \times 100\%$. This change in the demand for the product is felt by all the middlemen in

†An equally effective scheme for deciding whether or not the various parameters have changed can be based on the likelihood ratio function. (See Wilks, 1962, Chapter 13).

the market and they then "reoptimize" the prevailing price using the equation for the optimum price given in §13.3.3. If, in performing the parameter estimation the mean changes while the variance is constant then the middleman concludes that his price is out of line with the prevailing market price. He will then use the $\bar{p}$ equation of §13.3.3 to choose his new price. It can be shown easily that $\pi(\bar{p}) \geqslant 0$, enabling the middleman to stay in business. We could extend the framework of investigation further and determine an optimal Q to maximize $\pi(\bar{p})$. We omit the details since they are routine. Finally, we point out that disregarding the price elasticity of the customer arrival rate ρ'/ρ is consistent with our assumption that the middleman's price is not too far off from the "prevailing" price (see Appendix 13-D). If this assumption is dropped, we must retain the ρ'/ρ term.

APPENDIX 13-A: COMPUTATIONS OF $E(T_Q)$ AND var (T_Q)

We see that the event

$$\left(\sum_{i=1}^{r-1} q_i < Q \leqslant \sum_{i=1}^{r} q_i\right) = \left(\sum_{i=1}^{r-1} q_i < Q \quad \text{and} \quad Q \leqslant \sum_{i=1}^{r} q_i\right)$$

where

$$\left(\sum_{i=1}^{r-1} q_i < Q\right) = \left(\sum_{i=1}^{r-1} q_i < Q \quad \text{and} \quad \sum_{i=1}^{r} q_i < Q\right)$$

$$\cup \left(\sum_{i=1}^{r-1} q_i < Q \quad \text{and} \quad \sum_{i=1}^{r} q_i \geqslant Q\right)$$

These two events on the right-hand side are mutually exclusive, hence

$$P\left(\sum_{i=1}^{r-1} q_i < Q \leqslant \sum_{i=1}^{r} q_i\right) = K_{r-1}(Q) - K_r(Q) \tag{13.A-1}$$

where

$$K_r(Q) = P\left(\sum_{i=1}^{r} q_i < Q\right), \qquad r = 1, 2, \ldots$$

$$K_0(Q) = 1, \qquad Q \geqslant 0$$

Then

$$E(T_Q) = E\left[E\left(\sum_{i=1}^{N_Q} T_i | N_Q = r\right)\right]$$

$$= E\left(\sum_{i=1}^{r} T_i\right) \qquad (13.A\text{-}2)$$

$$= E(N_Q)/\rho$$

where

$$E(N_Q) = \sum_{r=0}^{\infty} rP[N_Q = r]$$

$$= \sum_{r=0}^{\infty} r(K_{r-1}(Q) - K_r(Q))$$

$$= \sum_{r=0}^{\infty} K_r(Q)$$

We assume the series is absolutely convergent. We have

$$E(T_Q) = \sum_{r=0}^{\infty} K_r(Q)/\rho \qquad (13.A\text{-}3)$$

From the assumption, q_i are independently and identically distributed with the probability density function $g(q; \theta)$. Denote its Laplace transform by $\hat{g}(s; \theta)$, $\mathfrak{L}(g(q; \theta)) = \hat{g}(s; \theta)$. The the Laplace transform of the density $\Sigma_{i=1}^{r} q_i$ is $[\hat{g}(s; \theta)]^r$. Hence

$$K_r(s) = \frac{1}{s}[\hat{g}(s; \theta)]^r$$

$$\mathfrak{L}(E(T_Q)) = \frac{1}{s\rho} \sum_{r=0}^{\infty} (\hat{g})^r \qquad (13.A\text{-}4)$$

$$= \frac{1}{\rho s(1 - \hat{g})}$$

Equation (13.A-3) shows that the mean reorder time is equal to the average number of customers between reorder $E(N_Q)$ times the mean customer interarrival time $1/\rho$.

When we specialize to the probability distribution function of Γ-density function, $\hat{g}$ is given by

$$\hat{g}(s) = \int_0^\infty \frac{\mu(\mu x)^{\alpha-1} e^{-\mu x}}{\Gamma(\alpha)} e^{-sx}\, dx \qquad (13.A\text{-}5)$$

$$= (\mu/(s+\mu))^\alpha \qquad (13.A\text{-}6)$$

and

$$K_r(s) = \frac{1}{s}\left(\frac{\mu}{s+\mu}\right)^{\alpha r}$$

From (13.A-4), then

$$\rho\mathcal{L}(E(T_Q)) = \frac{1}{s\left[1 - \left(1 + \frac{s}{\mu}\right)^{-\alpha}\right]}$$

We assume α to be a positive integer so that (13.A-6) is a rational function.

An asymptotic expression of the above for large Q can be obtained by expanding the above as a Laurent series expansion in s,

$$\rho\mathcal{L}[E(T_Q)] = \frac{\mu}{\alpha}\left[\frac{1}{s^2} + \frac{\alpha+1}{2\mu}\frac{1}{s} + o(1)\right]$$

or

$$\rho E(T_Q) = \frac{\mu}{\alpha}\left[Q + \frac{\alpha+1}{2\mu} + o(1)\right] \qquad (13.A\text{-}7)$$

It may be instructive to rewrite (13.A-7) in terms of the mean $m = E(q)$ and the variance $\sigma^2 = \text{var}(q)$ as

$$E(T_Q) = \frac{Q}{\rho}\left[\frac{1}{m} + \frac{1}{Q}\frac{\sigma^2 + m^2}{2m^2} + o\left(\frac{1}{Q}\right)\right] \qquad (13.A\text{-}8)$$

VARIANCE OF T_Q We have

$$E(T_Q^2) = E\left[E\left(\sum_{i=1}^{N_Q} T_i\right)^2 | N_Q = n\right]$$

where

$$E\left[\left(\sum_{i=1}^{N_Q} T_i\right)^2 \Big| N_Q = n\right] = n(n+1)/\rho^2 \qquad (13.A\text{-}9)$$

since $\sum_{i=1}^{n} T_i$ is Poisson with mean n/ρ and variance n/ρ^2. Thus from (13.A-1) and (13.A-7)

$$\begin{aligned}
\rho^2 E(T_Q^2) &= E[N_Q(N_Q+1)] \\
&= \sum_{r=0}^{\infty} r(r+1)[K_{r-1}(Q) - K_r(Q)] \\
&= 2\sum_{r=0}^{\infty} rK_r(Q) + \sum_{r=0}^{\infty} K_r(Q) \qquad (13.A\text{-}10)
\end{aligned}$$

Take the Laplace transform of the above with respect to Q to obtain

$$\rho^2 \mathfrak{L}(E(T_Q^2))/2 = \sum_{r=0}^{\infty} (r+1)(\hat{g})^2$$

In the special case of Γ-distribution, K_r is given by (13.A-6) and

$$\begin{aligned}
\rho^2 \mathfrak{L}[E(T_Q^2)]/2 &= 1/s\{1 - [1 + (s/\mu)]^{-\alpha}\}^2 \\
&= \frac{\mu^2}{\alpha^2 s^3}\left[1 + \frac{\alpha+1}{\mu}s + \frac{3(\alpha+1)(5\alpha+1)}{4\mu^2}s^2 + O(s^3)\right]
\end{aligned}$$

or

$$\frac{\rho^2 E(T_Q^2)}{2} = \frac{\mu^2}{\alpha^2}\left[\frac{Q^2}{2} + \frac{\alpha+1}{\mu}Q + \frac{(\alpha+1)(5\alpha+1)}{12\mu^2} + O(1)\right]$$

Hence

$$\begin{aligned}
\operatorname{var}(T_Q) &= E(T_Q^2) - E(T_Q)^2 \\
&= \frac{(\alpha+1)}{\alpha^2\rho^2}Q\left[1 + \frac{7\alpha-1}{12\mu Q} + o\left(\frac{1}{Q}\right)\right]
\end{aligned}$$

The variable $E(T_Q)$ and var (T_Q) are used to estimate ρ or to test the hypothesis that the subjective value of ρ held by the middleman is the correct ρ.

APPENDIX 13-B: DERIVATION OF THE BETA DISTRIBUTIONS

The random variable t_2/T_Q that appears in (13.3-3) has the form $t_2/[t_2 + (t_1 + t_3 + \cdots + t_r)]$, where t_2 is independent of $t_1 + t_3 + \cdots + t_r$. The variable t_2 has the density function $\rho e^{-\rho t_2}$ and $U = t_1 + t_3 + \cdots + t_r$ is distributed according to the density $[\rho(\rho u)^{r-2}/(r-2)!]e^{-\rho u}$.

We next show that variable $Z = t_2/(t_2 + U)$ has the Beta distribution $r(1-z)^{r-2}$, $0 \leqslant z \leqslant 1$. Its mean is $1/r$. Similarly, $t_3/T_q = t_3/[t_3 + (t_1 + t_2 + t_4 + \cdots + t_r)]$ has mean $1/r$.

We follow (Cramér, 1946) in deriving the probability density function for t_1/t_2, where t_1 and t_2 are independent and have Γ-distributions.

Let t_1 and t_2 be independently distributed random variables with the density functions

$$\frac{\rho(\rho t_1)^{\alpha-1}e^{-\rho t_1}}{\Gamma(\alpha)} \quad \text{and} \quad \frac{\rho(\rho t_2)^{\beta-1}e^{-\rho t_2}}{\Gamma(\beta)}$$

respectively.

Define a random variable Z by

$$Z = t_1/(t_1 + t_2) = (t_1/t_2)/[(t_1/t_2) + 1]$$

Its probability distribution function is

$$P(Z < x) = P\left(\frac{t_1}{t_2} < \frac{x}{1-x}\right) \tag{13.B-1}$$

We derive the probability density function for $P(t_1/t_2 < y)$ first. From the independence of t_1 and t_2, this probability is given as

$$P(t_1/t_2 < y) = \iint_{\substack{0 \leqslant \tau_1 \leqslant \tau_2 y \\ 0 < \tau_2}} g(\tau_1; \alpha)\, g(\tau_2; \beta)\, d\tau_1 d\tau_2 \tag{13.B-2}$$

Change variables from (τ_1, τ_2) to (u, v),

$$\tau_2 = v$$

$$\tau_1 = uv$$

Then $\partial(\tau_1, \tau_2)/\partial(u, v) = v$ and

$$P(t_1/t_2 < y) = \iint_{\substack{v \geqslant 0 \\ 0 \leqslant u \leqslant y}} g(uv; \alpha)\, g(v; \beta) v \, du dv$$

$$= \frac{\Gamma(\alpha + \beta)}{\Gamma(\alpha)\Gamma(\beta)} \int_{0 \leqslant u \leqslant y} \frac{u^{\alpha - 1}}{(u + 1)^{\alpha + \beta}} \, du$$

Hence the density function is

$$P(t_1/t_2 = y) = \frac{\Gamma(\alpha + \beta)}{\Gamma(\alpha)\Gamma(\beta)} y^{\alpha - 1}/(y + 1)^{\alpha + \beta}$$

From (13.B-1) and (13.B-2), the density function for Z is the Beta distribution

$$P(Z) = \frac{\Gamma(\alpha + \beta)}{\Gamma(\alpha)\Gamma(\beta)} x^{\alpha - 1}(1 - x)^{\beta - 1}$$

APPENDIX 13-C: COMPUTATION OF $E(S) = EV_1/ET_1$

Let

$$\bar{V} = EV_i$$

$$\bar{T} = ET_i, \qquad i = 1, \ldots, n$$

Then, I_N is expressible as

$$I_N = \frac{\bar{V} + \sum_{i=1}^{N} (V_i - \bar{V})/N}{\bar{T} + \sum_{i=1}^{N} (T_i - \bar{T})/N}$$

$$= \frac{\bar{V}}{\bar{T}} \frac{1 + \dfrac{1}{\bar{V}N} \sum_{i=1}^{N} (V_i - \bar{V})}{1 + \dfrac{1}{\bar{T}N} \sum_{i=1}^{N} (T_i - \bar{T})}$$

By the Kolmogorov's strong law of large numbers (Chung, 1968), we have

$$\frac{1}{N}\sum_{i=1}^{N}\left(V_i - \overline{V}\right) \to 0 \qquad \text{a.s.} \qquad \text{as } N \to \infty$$

$$\frac{1}{N}\sum_{i=1}^{N}\left(T_i - \overline{T}\right) \to 0 \qquad \text{a.s.} \qquad \text{as } N \to \infty$$

Thus

$$I_N \to \overline{V}/\overline{T} \qquad \text{a.s.}$$

From (13.3-1), we know that

$$\overline{T} = \frac{\mu}{\rho\alpha}\left[Q + \frac{\alpha+1}{2\mu} + o(1)\right]$$

we next compute $\overline{V}$. Suppose $N_Q = r$. Then the time integral of the inventory is

$$V = Qt_1 + (Q - q_1)t_2 + \cdots + \left(Q - \sum_{i=1}^{r-1} q_i\right)t_r$$

where t_i is the interval between the customer arrival. Take the expectation with respect to t_i first

$$\begin{aligned} E_t(V|N_Q = r) &= \frac{1}{\rho}\left[Q + (Q - q_1) + \cdots + \left(Q - \sum_{1}^{r-1} q_i\right)\right] \\ &= \frac{1}{\rho}\left[rQ - \left\{q_1 + (q_1 + q_2) + \cdots + \sum_{i=1}^{r-1} q_i\right\}\right] \end{aligned}$$

Assume $Q \gg 1$ so that

$$\begin{aligned} E(q_1|q_1 < Q) &\cong E(q_1) + o(1) \\ &= \alpha/\mu + o(1) \end{aligned}$$

$$E(q_1 + q_2|q_1 + q_2 < Q) \cong \frac{2\alpha}{\mu} + o\left(\frac{1}{Q}\right)$$

Then

$$E(V|N_Q = r) \cong \frac{1}{\rho}\left[rQ - \frac{\alpha}{\mu}(1 + 2 + \cdots + r - 1)\right] + o(1)$$

$$= \frac{1}{\rho}\left[rQ - \frac{\alpha}{\mu}\,\frac{r(r-1)}{2}\right] + o(1)$$

Thus

$$\bar{V} = E(V) \cong \frac{1}{\rho}\left[\bar{r}Q - \frac{\alpha}{\mu}\,\frac{\overline{r(r-1)}}{2}\right] + o(1)$$

We have computed in Appendix 13.A that

$$\overline{r(r-1)} = \left(\frac{\mu}{\alpha}\right)^2 Q^2 + \frac{2\mu}{\alpha^2}Q + \frac{1-\alpha^2}{6\alpha^2} + o(1)$$

$$\bar{r} = \left(\frac{\mu}{\alpha}\right)Q + \frac{\alpha+1}{2\alpha} + o(1)$$

Substituting these, we obtain

$$\rho\bar{V} = \frac{\mu}{2\alpha}Q^2 + \frac{\alpha-1}{2\alpha}Q - \frac{1-\alpha^2}{12\alpha\mu} + o(1)$$

From this and $\bar{T}$ obtained above we derive $E(S)$ as shown in (13.3-5).

APPENDIX 13-D: OPTIMAL PRICE FOR A GIVEN Q

Differentiate π of (13.3-9) with respect to p

$$0 = \frac{\partial\pi}{\partial p} = A'[p - c - (\delta/Q)] + A - ds'$$

where

$$A = \frac{\rho\alpha}{\mu}\,\frac{1 + (\alpha+1)/\mu Q}{1 + (\alpha+1)/2\mu Q}$$

or

$$p - c - (\delta/Q) = \frac{ds'}{A'} - \frac{A}{A'} \tag{13.D-1}$$

where

$$\frac{A'}{A} = \frac{\rho'}{\rho} + \frac{\alpha'}{\alpha} - \frac{\mu'}{\mu} + \frac{1}{2}\left(\frac{\alpha'}{\alpha}\,\frac{\alpha}{\mu Q} - \frac{\mu'}{\mu}\cdot\frac{\alpha+1}{\mu Q}\right)\cdot$$

$$\frac{1}{[1 + (\alpha+1)/\mu Q)][1 + (\alpha+1)/2\mu Q)]}$$

Under the assumption that

$$\frac{\alpha/\mu}{Q} \ll 1 \quad \text{and} \quad \rho' = 0$$

we see that

$$A \cong m\rho \quad \text{and} \quad A'/A \cong m'/m \tag{13.D-2}$$

where the symbol $'$ denotes differentiation with respect to p. By definition $s = (m/4) - (k/4\mu)$. Thus

$$s' = \frac{m'}{4} + \frac{k}{4\mu^2}\,\mu'$$

Combining the expressions for m' and μ', obtainable from (13.2-7), we see that

$$\frac{s'}{A'} \cong \frac{1}{4\rho}\cdot\left[1 - \frac{k(2\zeta - 1)}{\bar{\alpha}} + \frac{2k\zeta(2\zeta - 1)}{\bar{\alpha}}\,\epsilon(p - \bar{p})\right] \tag{13.D-3}$$

We also have

$$\frac{m(p)}{m'(p)} \cong \frac{1}{\epsilon'(0)} + (p - \bar{p})$$

Substituting (13.D-2)–(13.D-3) into (13.D-1), we obtain the equation for the middleman's optimal price when he is in equilibrium with the prevailing price,

$$\begin{aligned} \bar{p} - c - \frac{\delta}{Q} &= -\frac{m(\bar{p})}{m'(\bar{p})} + \frac{d}{4\rho}\left[1 - \frac{(2\zeta - 1)k}{\alpha(\bar{p})}\right] \\ &= -\frac{1}{\epsilon'(0)} + \frac{d}{4\rho}\left[1 - \frac{(2\zeta - 1)k}{\alpha(\bar{p})}\right] \end{aligned} \tag{13.D-4}$$

APPENDIX 13-E: MAXIMUM LIKELIHOOD ESTIMATES AND ASYMPTOTIC DISTRIBUTIONS

We summarize some useful facts on the maximum likelihood estimates for easy reference.

13.E.1 Arrival Rate ρ

Let N_i be the number of customers who arrived during the ith reorder cycle. The variable T_i is the length of the ith reorder cycle. Then

$$P(N(T_i) = N_i) = \frac{(\rho T_i)^{N_i} e^{-\rho T_i}}{N_i!}$$

For n reorder cycles the likelihood function is

$$L(N_1, T_i, \ldots, N_n, T_n) = \prod_{i=1}^{n} (\rho T_i)^{N_i} \frac{e^{-\rho T_i}}{N_i!}$$

Maximizing this expression with respect to ρ yields

$$\hat{\rho} = \frac{N}{T} = \frac{\sum_{i=1}^{n} N_i}{\sum_{i=1}^{n} T_i}$$

Clearly

$$E(\hat{\rho}|T) = \frac{1}{T} E(N|T) = \frac{\rho T}{T} = \rho$$

$$E(\rho^2|T) = \frac{1}{T^2} E(N^2|T) = \frac{1}{T^2} \left[T^2\rho^2 + \rho T\right] = \rho^2 + \frac{\rho}{T}$$

$$\text{Var}(\hat{\rho}|T) = \frac{\rho}{T}$$

13.E.2 Demand Shape, Mean and Standard Deviation

Assume the estimation is to be based on the observation of n customers. The demand density is assumed to have a Gamma distribution

$$f(q) = \frac{\mu^\alpha q^{\alpha-1} e^{-\mu q}}{\Gamma(\alpha)} \qquad \text{with } \alpha \text{ an integer}$$

Then if m is the mean and σ the standard deviation of the distribution

$$f(q) = \left(\frac{\alpha}{m}\right)^\alpha q^{\alpha-1} e^{-(\alpha/\mu)q} / \Gamma(\alpha)$$

$$= \left(\frac{\alpha^{1/2}}{\sigma}\right)^\alpha q^{\alpha-1} e^{-(\alpha^{1/2}/\sigma)q} / \Gamma(\alpha)$$

With these three equivalent versions of the demand density the following likelihood functions occur:

$$L(q, \mu, \alpha) = \mu^{n\alpha} \prod_{i=1}^{n} q_i^{\alpha-1} e^{-\mu \Sigma_{i=1}^{n} q_i} / \Gamma^n(\alpha)$$

$$L(q, m, \alpha) = \left(\frac{\alpha}{m}\right)^{\alpha} \prod_{i=1}^{n} q_i^{\alpha-1} e^{-(\alpha/m)\Sigma_{i=1}^{n} q_i} / \Gamma^n(\alpha)$$

$$L(q, \sigma, \alpha) = \left(\frac{\alpha^{1/2}}{\sigma}\right)^{\alpha} \prod_{i=1}^{n} q_i^{\alpha-1} e^{-[(\alpha^{1/2})/\sigma]\Sigma_{i=1}^{n} q_i} / \Gamma^n(\alpha)$$

Taking the natural logarithm of each of these expressions and maximizing them with respect to the desired parameters yields the maximum likelihood estimates,

$$\hat{\mu} = \hat{\alpha} \Big/ \frac{1}{n} \sum_{i=1}^{n} q_i$$

$\hat{\alpha}$ is the solution to

$$\log \hat{\alpha} - \psi(\hat{\alpha}) - \ln \frac{1}{n} \Sigma q_i + \frac{1}{n} \Sigma \ln q_i = 0$$

where

$$\psi(\hat{\alpha}) = \frac{d}{d\alpha} \ln \Gamma(\hat{\alpha}) = \frac{\Gamma'(\hat{\alpha})}{\Gamma(\hat{\alpha})}$$

and asymptotically

$$\psi(\hat{\alpha}) \approx \ln \hat{\alpha} - \frac{1}{2\hat{\alpha}} - \frac{1}{12\hat{\alpha}^2} + o\left(\frac{1}{\hat{\alpha}^2}\right)$$

If α is large enough, then $\hat{\alpha}$ is the solution to

$$\frac{1}{2\hat{\alpha}} - \ln \frac{\Sigma q_i}{n} + \frac{1}{n} \Sigma \ln q_i = 0$$

Epilogue

You undoubtedly would like to know what has been accomplished in this book and also what is likely to be in store for control and system theory in economics.

We wanted to make a case for control and system theory in economics: control and system theory is useful, and even crucial in some cases, to theoretical economics. Economic agents influence each other, and sectors of an economy are interconnected. There are indirect as well as direct transmission paths and/or feedback loops in economies. Some indirect paths are due to the dynamics of the systems and may not be so obvious; thus, counter-intuitive phenomena may occur because of the complexity of these feedback loops. Unless we treat economic systems as dynamic systems with the help of control and system theory, we cannot adequately handle some of the phenomena associated with dynamic feedback loops. We cannot be, nor have we been encyclopedic, in our choice of examples to make our case in support of control and system theory. There are many topics that we have left out and that would have served our purpose just as well. For example, we have hardly touched on coordination, except briefly in Chapter 6, a topic of major interest in welfare economics, the coordination in interdependent economies in international economics, the coordination of decisions in the market places by firms and households, and so on.

In Part I, we introduced the state-space representations as an alternative to more traditional representations in economics, such as simultaneous or final forms, as more useful representations to deal with dynamic economic phenomena. We have established their equivalence to traditional forms and also discussed their properties. We used the state-space representations to introduce and elaborate on the three basic properties of dynamic systems: stability, controllability and observability. We have illustrated the importance of these concepts in delimiting potential performance of dynamic models throughout this book, and how they naturally arise as technical conditions in such questions as existence of stabilization policies, optimal control rules, and so on. We have discussed transfer functions as a dynamic generalization of multipliers in economics in Chapter 2. We have introduced bilinear dynamic systems in Chapter 2 and developed them further in Chapters 3 and 11 as potentially more useful approximations to nonlinear dynamic economic models.

The concepts and tools of state-space representations are used to analyze stability of macroeconomic systems and microeconomic models with no exogenous random disturbances in Part II. We have established connection of the assignment problem in economics with the decoupling problems in control. As a byproduct, we have devised a useful analytical tool for dealing with rationing and short-run and long-run analysis in Chapter 7. Autoregressive moving average models are shown to be equivalent to the state-state models of Chapter 2. Prediction of endogenous variables are tied to the Kalman filter and their asymptotic behavior shown to depend on controllability and observability of the models in Chapters 9 and 10.

Chapters 11 through 13 deal with new results in economic models of imperfect information in which control and system theoretic tools are applied to examine decision-making processes of economic agents without perfect information assumption. These chapters best reflect our view of the potential benefits of system theoretic analysis in economics: first, in microeconomic models of markets, and later in macroeconomics.

To sum up, because there are indirect as well as direct transmission and feedback paths in the economy which are not so obvious, some due to dynamics and some due to feedback paths, counter-intuitive phenomena may arise because of the complexities of the feedback loops. Unless we treat economies as dynamic systems with feedbacks with the help of control and system theory we cannot adequately treat phenomena arising from dynamic feedback loops. For example, by examining controllability and observability properties of dynamic systems, we can decide what kinds of instruments are needed to better stabilize the system or what new signals are desirable to produce more stable market behavior and so on.

If we have been successful in our efforts, we should have convinced you of the usefulness of state-space representations in dynamic economic analysis. If you are tempted to apply some of the theoretical concepts or tools introduced in this book, or if you now have a better intuitive grasp of dynamic systems, then we feel that this book has accomplished its purpose.

REFERENCES

Aitchison, J., and J. A. C. Brown, *The Lognormal Distribution*, Cambridge University Press, London and New York (1957).

Akaike, H., "Stochastic Theory of Minimal Realization," *IEEE Trans Aut. Control* **19** (1974), pp. 667–74.

Akaike, H., "Markovian Representation of Stochastic Processes by Canonical Variables," *SIAM J. Control* **13** (1975), pp. 162–73.

Akaike, H., "Markovian Representation of Stochastic Processes and Its Application to Analysis of Autoregressive Moving Average Processes," To appear in *Ann. Inst. Stat. Math.*

Albert, A. E. and L. A. Gardner, Jr., *Stochastic Approximation and Nonlinear Regression*, The M.I.T. Press, Cambridge, Mass. (1967).

Alchian, A. A., "Information Costs, Pricing and Resource Unemployment," in Phelps *et al.*, *Micro-economic Foundations of Employment and Inflation Theory*, W. W. Norton and Company, Inc., New York (1970), pp. 27–52.

Allen, R. G. D., *Macro-Economic Theory, a Mathematical Treatment*, Macmillan & Company, London (1967).

Alspach, D. L., "A Dual Control for Linear Systems with Control Dependent Plant and Measurement Noise," *Proc. IEEE Conf. Decision and Control* (December 1973), pp. 681–689.

Andersen, L. C., "A Study of Factors Affecting the Money Stock," *Rev. Fed. Res. Bulletin* **51** (October 1965), p. 1379.

Anderson, T. W., *An Introduction to Multivariate Statistical Analysis*, John Wiley & Sons, New York (1968).

Annals of Economic and Social Measurement **1** (4), (1972), and **3** (1), (1974).

Aoki, M., *Optimization of Stochastic Systems*, Academic Press, New York (1967).

Aoki, M., "Aggregation," Chapter 5 in *Optimization Methods for Large-Scale Systems . . . with Applications*, D. A. Wismer (ed.), McGraw-Hill Book Company, Inc., New York (1971).

Aoki, M., "On Feedback Stabilizability of Decentralized Dynamic Systems," *Automatica* **8** (1972), pp. 163–173.

Aoki, M., "On Sufficient Conditions for Optimal Stabilization Policies," *Rev. Econ. Stud.* **40** (1), (Jan. 1973a), pp. 131–138.

Aoki, M., "On a Dual Control Approach to Pricing Policies of a Trading Specialist," Lecture Notes in Computer Science No. 4, *5th Conference on Optimization Techniques, Part II* (1973b), G. Goos (ed.), Springer Verlag, pp. 272–282.

Aoki, M., "Local Controllability of a Decentralized Economic System," *Rev. Econ. Stud.* **41**, (1), No. 125, (Jan. 1974a), pp. 51–64.

Aoki, M., "On Some Price Adjustment Schemes," *Ann. Econ. Soc. Measurements* **3** (Jan. 1974b), pp. 95–115.

Aoki, M., "Noninteracting Control of Macroeconomic Variables: Implications on Policy Mix Considerations," *J. Econometrics* **2**, No. 4 (1974c), pp. 261–281.

Aoki, M., "A Model of Quantity Adjustment by a Firm: A Kiefer-Wolfowitz Like Stochastic Approximation Algorithm as a Bounded Rationality Algorithm," Presented at the North American Meeting, the Econometric Society, San Francisco (Dec. 1974d).

Aoki, M., "On A Generalization of Tinbergen's Condition in the Theory of Policy to Dynamic Models," *Rev. Econ. Stud.* **42** (1975a), pp. 293–296.

Aoki, M., "Control of Linear Discrete-Time Stochastic Dynamic Systems with Multiplicative Disturbances," *IEEE Trans. Aut. Control*, **AC-20** (June 1975b).

Aoki, M., "Customer Arrival Rate as a Signal to a Middleman," Presented at 1975 Econometric Society World Congress (Aug. 1975c).

Aoki, M., "Output Decisions by a Firm: An Example of Dual Control Problem with Information Externality" in *Adaptive Economic Models* (R. H. Day and T. Groves, eds.), Academic Press, Inc., New York (1975d).

Aoki, M., and J. R. Huddle, "Estimation of the State Vector of a Linear Stochastic System with a Constrained Estimator," *IEEE Trans. Aut. Control*, **AC-12** (Aug. 1967), pp. 432–433.

Aoki, M. and M. T. Li, "Partial Reconstruction of State Vectors in Decentralized Dynamic Systems," *IEEE Trans. Aut. Control*, **AC-18** (June 1973), pp. 289–292.

Archibald, G. C., and R. G. Lipsey, "Value and Monetary Theory: Temporary Versus Full Equilibrium," in R. W. Clower (ed.), *Monetary Theory*, Penguin Books, Baltimore (1969).

Arrow, K., "Toward a Theory of Price Adjustment," in M. Abramovitz *et al.*, (eds.), *The Allocation of Economic Resources*, Stanford University Press, Stanford (1959).

Arrow, K., and F. H. Hahn, *General Competitive Analysis*, Holden-Day Inc., San Francisco (1971).

Arrow, K., and M. Kurz, *Public Investment, the Rate of Return, and Optimal Fiscal Policy*, John Hopkins Press, Baltimore (1970).

Athans, M., and D. Kendrick, "Control Theory and Economics: A Survey, Forecast, and Speculations," Report 73–74, Department of Economics, University of Texas, Austin (Oct. 1973).

Barrett, J. F., *et al.*, "Macroeconomic Modeling: A Critical Appraisal," *Proc. IFAC/IFOSRS International Conference on Dynamic Modeling and Control of National Economies*, IEEE Conf. Publication No. 101 (1973), pp. 60–80.

Barro, R. J., "A Theory of Monopolistic Price Adjustment," *Rev. Econ. Stud.*, **34**, (Jan. 1972), pp. 17–26.

Barro, R. J., and H. I. Grossman, "A General Disequilibrium Model of Income and Employment," *American Economic Review* (March 1971), pp. 82–93.

Bar-Shalom, Y., and R. Sivan, "On the Optimal Control of Discrete-Time Linear

Systems with Random Parameters," *IEEE Trans. Aut. Control*, **AC-14**, No. 1 (Feb. 1969), pp. 3–8.

Basile, G., and G. Marro, "On the Perfect Output Controllability of Linear Dynamic Systems," *Ricerche di Automatica* **2**, (1971), pp. 1–10.

Basmann, R. L., "A Note on the Statistical Testability of 'Explicit Causal Chains' Against the Class of 'Interdependent' Models," *J. Am. Stat. Assoc.* **60** (Dec. 1965), pp. 1080–1093.

Baumol, W. J., *Economic Dynamics, an Introduction* (3rd ed.), Macmillan Co., New York (1970).

Bellman, R., *Stability Theory of Differential Equations*, McGraw-Hill Company, Inc., New York, (1953).

Bellman, R., *Introduction to Matrix Theory*, McGraw-Hill Company, Inc., New York (1960).

Benavie, A., *Mathematical Techniques for Economic Analysis*, Prentice Hall, Englewood Cliffs, N.J. (1972).

Bergstrom, A. R., Chapter 5 in *The Construction and Use of Economic Models*, The English Universities Press, Ltd., London (1967).

Beutler, F. J., "On Two Discrete-Time System Stability Concepts and Supermartingales," *J. Math. Anal. App.* **44** (1973), pp. 464–471.

Bhattacharyya, S. P., J. B. Pearson, and W. M. Wonham, "On Zeroing the Output of a Linear System," *Inf. Control* **20** (1972), pp. 135–142.

Blin, J. M., *Patterns and Configurations in Economic Science*, D. Reidel Pub. Co., Boston, Mass., (1973).

Brainard, W., "Uncertainty and the Effectiveness of Policy," *Am. Econ. Rev.* **57**, (May 1967), pp. 411–433.

Bray, J., "Predictive Control of A Stochastic Model of the U. K. Economy, Simulating Present Policy Making Practice by the U. K. Government," *Ann. Econ. Soc. Measurements* **3** (Jan. 1974), pp. 239–256.

Brent, R. P., *Algorithms for Minimization Without Derivatives*, Prentice-Hall, Inc., Englewood Cliffs, N.J. (1973).

Brito, D. L. "Estimation, Prediction and Economic Control," *International Economic Review* **14**, No. 3 (Oct. 1973), pp. 222–231.

Brockett, R. W., and M. Mesarovic, "The Reproducibility of Multivariable Systems," *J. Math. Anal. Appl.* **11** (1965), pp. 548–563.

Brown, G. W., "Iterative Solution of Games by Fictitious Play" in T. C. Koopmans (ed.), *Activity Analysis of Production and Allocation*, John Wiley & Sons, Inc., New York (1951).

Brunner, K. (ed.), *Targets and Indicators of Monetary Policy*, Chandler Pub. Co., San Francisco, Calif. (1969).

Burger, A. E., *The Money Supply Process*, Wadsworth Publishing Co., Inc., Belmont, California (1971).

Chipman, J. S., "The Aggregation Problem in Econometrics," To appear in *Advances in Applied Probability* (University of Alberta).

Chow, G. C., "Problems of Economic Policy from the Viewpoint of Optimal Control," *Am. Econ. Rev.* **43**, No. 5 (Dec. 1973), pp. 825–837.

Chow, G. C., "A Solution to Optimal Control of Linear Systems with Unknown

Parameters," presented at the third NBER Stochatic Control Conference, Washington D. C., (May 1974).

Chung, K. L., "On Stochastic Approximation Method," *Ann. Math. Stat.* **25** (1954), pp. 463–483.

Chung, K. L., *A Course in Probability Theory*, Harcourt, Brace and World, Inc., New York (1968).

Clower, R. W., "Competition, Monopoly, and the Theory of Price," *Pakistan Economic Journal* (Sept. 1955), pp. 219–226.

Clower, R. W., "Oligopoly Theory: A Dynamic Approach," *Proc. 34th Ann. Conf. Western Economic Association* (1959a).

Clower, R. W., "Some Theory of an Ignorant Monopolist," *Economic Journal* (Dec. 1959b), pp. 705–716.

Connors, M. M., "Controllability of Discrete, Linear Random Dynamical Systems," *SIAM J. Control* **5** (May 1967), pp. 183–210.

Cooper, J. P., *Development of the Monetary Sector, Prediction and Policy Analysis—The FRB-MIT-Penn Model.* D.C. Heath & Co., Lexington, Mass., (1974).

Cooper, P., and S. Fischer, "A Method for Stochastic Control of Nonlinear Econometric Models and an Application," *Econometrica* **43** (1975), pp. 147–162.

Cooper, R. N., *Economics of Interdependence*, McGraw-Hill Book Company, Inc., New York (1968).

Corfmat, J. P., and A. S. Morse, "Stabilization with Decentralized Feedback Control," *IEEE Trans. Aut. Control*, **AC-18** (Dec. 1973), pp. 679–682.

Cox, D. R., *Renewal Theory*, John Wiley & Sons, Inc., New York (1962).

Cox, D. R., and P. A. W. Lewis, *The Statistical Analysis of Series of Events*, Methuen and Company, Ltd., London (1966).

Cramér, H., *Mathematical Methods of Statistic*, Princeton University Press, Princeton, New Jersey (1946).

Curtain, R. F. (ed.), *Stability of Stochastic Dynamical Systems*, Lecture Notes in Mathematics, Springer-Verlag, No. 294 New York (1972).

Debreu, G., *Theory of Value an Axiomatic Analysis of Economic Equilibrium*, Cowles Foundation Monograph **17**, John Wiley & Sons, New York (1959).

Derman, C., "An Application of Chung's Lemma to the Kiefer-Wolfowitz Stochastic Approximation Procedure," *Ann. Math. Stat.* **27** (1956), pp. 532–536.

Desoer, C. A., *Notes for a Second Course on Linear Systems*, Van Nostrand Reinhold Co., New York (1970).

Dorato, P., "On the Inverse of Linear Dynamical Systems," *IEEE Trans. Systems Science and Cybernetics* **SSC-5** (Jan. 1969), pp. 43–48.

Dreyfus, S. E., "Some Types of Optimal Control of Stochastic Systems," *SIAM J. Control* **2** (Jan. 1964), pp. 120–134.

Dupač, C., "On the Kiefer-Wolfowitz Approximation Method," MIT Lincoln Lab. Translation 22G-0008, Lexington, MA (Feb. 1960).

Elliott, D. F., and D. D. Sworder, "Applications of a Simplified Multidimensional

Stochastic Approximation Algorithms," *IEEE Trans. Aut. Control*, **AC-15** (1970), pp. 101–104.

Elliott, D. L., "Controllable Nonlinear Systems Driven by White Noise." Ph.D. dissertation, School of Engineering and Applied Science, Univ. of Calif. at Los Angeles (1969).

Fadeeva, V. N., *Computational Methods in Linear Algebra*, Dover Publications, Inc., New York (1955).

Falb, P. L., and W. A. Wolovich, "Decoupling in the Design and Synthesis of Multivariable Control," *IEEE Trans. Aut. Control*, **AC-12** (Dec. 1967), pp. 651–659.

Federal Reserve Bank of Boston, *Controlling Monetary Aggregates*, Conference Series No. 1 (1970).

Fel'dbaum, A. A., "Theory of Dual Control, I–IV," *Automat. Remote Control* **21** (1960), pp. 1240–1249, 1453–1464; **22** (1961), pp. 3–16, 129–143.

Fisher, F. M., "The Stability of the Cournot Oligopoly Solution: The Effects of the Speeds of Adjustment and Increasing Marginal Costs," *Rev. Econ. Stud.* **28** (1961), pp. 125–135.

Fisher, W. D., *Clustering and Aggregation in Economics*, John Hopkins Press, Baltimore, Md. (1969).

Fitts, J. M., "On the Observability of Nonlinear Systems with Applications to Nonlinear Regression Analysis," *Information Sciences* **4** (1972), pp. 129–156.

Fleming, J. M., "Targets and Instruments," *IFM Staff Papers* **15** (Nov. 1968), pp. 387–402.

Friedman, B. M., "Optimal Economic Stabilization Policy: An Extended Framework," *J. P. Econ.* **80** (1972), pp. 1002–1022.

Friedman, B. M., *Methods in Optimization for Economic Stabilization Policies*, North Holland Publ. Co., Amsterdam (1974).

Frisch, R., "On the Motion of Equilibrium and Disequilibrium," *Rev. Econ. Stud.* **3** (1935–36), pp. 100–106.

Gantmacher, F. R., *The Theory of Matrices*, Vol. 1, Chelsea Pub. Company, New York (1959).

Gilbert, E. G., "The Decoupling of Multivariable Systems by State Feedback," *SIAM J. on Control* **7** (1969), pp. 50–63.

Goka, T., T. J. Tarn, and D. L. Elliott, "On the Controllability of Discrete Bilinear Systems," *Automatica* **9** (1973), pp. 615–622.

Gould, J. P., and C. R. Nelson, "The Stochastic Structure of the Monetary and Macroeconomic Theory," *Am. Econ. Rev.* **64** (1974), pp. 405–418.

Gould, J. R., and S. G. B. Henry, "The Effects of Price Control on a Related Market," *Economic J.*, (Feb. 1967), pp. 42–49.

Grandmont, J-M., and Y. Younes, "On the Role of Money and the Existence of a Monetary Equilibrium," *Rev. Econ. Stud.* **34** (July 1972), pp. 355–372.

Gramlich, E. M., "The Usefulness of Monetary and Fiscal Policy as Discretionary Stabilization Tools," *J. Money, Credit and Banking* **3** (1971), pp. 506–532.

Grossman, H. I., "Theories of Markets without Recontracting," *J. Econ. Theory* **1** (Dec. 1969), pp. 476–479.

Grossman, H. I., "Money, Interest and Prices in Market Disequilibrium," *J.P.E.* (Sept./Oct. 1971), pp. 943–961.

Gunckel, T. L. II, and G. F. Franklin, "A General Solution for Linear Sample-data Control," *Trans. ASME, J. Basic Eng., Ser. D.*, **85** (June 1963), pp. 197–203.

Haavelmo, T., "What can Static Equilibrium Models Tell Us," *West. Econ. Jou.* **12** (March 1974), pp. 27–34.

Hadley, G., and M. C. Kemp, *Variational Methods in Economics*, North Holland Publ. Co., Amsterdam, (1971).

Haynes, G. W., and H. Hermes, "Nonlinear Controllability via Lie Theory," *SIAM J. Control* **8** No. 4, (Nov. 1970), pp. 450–460.

Henderson, D. W., and S. J. Turnovsky, "Optimal Macroeconomic Policy Adjustment Under Condition of Risk," *Jou. Econ. Theory* **4** (1972), pp. 58–71.

Hermes, H. and J. P. LaSalle, *Functional Analysis and Time Optimal Control*, Academic Press, New York (1969).

Heymann, M., "Comments on Pole Assignemnt in Multi-Input Controllable Linear Systems," *IEEE Trans. Aut. Control*, **AC-13** (Dec. 1968), pp. 748–749.

Hodges, J. L., and E. L. Lehmann, "Two Approximations to the Robins-Monro Process," *Proc. 3rd Berkeley Symposium on Math. Statistics and Probability*, University of Calif. Press, Berkeley (1956), pp. 95–104.

Holbrook, R. S., "Optimal Economic Policy and the Problem of Instrument Stability," *Am. Econ. Rev.* **62** (March 1972), pp. 57–65.

Holt, C. C., "Linear Decision Rules for Economic Stabilization and Growth," *Q. J. Econ.* **76**, No. 1 (Feb. 1962), pp. 20–45.

Holt, C. C., F. Modiglinani, J. F. Muth, and H. A. Simon, *Planning, Production, Inventories and Work Force*, Prentice-Hall, Englewood Cliffs, N. J. (1960).

Householder, A. S., *The Theory of Matrices in Numerical Analysis*, Blaisdell Publishing Co., New York (1964).

Howitt, P., "The Short Run Dynamics of Monetary Exchange," Chapter 3, unpublished Ph.D. dissertation, Northwestern University (1973).

IFAC/IFORS International Conference on Dynamic Modeling and Control of National Economics, IEEE Conf. Pub. No. 101, London, England (1973).

Intriligator, M. D., *Mathematical Optimization and Economic Theory*, Sec. 11.3, Prentice-Hall, Englewood Cliffs, N. J. (1971).

Joseph, P. D., and J. T. Tou, "On Linear Control Theory," *AIEE Trans. (Appl. Ind.)* **80** (Sept. 1961), pp. 193–196.

Kalman, R., "A Contribution to the Theory of Optimal Control," *Bol. Socied. Mat. Mexicana* **5** (1960), pp. 102–119.

Kaplan, W., *Operational Methods for Linear Systems,* Addison-Wesley Pub. Co., Reading, Mass. (1962).

Karlin, S., *Mathematical Methods and Theory in Games, Programming, and Economics*, Addison-Wesley, Reading, Massachusetts (1959).

Kendrick, D. A., "On the Leontief Dynamic Inverse," *Quart. J. Economics* (1972), **86** pp. 653–696.

Kendrick, D. A., and J. Mayors, "Stochastic Control with Uncertain Macroeconomic Parameters," Presented at 2nd Stochastic Control Conf. NBER, University of Chicago, Chicago, Ill. (May 1973).

Kiefer, J., and J. Wolfowitz, "Stochastic Estimation of the Maximum of a Regression Function," *Ann. Math. Stat.* **23** (1952), pp. 462–466.

Kleinman, D. L., "Optimal Stationary Control of Linear Systems with Control-dependent Noise," *IEEE Trans. Aut. Control, AC*-14 (Dec. 1969) pp. 673–677.

Kmenta, J., and P. E. Smith, "Automonomous Expenditure Versus Money Supply: An application of Dynamic Multiplier," *Rev. Econ. Stat.* **55**, (Aug. 1973), pp. 299–307.

Knopp, K., *Infinite Sequences and Series*, Dover Publications, Inc., New York (1956).

Koopmans, T. C., H. Rubin and R. B. Leipnik, "Measuring the Equation Systems of Dynamic Economics," Chapter II in Koopmans, T. C. (ed.) *Statistical Inference in Dynamic Economics*, John Wiley & Sons, New York (1950).

Kou, S. R., D. L. Elliott, and T. J. Tarn, "Observability of Nonlinear Systems," *Information and Control* **22** No. 1, (1973), pp. 89–99.

Kučera, V., "A Contribution to Matrix Quadratic Equations," *IEEE Trans. Aut. Control*, **AC-17** (June 1972), pp. 344–346.

Kushner, H. J., *Stochastic Stability and Control*, Academic Press, New York, (1967).

Kushner, H. J., "Stochastic Approximation Algorithms for Local Optimization of Functions with Nonunique Stationary Points," *IEEE Trans. Aut. Control*, **AC-17** (Oct. 1972), pp. 646–654.

Kushner, H. J., "Some Basic Ideas in Stochastic Stability," *Ann. Econ. Soc. Meas.* **3**, No. 1 (Jan. 1974), pp. 85–90.

Lakshmikantham, V. and Leela, S., *Differential and Integral Inequalities* (Vols. 1 and 2), Academic Press, New York (1969).

LaSalle, J. and Lefchetz, S., *Stability by Liapunov's Direct Method*, Academic Press, New York (1961).

Lefschetz, S., *Differential Equations: Geometric Theory*, Interscience Publishers, New York (1962).

Leijonhufvud, A., "Effective Demand Failure," *Swed. Econ. J.* **75** (March 1973), pp. 27–48.

Leijonhufvud, A., "The Varieties of Price Theory: What Microfoundations for Macrotheory," Discussion Paper No. 44, Department of Economics, UCLA, Los Angeles (January 1974).

Leondes, C. T., and L. Novak, "Reduced-Order Observers for Linear Discrete-Time Systems," *Proc. IEEE Conference on Decision and Control* (1972), pp. 206–210.

Livesey, D. A., "Control Theory and Input-Output Analysis," *Int'l. J. System Science* **2** (1971), pp. 307–318.

Livesey, D. A., "Can Macro-Economic Planning Problems Ever be Treated as a Quadratic Regulator Problem?," *Proc. IFAC/IFORS International Conference on Dynamic Modeling and Control of National Economies* (Aug. 1973), IEE Conf. Pub. No. 101, London, England, pp. 1–14.

Lucas, R. E., Jr., "Some International Evidence on Output-Inflation Trade-offs." *Amer. Econ. Rev.* **63** (1973a), pp. 326–34.

Lucas, R. E., Jr., "Econometric Policy Evaluation: A Critique," Graduate School of Industrial Administration, Carnegie-Mellon University (May 1973b).

Lucas, R. E., Jr., and Edward C. Prescott, "Equilibrium Search and Unemployment," *J. Econ. Theory* **7** (1974), pp. 188–209.

Luenberger, D. G., "An Introduction to Observers," *IEEE Trans. Aut. Control*, **AC-16**, (Dec. 1971), pp. 596–602.

McFadden, D., "On the Controllability of Decentralized Macroeconomic Systems: The Assignment Problem," in *Mathematical Systems Theory and Economics I*, H. W. Kuhn and G. P. Szego (eds.), Springer-Verlag, New York (1969).

MacLane, P. J., "Optimal Stochastic Control of Linear Systems with State and Control-dependent Disturbances," *IEEE Trans. Aut. Control*, **AC-16** (Dec. 1971), pp. 793–798.

Malinvaud, E., "First Order Certainty Equivalence." *Econometrica* **37** (Oct. 1969), pp. 706–18.

Marschak, J., and R. Radner, *Team Theory*, Yale University Press, New Haven, Conn. (1971).

Matrosov, V. M., "Method of Lyapunov Vector Functions in Feedback Systems," *Automation and Remote Control* **33** (1972), pp. 1458–1468.

Mehra, R. K., "On-Line Identification of Linear Dynamic Systems with Applications to Kalman Filtering," *IEEE Trans.*, **AC-16**, No. 1, (Feb. 1971), pp. 12–21.

Mehra, R. K., "Identification in Control and Econometrics; Similarities and Differences," Tech. Rept. Div. Engineering and Applied Physics, Harvard University, Cambridge, Mass. (1973).

Michel, A. N., "Stability Analysis of Interconnected Systems," *SIAM J. Control* **12** (Aug. 1974), pp. 554–579.

Mortensen, D. T., "Rational Price Dispersion, Search and Adjustment" Presented at the North American Econometric Society Meeting, San Francisco, (Dec. 1974).

Muench, T. and N. Wallace, "On Stabilization Policy: Goals and Models," *AER* **64** (1974), pp. 330–337.

Miller, K. S., *Linear Difference Equation*, W. A. Benjamin, New York (1968).

Morse, A. S., and W. M. Wonham, "Status of Noninteracting Control," *IEEE Trans. Aut. Control*, **AC-16**, (Dec. 1971), pp. 568–581.

Mundell, R. A., "The Monetary Dynamics of International Adjustment Under Fixed or Flexible Exchange Rates," *Quart. Jou. Economics* **74** (1960), pp. 227–257.

Mundell, R. A., "Appropriate Use of Monetary and Fiscal Policy for Internal and External Stability," *IMF Staff Papers* **9** (1962), pp. 70–79.

Mundell, R. A., "Hicksian Stability, Currency Markets, and the Theory of Economic Policy," in *Value, Capital and Growth*, J. N. Wolfe (ed.), Edinburgh University Press, Edinburgh, Scotland (1968).

Murphy, W. J., "Optimal Stochastic Control of Discrete Linear Systems with Unknown Gain," *IEEE Trans. Aut. Control*, **AC-13**, No. 4 (Aug. 1968), pp. 338–344.

Muth, J. "Rational Expectation and the Theory of Price Movements," *Econometrica* **29** (July 1961), pp. 315–335.

Negishi, T., "The Stability of Competitive Economy: A Survey Article," *Econometrics* **30** (1962), pp. 635–669.

Nerlove, M., "Lags in Economic Behavior," *Econometrica* **40** (1972), pp. 221–252.

Newman, P., "Some Notes on Stability Conditions," *Rev. Econ. Stud.* **27** (1959–60), pp. 1–9.

Newman, P., "Approaches to Stability Analysis," *Economica*, Series 2, **28** (1961), pp. 12–29.

Nikaido, H., Chapter VII, *Convex Structures and Economic Theory*, Academic Press, New York (1968).

Norman, A. L., "On the Relationship between Linear Feedback Control and First Period Certainty Equivalence," *Int'l. Econ. Rev.* **15** (1974), pp. 209–215.

Patrick, J. D., "Establishing Convergent Decentralized Policy Assignment," *J. Int'l. Economic* **3** (1973), pp. 37–52.

Payne, H. J. and L. M. Silverman, "Matrix Riccati Equations and System Structure," *Preprint: IEEE Conference on Decision and Control* (1973), pp. 558–563.

Peng, T. K. C., "Invariance and Stability for Bounded Uncertain Systems," *SIAM J. Control* **10** (1972), pp. 679–690.

Peston, M., "Econometrics and Control: Some General Comments," *Proc. IFAC/IFORS International Conference on Dynamic Modeling and Control of National Economics* (July 1973), pp. 9–12; IEEE Conf. Pub. No. 101 (1973), London, England, pp. 15–30.

Phelps, E. S., *et al.*, *Microeconomic Foundations of Employment and Inflation Theory*, W. W. Norton and Company Inc., New York (1970).

Phillips, A. W., "Stabilization Policy in a Closed Economy," *Econ. J.* **64** (June 1954), pp. 290–323.

Pindyck, R. S., *Optimal Planing for Economic Stabilization*, North Holland Publ. Co., Amsterdam (1973).

Pindyck, R. S., and A. Martens, "Optimal Control Theory and Development Planning," *Proc. IEEE Conference on Decision and Control*, IEEE Catalog No. 72 CHO 705-4 SCS, (1972), pp. 31–36.

Polak, E., and E. Wong, *Notes for a First Course on Linear Systems*, Van Nostrand Reinhold, New York (1970).

Pontryagin, L. S., *et al.*, *The Mathematical Theory of Optimal Processes*, Interscience Publishers, New York (1962).

Prescott, E. C., "Adaptive Decision Rules for Macroeconomic Planning," *Western Econ. J.* **9** (Dec. 1971), pp. 369–78.

Prescott, E. C., "The Multi-Period Control Problem under Uncertainty," *Econometrica* **40** (Nov. 1972), pp. 1043–58.

Prescott, E. C., "Market Structure and Monopoly Profits: A Dynamic Theory," *J. Econ. Theory* **6** (1973), pp. 546–557.

Preston, A. J., "A Dynamic Generalization of Tinbergen's Theory of Policy," *Rev. Econ. Stud.* **41** (1974), pp. 65–74.

Preston, A. J., and K. D. Wall, "Some Aspects of the Use of State Space Models in Econometrics," in *Proc. IFAC/IFORS International Conference on Dynamic Modeling and Control of National Economies* (July 1973), IEE Conf. Pub. No.

101, London, England (1973), pp. 226–239.
Rausser, G. C., and J. W. Freebairn, "Approximate Adaptive Control Solutions to U. S. Beef Trade Policy," *Ann. Econ. Soc. Measurement* **3** (Jan. 1974), pp. 177–203.
Rink, R. E., and R. R. Mohler, "Completely Controllable Bilinear Systems," *SIAM J. Control* **6** (1968), pp. 477–486.
Rissanen, J., "An Algebraic Approach to the Problems of Linear Prediction and Identification," IBM Research Report RJ468 (Oct. 1967).
Robbins, H., and S. Monro, "A Stochastic Approximation Method," *Ann. Math. Stat.* **22** (1951), pp. 400–407.
Rothschild, M., "Models of Market Organization with Imperfect Information: A Survey," *Jou. Pol. Econ.*, **81**, No. 6, (Nov./Dec., 1973), pp. 1283–1308.
Sain, M. K. and J. L. Massey, "Invertibility of Linear Time-Invariant Dynamical Systems," *IEEE Trans. Aut. Control*, **AC-14** (April 1969), pp. 141–149.
Samuelson, P. A., "Interactions Between the Multiplier Analysis and the Principle of Acceleration," *Rev. Econ. Stud.* **21** No. 2, (May 1939), pp. 75–78.
Samuelson, P. A., *Foundations of Economic Analysis*, Harvard University Press Cambridge, Massachusetts (1947).
Samuelson, P. A., "Mathematics of Speculative Price," in *Mathematical Topics in Economic Theory and Computation*, Richard H. Day and Stephen M. Robinson (eds.), SIAM, Philadelphia, Penn. (1972), pp. 1–42.
Saperstone, S. H. and Yorke, J. A., "Controllability of Linear Oscillatory Systems using Positive Controls" in J. A. Yorke (ed.), *Seminar on Differential Equations and Dynamical Systems, II*, Lecture Notes in Mathematics, Springer-Verlag, New York (1970).
Sargent, T. J. and N. Wallace, "Rational Expectations, the Optimal Monetary Instrument and the Optimal Money Supply Rule," *J. Pol. Econ.* **82** (April 1975), pp. 241–254.
Savant, C. J., *Basic Feedback Control System Design*, McGraw-Hill Book Company, Inc., New York (1958).
Scarf, H. E., D. M. Gilford, and M. W. Shelly (eds.), *Multistage Inventory Models and Techniques*, Stanford University Press, Stanford, Calif. (1963).
Schmidt, P. and R. N. Waud, "The Almon Lag Technique and the Monetary Versus Fiscal Policy Debate," *J. Amer. Stat. Assoc.* **68** (March 1973), pp. 11–19.
Shell, K., *Essays on the Theory of Optimal Economic Growth*, MIT Press, Cambridge, Mass. (1966).
Shupp, F. R., "Optimal Control, Uncertainty and a Temporary Income Policy," *Proc. IEEE Conf. on Decision and Control*, New Orleans, La. (1972), pp. 21–25.
Siljak, D. D., "Stability of Large-Scale Systems Under Structural Perturbations," *IEEE Transaction SMC*-3, (1973), pp. 415–417.
Silverman, L. M., and H. E. Meadows, "Controllability and Observability in Time-Variable Linear Systems," *J. SIAM Control* **5** (1969), pp. 64–73.
Silverman, L. M., and H. J. Payne, "Input-Output Structure of Linear Systems with Applications to the Decoupling Problem," *SIAM J. Control* **9** (May 1971), pp. 199–233.

Simon, H. A., "Dynamic Programming under Uncertainty with a Quadratic Criterion Function," *Econometrica* **24** (Jan. 1956), pp. 74–81.

Simon, H. A. and A. Ando, "Aggregation of Variables in Dynamic Systems," *Econometrica* **29** (April 1961), pp. 111–138.

Simons, H. C., "Rules vs. Authorities in Monetary Policy," *Journal of Political Economy* **44** (Feb. 1936), pp. 1–30.

Swoboda, A. K., "On Limited Information and the Assignment Problem," in Claassen, E., and P. Salin (eds.), *Stabilization Policies in Interdependent Economies*, North Holland Pub. Co., Amsterdam (1972).

Sworder, D. D., *Optimal Adaptive Control System*, Academic Press, New York (1966).

Takahashi, T., *Mathematics of Automatic Control*, Holt, Rinehart and Winston Inc., New York (1966).

Tarn, R. J., D. L. Elliott, and T. Goka, "Controllability of Discrete Bilinear Systems with Bounded Control," *IEEE Trans. Aut. Control*, **AC-18**, (June 1973), pp. 298–301.

Taylor, J. B., "Asymptotic Properties of Multiperiod Control Rules in the Linear Regression Model," Tech. Report No. 79, Institute for Mathematical Studies in the Social Sciences, Stanford University, Stanford, Calif. (Dec. 1972).

Taylor, J. B., "A Criterion for Multiperiod Controls in Economic Models with Unknown Parameters," Presented at 2nd Stochastic Control Conf. NBER, University of Chicago, Chicago, Ill. (May 1973).

Theil, H., "Linear Decision Rules for Macrodynamic Policy Problems," in *Quantitative Planning of Economic Policy*, B. G. Hickman (ed.), The Brookings Institute, Washington, D. C. (1965).

Theil, H., "A Note on Certainty Equivalence in Dynamic Planning," *Econometrica* **25** (April 1957), pp. 346–349.

Theil, H., and T. Kloek, "The Operational Implications of Imperfect Models," in *Math. Methods in the Social Sciences*, K. J. Arrow *et al.*, (ed.), Stanford University Press, Stanford, Calif. (1960).

Tinbergen, J., *On the Theory of Economic Policy* (2nd ed.), North Holland Publ. Co., Amsterdam (1955).

Tinsley, P., *et al.*, "On NERFF Solutions of Macroeconomic Tracking Problems," FRB Special Studies Paper No. 48 (1974).

Truxal, J. G., *Automatic Feedback Control System Synthesis*, McGraw-Hill Book Company, Inc., New York (1955).

Tse, E., and M. Athans, "Adaptive Stochastic Control for a Class of Linear Systems," *IEEE Trans. Aut. Control*, **AC-17**, No. 1 (Feb. 1972), pp. 38–52.

Tse, E., Y. Bar-Shalom and L. Meier, III, "Wide-Sense Adaptive Dual Control for Nonlinear Stochastic Systems," *IEEE Trans. Aut. Control*, **AC-18** (April 1973), pp. 109–117.

Turnovsky, S. J., "The Stability Properties of Optimal Economic Policies," *Am. Econ. Rev.* **64**, No. 1 (March 1974), pp. 136–148.

Tustin, A., *The Mechanism of Economic Systems; an Approach to the Problem of Economic Stabilization from the Point of View of Control-System Engineering*,

Cambridge, Harvard University Press, Cambridge, Mass. (1953).

Wasan, M. T., *Stochastic Approximation*, Cambridge University Press, Cambridge, England (1969).

Waud, R., "Proximate Targets and Monetary Policy," *Economic J.*, **83** (March 1973), pp. 1–20.

Weiss, L., "Controllability for Various Linear and Nonlinear System Models," in J. A. York (ed.), *Seminar on Differential Equations and Dynamical Systems, II*, No. 144, Lecture Notes in Mathematics, Springer-Verlag, New York (1970), pp. 250–261.

Wilkinson, J. H., *The Algebraic Eigenvalue Problem*, Oxford Clarendon Press, Oxford, England (1965).

Wilks, S. S., *Mathematical Statistics*, John Wiley & Sons, Inc., (1962).

Wolovich, W. A., "Static Decoupling," *IEEE Trans. Aut. Control*, **AC-18**, (1973), pp. 536–537.

Wolovich, W. A., *Linear Multivariable Systems*, Springer-Verlag, New York (1974).

Wonham, W. M., "Optimal Stationary Control of A Linear System with State-dependent Noise," *SIAM J. Control* **5** (1967), pp. 486–500.

Wonham, W. M., "On Pole Assignment in Multi-Input Controllability Linear Systems," *IEEE Trans. Aut. Control*, **AC-12** (1967), pp. 660–665.

Wonham, W. M., and A. S. Morse, "Decoupling and Pole Assignment in Linear Multivariable Systems: A Geometric Approach," *SIAM J. Control* **8** No. 1, (Feb. 1970), pp. 1–18.

Woodside, C. M., "Uncertainty in Policy Optimization Experiments on a Large Econometric Model," in *Proc. IFAC/IFORS International Conference on Dynamic Modeling and Control of National Economies* (July 1973), IEE Conf. Pub. No. 101, London, England (1973), pp. 418–425.

Yüksel, Y. O., and J. J. Bongiorno, Jr., "Observers for Linear Multivariable Systems with Applications," *IEEE Trans. Aut. Control*, **AC-16**, (Dec. 1971), pp. 603–613.

Zadeh, Lotfi A., and Charles A. Desoer, *Linear Systems Theory: The State Space Approach*, McGraw-Hill Book Company, Inc., New York (1963).

Appendix A

Functions of Matrices

We formally define $f(A)$ where A is an $n \times n$ matrix for any $f(\cdot)$ having a Taylor series expansion of the form

$$f(\lambda) = \alpha_0 + \alpha_1\lambda + \alpha_2\lambda^2 + \cdots$$

to be equal to

$$f(A) = \alpha_0 I + \alpha_1 A + \alpha_2 A^2 + \cdots$$

provided the series converges, i.e., n^2 series of all n^2 elements of the $n \times n$ matrices converge.

For example $e^{at} = \Sigma_{n=0}^{\infty}(at)^n/n!$ converges absolutely for any finite t. Then, from $\|A^n\| < \|A\|^n$, we see that

$$\|I + At + (At)^2/2! + \cdots \| \leqslant \|I\| + \|At\| + \|At\|^2/2! + \cdots$$

$$= \exp \|At\|$$

where we define $\|A\|^2$ = maximum eigenvalue of $A^{t}A$. Thus, we can define $\exp At$ by $\Sigma_{n=0}^{\infty}(At)^n/n!$.

In computing $\exp At$ and in other operations involving A, the next theorem is very useful.

Cayley–Hamilton Theorem *Every matrix satisfies its own characteristic equation considered as the matrix polynomial.*

In other words, let A be any $n \times n$ matrix. Its characteristic equation is defined to be

$$\rho(\lambda) = |\lambda I - A|$$

where $|\ |$ denotes determinant. It is an nth order polynomial in λ,

$$\rho(\lambda) = \lambda^n + \alpha_1\lambda^{n-1} + \cdots + \alpha_n$$

The corresponding matrix polynominal in A is

$$\rho(A) = A^n + \alpha_1 A^{n-1} + \cdots + \alpha_n I$$

The Cayley-Hamilton Theorem states that

$$\rho(A) = 0$$

Some special cases are easily verified. For example, suppose A has n distinct eigenvalues so that there exists a nonsingular matrix T such that

$$T^{-1}AT = \text{diag}(\lambda_1, \ldots, \lambda_n)$$

Then

$$\begin{aligned} T^{-1}\rho(A)T &= \rho(T^{-1}AT) \\ &= \text{diag}(\rho(\lambda_1), \ldots, \rho(\lambda_n)) \end{aligned}$$

or

$$\rho(A) = T\,\text{diag}(\rho(\lambda_1), \ldots, \rho(\lambda_n))T^{-1}$$

since

$$T^{-1}A^kT = (T^{-1}AT)(T^{-1}AT) \cdots (T^{-1}AT)$$

k-fold multiplication with $k = 1, 2, \ldots$

Now

$$\rho(\lambda_i) = 0, \qquad i = 1, \ldots, n$$

since λ's are the roots of the characteristic equation. Therefore $\rho(A) = 0$. For the proof of the general case, see Bellman (1960).

One useful application of the Cayley-Hamilton Theorem is to compute high order polynomials in A or functions defined by infinite series such as $\exp At$. From the theorem, we have

$$A^n = -\left(\alpha_1 A^{n-1} + \cdots + \alpha_n I\right)$$

Multiplying both sides by A,

$$\begin{aligned} A^{n+1} &= -\alpha_1 A^n - \left(\alpha_2 A^{n-1} + \cdots + \alpha_n A\right) \\ &= -\alpha_1\left(\alpha_1 A^{n-1} + \cdots + \alpha_n I\right) - \left(\alpha_2 A^{n-1} + \cdots + \alpha_n A\right) \end{aligned}$$

showing that A^n and A^{n+1} are expressible as polynomials in A of *at most* degree $n - 1$. We see easily that this is true for A^k, $k \geqslant n$. This is a special case of the next fact.

Fact *Let $p(\lambda)$ be a polynomial in λ of at least degree n. Then $p(A) = r(A)$, where $r(\lambda)$ is the remainder when $p(\lambda)$ is divided by the characteristic polynomial $\rho(\lambda)$.*

This fact becomes obvious when we perform the long division

$$p(\lambda) = \rho(\lambda)q(\lambda) + r(r)$$

and note that $\rho(A) = 0$ by the Cayley-Hamilton Theorem.

We next demonstrate the application of the theorem to evaluate $\exp At$.

We have by definition $\exp At = I + At + (At)^2/2! + \cdots$. For any $k \geqslant n$, we have

$$(At)^k/k! = \frac{t^k}{k!}A^k$$

$$= \frac{t^k}{k!}r_k(A)$$

where the Fact is used

$$A^k = r_k(A), \qquad k \geqslant n$$

where $r_k(\lambda)$ is the remainder term after dividing λ^k by $\rho(\lambda)$. Thus

$$\exp At = I + At + \cdots + (At)^{n-1}/(n-1)!$$

$$+ \frac{t^n}{n!}r_n(A) + \frac{t^{n+1}}{(n+1)!}r_{n+1}(A) + \cdots$$

where

$$r_k(\lambda) = \beta_0{}^k + \beta_1{}^k\lambda + \cdots + \beta_{n-1}{}^k\lambda^{n-1}$$

Thus

$$\exp At = \left(\sum_{k=0}^{\infty} \beta_0{}^k\right)I + \left(\sum_{k=0}^{\infty} \beta_1{}^k\right)A + \cdots + \left(\sum_{k=0}^{\infty} \beta_{n-1}{}^k\right)A^{n-1} \qquad \text{(A-1)}$$

where

$$\beta_i^k = \begin{cases} t^i/i!, & k = i \\ 0, & k \neq i, \quad k < n \end{cases}$$

Since the original series converges, the coefficients in (A.1) are all finite.

Equation (A.1) shows that $\exp At$ is a polynomial of degree $(n - 1)$ in A,

$$\exp At = \alpha_0 I + \alpha_1 A + \cdots + \alpha_{n-1} A^{n-1}$$

To determine the coefficients if A has n distinct eigenvalues we can use

$$e^{\lambda_i t} = \alpha_0 + \alpha_1 \lambda_i + \cdots + \alpha_{n-1} \lambda_i^{n-1}, \qquad i = 1, \ldots, n$$

to solve for α's.

If some λ's are repeated m times we obtain m independent equations by evaluating $d_j / d\lambda^j_i \, e^{\lambda_i t}, j = 1, 2, \ldots, m - 1$. Again, we can solve for α's. See Polak-Wong (1970) or Zadeh-Desoer (1963) for further details.

Appendix B

Some Useful Relations

Lemma 1 *Given a matrix with compatible submatrices A, B, C and D*

$$\begin{pmatrix} A & B \\ C & D \end{pmatrix}$$

its determinant is given by

$$\det\begin{pmatrix} A & B \\ C & D \end{pmatrix} = |A| \cdot |D - CA^{-1}B|, \qquad \text{if } |A| \neq 0 \tag{B-1}$$

$$\det\begin{pmatrix} A & B \\ C & D \end{pmatrix} = |D| \cdot |A - BD^{-1}C|, \qquad \text{if } |D| \neq 0 \tag{B-2}$$

This fact is a straightforward generalization of the fact that addition of a multiple of a row to another row does not change the determinant value. For a proof, see Gantmacher (1959, page 45).

Lemma 2 *Let $\boldsymbol{a}$ and $\boldsymbol{b}$ be any n-dimensional vectors. Then*

$$|I + \boldsymbol{a}\boldsymbol{b}^{\mathrm{t}}| = 1 + \boldsymbol{b}^{\mathrm{t}}\boldsymbol{a} \tag{B-3}$$

$$(I + \boldsymbol{a}\boldsymbol{b}^{\mathrm{t}})^{-1} = I - \frac{\boldsymbol{a}\boldsymbol{b}^{\mathrm{t}}}{1 + \boldsymbol{b}^{\mathrm{t}}\boldsymbol{a}} \tag{B-4}$$

Proof The relation (B-4) can be verified by direct computation.

The relation (B-3) may be established by induction on the dimension n. Clearly, it is true for $n = 1$. Suppose that (B-3) is true for n. We show that it is true also for $n + 1$.

Let I_n be the $n \times n$ identity matrix. Let α and β be scalars.

$$\left| I_{n+1} + \begin{pmatrix} \boldsymbol{a} \\ \alpha \end{pmatrix} (\boldsymbol{b}^{\mathrm{t}} \; \beta) \right| = \begin{vmatrix} I_n + \boldsymbol{a}\boldsymbol{b}^{\mathrm{t}} & \boldsymbol{a}\beta \\ \alpha\boldsymbol{b}^{\mathrm{t}} & 1 + \alpha\beta \end{vmatrix}$$

$$= |I_n + \boldsymbol{a}\boldsymbol{b}^{\mathrm{t}}| \left[1 + \alpha\beta - \alpha\beta\boldsymbol{b}^{\mathrm{t}}(I_n + \boldsymbol{a}\boldsymbol{b}^{\mathrm{t}})^{-1}\boldsymbol{a} \right]$$

where Lemma 1 is used.

Apply (B·4) to the second factor of the above expression. Then the right-hand side becomes

$$|I_n + \boldsymbol{a}\boldsymbol{b}^{\mathrm{t}}|\left(1 + \frac{\alpha\beta}{1 + \boldsymbol{b}^{\mathrm{t}}\boldsymbol{a}}\right) = 1 + \boldsymbol{b}^{\mathrm{t}}\boldsymbol{a} + \alpha\beta$$

$$= 1 + (\boldsymbol{b}^{\mathrm{t}} \quad \beta)\binom{\boldsymbol{a}}{\alpha}$$

as required.

Corollary

$$|D + \boldsymbol{a}\boldsymbol{b}^{\mathrm{t}}| = |D|(1 + \boldsymbol{b}^{\mathrm{t}}D^{-1}\boldsymbol{a})$$

where D is a diagonal matrix.

$$(A + \boldsymbol{a}\boldsymbol{b}^{\mathrm{t}})^{-1} = A^{-1} - \frac{A^{-1}\boldsymbol{a}\boldsymbol{b}^{\mathrm{t}}A^{-1}}{1 + \boldsymbol{b}^{\mathrm{t}}A^{-1}\boldsymbol{a}}$$

when A^{-1} exists.

Appendix C

Approximate Determination of Roots of a Polynomial

Suppose a set of n distinct negative real numbers $\lambda_1 < \lambda_2 < \ldots < \lambda_n < 0$ and another set of n numbers $\theta_1, \ldots, \theta_n$ are used to define a polynomial of degree n

$$g(\lambda) = \pi(\lambda)\left\{1 - \sum_{i=1}^{n} \frac{\theta_i}{\lambda - \lambda_i}\right\} \tag{C-1}$$

with

$$\pi(\lambda) = \prod_{i=1}^{n} (\lambda - \lambda_i).$$

Here we assume θ_i's are all positive. The cases where θ_i's are all negative or of indeterminate signs can be treated analogously.

If $\Sigma_{i=1}^{n}\theta_i/\lambda_i = 1$, then clearly $g(0) = 0$. We carry out our consideration for this case. The other possibilities $g(0) > 0$ or $g(0) < 0$ can be discussed analogously (see also Wilkinson, 1965).

We see that

$$\begin{aligned} g(\lambda_i) &= -\pi'(\lambda_i)\theta_i \\ &= -\theta_i \prod_{j \neq i} (\lambda_i - \lambda_j) \end{aligned} \tag{C-2}$$

By the assumption on the ordering of λ_i's, we see that the sign of $g(\lambda)$ changes as λ increases from $\lambda < \lambda_1$ to $\lambda_1 < \lambda < \lambda_2, \ldots,$ to $\lambda_n < \lambda$ with $g(\lambda_n) < 0$. Thus we have the following Lemma.

Lemma 3 *The polynomial* $g(\lambda)$ *defined by* (C-1) *has* n *distinct roots* $\lambda_1 < \mu_1 < \lambda_2, \lambda_2 < \mu_2 < \lambda_3, \ldots, \lambda_{n-1} < \mu_{n-1} < \lambda_n, \mu_n = 0$.

The first $n - 1$ roots of $g(\lambda)$ may be determined approximately by the Newton's method using $(\lambda_i + \lambda_{i+1})/2$ as the starting value, for example.

When more accurate approximation is desired, Newton's method may be applied successively, or parabolic or higher order methods may be applied successively (see, for example, Brent, 1973).

Appendix D

Root-Locus Method of Stability Analysis

Very often, only one or two parameters are free to be specified in stabilizing a control system. In such a case graphical techniques developed to analyze stability of a servomechanism are quite effective in determining the range of the parameter values for which the system is stable.

We now describe one such method, called the root-locus method. The method applies primarily to a single-input–single-output system, i.e., to a control system with a scalar instrument and a scalar target variable.

Suppose the input-output relation of a control system is described by a transfer function

$$\hat{y}(s)/\hat{x}(s) = h(s)$$

$$= \frac{kG(s)}{1 + kG(s)} \tag{D-1}$$

where k is a scalar parameter. We take $k > 0$ for the sake of definiteness.

For example, a control system described by

$$\dot{z} = Az + \boldsymbol{b}\boldsymbol{q}$$

$$y = \boldsymbol{c}^{\mathrm{t}}z$$

can be put into the above form with

$$G(s) = \boldsymbol{c}^{\mathrm{t}}(sI - A)\boldsymbol{b}$$

when $q = k(x - y)$, or when the loop is closed as shown in Fig. D-1. When A is of finite dimensional matrix, $G(s)$ is a rational function in s, i.e.,

$$G(s) = \frac{N(s)}{D(s)}$$

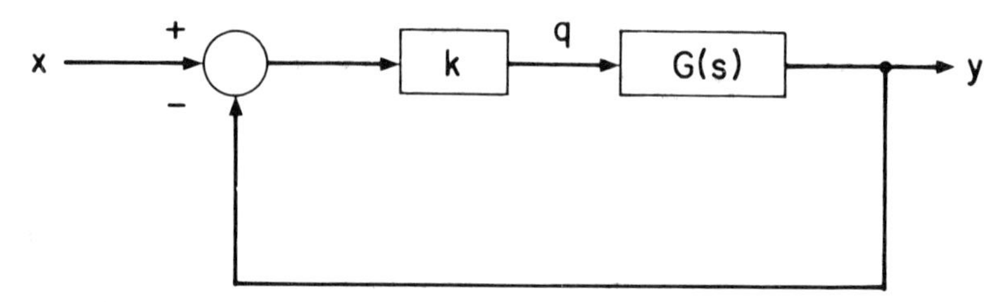

Fig. D-1 A Closed-loop System

where $N(s)$ and $D(s)$ are finite degree polynomials in s. The system (D-1) is asymptotically stable if and only if the poles, i.e., the roots of $1 + kG(s) = 0$, lie in the open left-half plane of the complex plane, or if $\text{Re}(s) < 0$ for s satisfying $1 + kG(s) = 0$. Since the roots of this equation depend on k continuously, we can obtain loci parametrized by k of the roots. These are the root-loci. The root-loci, if drawn, then give a graphical means of determining what the time response of (D-1) is for any chosen k value.

We now summarize some useful facts that characterize the root-loci (see most introductory textbooks on servomechanism or differential calculus for further details, for example, (Kaplan 1962, Takahashi 1966, Truxal 1955). Suppose that we know $G(s)$ in the factored form as

$$G(s) = \frac{\prod_{i=1}^{m} (s - z_i)}{\prod_{j=1}^{n} (s - p_j)}, \qquad m \leqslant n$$

The roots of $1 + kG(s)$ are such that $kG(s) = -1$. They satisfy

$$\angle G(s) = \sum_{i=1}^{m} \arg(s - z_i) - \sum_{j=1}^{n} \arg(s - p_j) = a\pi, \qquad a \text{ an odd integer}$$

and

$$|G(s)| = \frac{\prod_i |s - z_i|}{\prod_j |s - p_j|} = 1/k$$

Note that for $|s|$ sufficiently large, $G(s) \sim 1/s^{n-m}$. Thus, we obtain

(1) The root loci start at the poles of $G(s)$.
(2) There are $(n - m)$ straight line asymptotes with $a\pi/(n - m)$, $a = 1, 3, 5, \ldots$ degrees. The m branches of the root loci always terminate on the zeros of $G(s)$ and $n - m$ branches terminate at infinity.

For $|s|$ large,

$$G(s) = \frac{1}{s^{n-m} - (\Sigma_j p_j - \Sigma_i z_i)s^{n-m-1} + \cdots}$$

Therefore, we have:

(3) The hub of the asymptotes is given by

$$c = (\Sigma_j p_j - \Sigma_i z_i)/(n - m)$$

For k small, the loci are near the poles and for k large, the loci are close to the zeros. Thus:

(4) Let θ_l be the angle of departure of a locus from p_l. It is given by

$$\theta_l = a\pi - \sum_{j \neq l} (p_l - p_j) + \sum_{i=1}^{m} (p_l - z_i), \qquad a = 1, 3, 5, \ldots$$

Other rules exist such as those for determining the breakaway point of a locus from the real axis, and so on. We do not make use of these other rules in this book.

Example The root-loci with $G(s) = (s + 4)/[(s + 1)(s + 2)(s + 5)]$ are sketched in Fig. D-2.

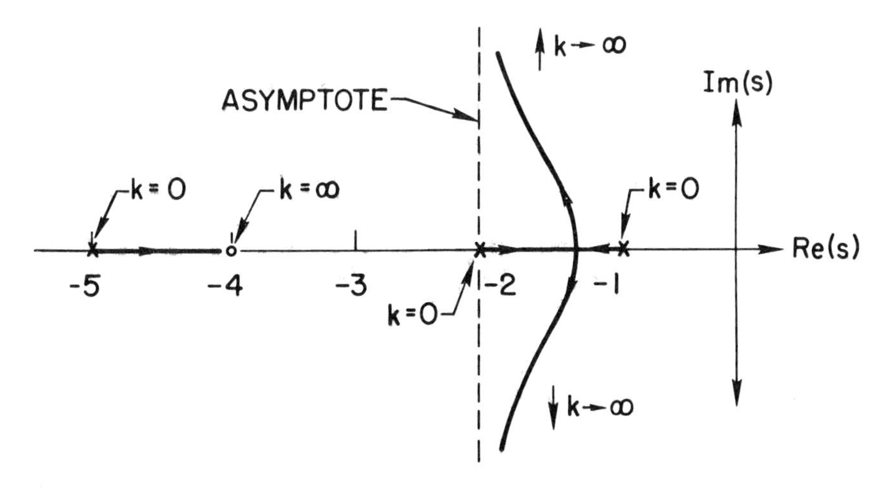

Fig. 10-2

INDEX

MASANAO AOKI

Masanao Aoki is Professor of Economics and Electrical Engineering at the University of Illinois and Professor of Engineering and Applied Science at UCLA. He is a graduate of the University of Tokyo and holds a Ph.D. in Control from the Tokyo Institute of Technology and a Ph.D. in Engineering from UCLA.

Professor Aoki is a member of the American Economic Association, the Society of Industrial and Applied Mathematics, the Institute of Electrical and Electronics Engineers, and the Econometrics Society.

In addition to *Dynamic Economics*, Professor Aoki has written three other books. They are *Optimization of Stochastic Systems* (Academic Press, 1967), *Introduction to Optimization Techniques* (Macmillan Book Co., 1971), and one book in Japanese.